Thinking Peaceful Change

Syracuse Studies on Peace and Conflict Resolution
Louis Kriesberg, *Series Editor*

Frank Möller

Thinking Peaceful Change

Baltic Security Policies and Security Community Building

SYRACUSE UNIVERSITY PRESS

The paper used in this publication meets the minimum requirements of American
National Standard for Information Sciences—Permanence of Paper for Printed
Library Materials, ANSI Z39.48-1984∞™

For a listing of books published and distributed by Syracuse University Press,
visit our Web site at SyracuseUniversityPress.syr.edu.

ISBN-13: 978-0-8156-3108-8
ISBN-10: 0-8156-3108-1

Library of Congress Cataloging-in-Publication Data
Möller, Frank
Thinking peaceful change : Baltic security policies and security community building /
Frank Möller.—1st ed. 2006.
p. cm.—(Syracuse studies on peace and conflict resolution)
Includes bibliographical references and index.
ISBN 0–8156–3108–1 (cloth : alk. paper)
1. National security—Baltic States. 2. Baltic States—Defenses.
3. Peace-building—International cooperation. I. Title.
UA646.53.M66 2006
355'.0335479—dc22   2006031892

To my parents,
Gertrud and Hans Möller

Unless we see "others" as having something in common with us—something that requires recognition—how can we respect their "otherness" except in parody form?

— NICK COULDRY,
Inside Culture: Re-imagining the Method of Cultural Studies

FRANK MÖLLER is a research fellow at the Tampere Peace Research Institute, University of Tampere, Finland.

Contents

Acknowledgments

IT IS ONE OF THE PLEASURES of writing a book to express gratitude to those who helped make it possible (without assigning to them the responsibility for any errors it may still have). Throughout the whole period of time I spent on writing this book, I was fortunate to benefit from the hospitality and generosity of a large number of people and institutions.

First, I thank my friends and colleagues at the Tampere Peace Research Institute (TAPRI) for providing the stimulating and challenging intellectual atmosphere within which the present book was written. TAPRI is an island of independent, quality-oriented research in an academic landscape that is increasingly exposed to political pressure and dominated by dogmatic rather than critical knowledge production. I am especially grateful to the former research director, Tuomo Melasuo, for inviting me to Tampere and for unconditionally supporting my work. I also thank his successor, Tarja Väyrynen, for believing in the quality of my research and insisting on publishing internationally. I am grateful to Unto Vesa, whose enthusiasm for and keen knowledge of the work of Karl Deutsch were invaluable inspirations during the whole project. In what turned out to be the first of countless conversations and exchanges of ideas, Samu Pehkonen, one of the most discreet and helpful people I have ever met, in the spring of the year 2000 directed my attention to collective memory and implanted in my thinking the idea that memory is indeed an important subject of peace research. Regarding chapter 12 of the present book, I acknowledge my intellectual debt to him. Unto Vesa and Matti Jutila thoroughly read and thoughtfully commented on the text. Roland Caldbeck checked the language of many drafts of the manuscript and pointed to several gaps and omissions. Aino Heiskanen dealt with all kinds of computer problems, and Anja Reini dealt with all sorts of administrative problems. They always found solutions in seemingly hopeless situations. Thank you.

At the very early stages of the project, Peter Wallensteen invited me to be a guest researcher to the Department of Peace and Conflict Research at Uppsala University. Erik Noreen and Björn Hagelin, whom I failed to convince of some admittedly immature ideas, helped me to reshape and specify my approach. They also arranged several interviews with colleagues in Stockholm that, although not being referred to in this book, proved very helpful for the organization of the research process. At the Schleswig-Holstein Institute for Peace Research, Hanne-Margret Birckenbach, Christian Wellmann, Walter Westphal, and Arend Wellmann thoroughly dissected my texts and provided a forum for critical discussion to which I certainly did not measure up at that time. Margitta Matthies gave me crucial information on the TAPRI grant for the year 2000, without which I would probably not have moved to Finland. Manfred Kerner invited me as a guest lecturer to the German-Latvian Social Science Center at the University of Latvia in Riga, thus giving me the chance not only to work with Latvian students, but also to try my ideas on a Baltic academic audience. Ulrich Albrecht and Manfred Kerner supervised my doctoral thesis at the Free University of Berlin, from which the present book is derived. Bernd Henningsen and Ruth Stanley served on the evaluation committee.

Björn Hagelin, Pertti Joenniemi, Joanna Kidd, Merje Kuus Feldman, Erik Mannik, Grazina Miniotaite, Romano Prodi, Steven Sawhill, David Smith, and Conrad Tribble generously shared with me their insider knowledge when otherwise available sources did not reveal what I needed to know. Burkhard Auffermann and Andreas Heinemann-Grüder read an earlier draft of the manuscript, pointed at gaps in the argumentation, raised many critical issues and made valuable and greatly appreciated suggestions as to the coherence of the text.

Tuomo Melasuo and Peter Gran brought me to Syracuse University Press. Louis Kriesberg accepted my manuscript for the Peace and Conflict Resolution series. Glenn Wright, the acquisitions editor, and Michael P. O'Connor, the editorial assistant for acquisitions, were instrumental in starting the publication process. Annie Barva was the copyeditor, and Pat Rimmer put together the index. Two anonymous referees for Syracuse University Press offered exquisite comments on both the structure and substance of earlier drafts of the manuscript. They helped to transform a highly unreadable doctoral thesis into a legible and coherent text and to increase the overall quality of the book. I thank all of these people for making the book possible.

As I worked on this project, my parents, Gertrud and Hans Möller, supported me without reservation. I suspect that it occasionally occurred to them that it was a slightly bizarre idea to try to make a living as a peace researcher rather than as a doctor or a judge, a choice they probably would have preferred especially in situations when one contract ended with no new contract in sight. Without their patience, moral encouragement, and financial support during the more difficult parts of the project, this book would not have been written.

A Note on Translation and Names: All translations of non-English source material are mine, except where indicated in the works cited list. The presentation of Latvian and Lithuanian names was a difficult issue because the use of special characters in these names has resulted in different spellings in various texts, so I have simply not used many of these characters in order to offer consistent spelling.

Abbreviations

ACVs	armored combat vehicles
AMEC	Arctic Military Environmental Cooperation
BaltBat	Baltic Peacekeeping Battalion
BALTNET	Baltic Regional Airspace Surveillance Network
BALTSEA	Baltic Security Assistance Forum
BEAR	Barents Euro-Arctic Region
CBSS	Council of the Baltic Sea States
CFE Treaty	Treaty on Conventional Armed Forces in Europe
CIS	Commonwealth of Independent States
CPSU	Communist Party of the Soviet Union
CSCE	Conference on Security and Cooperation in Europe
CTR	Cooperative Threat Reduction Program
e-PINE	Enhanced Partnership in Northern Europe
EU	European Union
IISS	International Institute of Strategic Studies, London
JCG	Joint Consultative Group
KFOR	NATO Kosovo Force
LDDP	Lithuanian Democratic Labor Party
MD	military district
NATO	North Atlantic Treaty Organization
NEI	United States Northern Europe Initiative
NGO	nongovernmental organization
OSCE	Organization for Security and Cooperation in Europe
PfP	Partnership for Peace
SAC	Strategic Air Command
SFOR	NATO Stabilization Force

TLE	treaty-limited equipment
UN	United Nations
USSR	Union of Soviet Socialist Republics
WEU	Western European Union
WTO	Warsaw Treaty Organization

Thinking Peaceful Change

1

Thinking Peaceful Change

WHY HAS PEACEFUL CHANGE in the Baltic Sea region not yet been translated into expectations of peaceful change? Why do the security policies of the Baltic states still reflect thinking of security as military security rather than as a security community? Why did it take ten years after Lithuania's application for North Atlantic Treaty Organization (NATO) membership for Baltic membership in the alliance to be realized, even though the United States has both energetically pushed forward enlargement and been an ardent rhetorical supporter of Baltic NATO membership? Why is it still too early to refer to all of the Baltic Sea states as a security community?

The explanation suggested in this book, which focuses on Estonia, Latvia, and Lithuania, is that security policy in these states, especially in the 1990s, has had often more to do with the construction of the nation-state than with security. The construction of the nation-state, in turn, has resulted in security policies that are suboptimal from the point of view of security community building. This may appear to be a rather unoriginal statement. After all, I am hardly the first to claim that there is an intimate relationship between the state and security. However, because of the frequent assessments that the nation-state is withering away in an era of globalization and that, to the extent that it still exists, it is entering a post-Westphalian phase with a decrease in responsibility, jurisdiction, and influence of state actors and an increase in the permeability of borders, material and otherwise, it may still be worthwhile to analyze such assessments empirically. As this book shows, thinking of security in terms of national security and equating national security with the security of the state and the capability to defend it militarily—both approaches increasingly marginalized in many theoretical writings on security—still enjoy much popularity among decision makers and prevents both a critical interrogation of de facto security policies and

1

their change from military and alliance-based conceptions to civilian and community-based conceptions.

In the Baltic Sea region in the early 1990s, the starting conditions for the transition to civilian and community-based conceptions of security were very promising. With the coming to power of Mikhail Gorbachev in the Soviet Union, new thinking had paved the way to a less antagonistic and more peaceful foreign and security policy. With the termination of the validity of the 1955 Treaty of Friendship, Cooperation, and Mutual Assistance on 1 July 1991 and the resulting dissolution of the Warsaw Treaty Organization (WTO), the primary means of Soviet military domination over parts of central, eastern, and Baltic Europe finally disappeared. It disappeared roughly eight months after the accession of the German Democratic Republic to the purview of the Federal Republic's Grundgesetz (Basic Law) on 3 October 1990. What is commonly referred to as German unification would not have been possible without civilian mass resistance to Soviet dominance in the German Democratic Republic in the late 1980s. The role taken by the military, especially by the Soviet forces stationed in the German Democratic Republic, during the mass demonstrations was mainly one of noninvolved bystanders, apparently based on instructions from Moscow.

The international accords in connection with the German unification resulted in "the biggest reductions in armed forces on a given territory within a given brief period" (Albrecht 1995b, 169). About one million troops were either disbanded or withdrawn from Germany. Likewise, the Soviet troops stationed in Estonia, Latvia, and Lithuania did not usually respond violently to the popular movements' striving for independence in the late 1980s and early 1990s. The movements, for their part, were strictly nonviolent, which indeed has to be seen as the condition for the possibility of their success in early autumn 1991. The dissolution of the Soviet Union and the independence of the Russian Federation also did not trigger violent actions in the Baltic Sea region.

The withdrawal of the Russian—former Soviet—troops from the territories of Poland, the former German Democratic Republic, Estonia, Latvia, and Lithuania has aptly been called "one of the greatest strategic retreats in human history" (Lieven 1996, 175). It was not only one of the greatest, but also one of the most peaceful military retreats in history, resulting in the withdrawal of altogether thirty-seven divisions from central Europe and the Baltic states. After the dissolution of the Soviet Union, the Russian troop pullout from Estonia, Latvia, and Lithuania amounted to 120,000 servicemen. The Russian authorities did not call into question the principle of withdrawal. Its implementation, however, caused dissatisfaction among them, as indicated by various official and semiofficial statements reflecting, among other things, the military value attached to the Baltic republics in Soviet times, political tensions between the Baltic states (as their status now was) and the Russian Federation, and unrealistic demands on the part of some Baltic decision makers as to the scope and speed of the troop pullout. Furthermore, technical problems caused delays and temporary interruptions of the troop withdrawal, thus intensifying anxieties in the Baltic states as to Russia's willingness to fulfill its international obligations.

The troop pullout, however, progressed quite smoothly at the operational level and was completed in Lithuania on 31 August 1993 and in Estonia and Latvia one year later. The Russian troop pullout from Poland and Germany was completed in October 1992 and August 1994, respectively. Altogether, 4,000 Russian servicemen were allowed to stay in Poland until 17 September 1993 to coordinate the transfer of troops withdrawn from Germany (Lachowski 1994, 577). Russia was also permitted to use the ballistic missile early-warning station in Skrunda, Latvia, up to August 1998. The subsequent demolition proceeded according to schedule. In compliance with the Russian troop withdrawal agreement, the dismantlement of the former Soviet nuclear submarine training base in Paldiski, Estonia, including two small nuclear test reactors, was completed in September 1995. This survey appears to be indicative of a rather smooth process of politicomilitary developments in the Baltic Sea region during the 1990s. However, appearances are deceptive, and although it is true that the tensions in the region did not escalate to violence, most political and military decision makers in the Baltic states did not at the time take peaceful change for granted (Möller 2002a). It is one of the aims of this book to show why this was so.

Without doubt, a decisive moment in the process of Baltic reindependence was the decision of the then chairman of the Parliament of the Russian Republic, Boris Yeltsin, to use the dramatic events in January 1991 as both an opportunity for his personal reckoning with Gorbachev and as a vehicle with which to help the Russian Republic replace the central Soviet authorities. Like the jurisdictional struggle of the Russian Republic's authorities against the Soviet center, Yeltsin's personal approach was "tactically ori-

4 • THINKING PEACEFUL CHANGE

ented to *weakening* the center and distributing its powers to the national re-publics, but strategically oriented to *capturing* the center and taking over its powers" (Brubaker 1996, 42, emphasis in original). The depiction of Yeltsin as "an embodiment of solidarity between the Russians and the Baltic nations in their fight for independence and democracy" (Vares 1995, 54) may reflect the popular impression at that time and the solidarity among democratic and pro-independence movements in several Soviet republics that no doubt existed (see Mihalisko 1991; Nahaylo 1991; Karklins 1994). Yet it probably underestimates Yeltsin's tactical use of the Baltic popular movements' con-flict with the Soviet central authorities, among other things, in order to rec-ommend himself to Western governments as a reliable and trustworthy partner and a potential successor of Gorbachev, who increasingly seemed to suffer from a lack of control over political and military developments in the Soviet Union.

When in January 1991 Soviet Ministry of the Interior troops attacked peaceful demonstrators in Vilnius, Lithuania, and Riga, Latvia, Gorbachev stayed in Moscow and kept silent while Yeltsin visited Tallinn, Estonia, at his own risk.[1] There, Yeltsin signed the Treaty on Fundamentals of Interstate Relations Between Estonia and Russia, in which both sides recognized the other side's right to *"realize its sovereignty in whatever form it chooses"* (Müller-son 1994, 119, emphasis in original). Furthermore, they recognized each other as sovereign states and entities under international law and declared their readiness to assist one another in case of a threat to sovereignty (Jonson 1992, 85). Moreover, Yeltsin called upon Russian soldiers not to use force against civilians because "violence against lawfulness and against the peo-ple of the Baltics will bring about a serious crisis, both in Russia itself and in the situation of the Russians who live in other Republics, including the Baltic Republics."[2] This statement was reportedly given "the widest possi-

1. A former advisor to Gorbachev writes that he had convinced Gorbachev to visit Vilnius the day after the military intervention. Gorbachev, however, changed his mind after a conver-sation with Vladimir Kryuchkov, then KGB chief, who argued that he could not guarantee Gor-bachev's security (Jakowlew 2003, 594–95).

2. Yeltsin, "Appeal to Russian Soldiers" (as cited in J. Trapans 1991a, 433). Yeltsin contin-ued by reminding the soldiers that the "sending of servicemen—called up for military service from the Russian Federation—outside the Republic in order to participate in carrying out du-ties not stipulated in article 29 of the RSFSR [Russian Soviet Federated Socialist Republic] Con-

ble publicity" by the Baltic governments at that time (Lieven 1994, 305). To-gether with Baltic leaders, Yeltsin also appealed to the United Nations (UN) to organize a conference in order to solve the Baltic issue by peaceful means (Kirby 1995, 434). In addition, the Presidium of the Russian Supreme Soviet issued a declaration declaring illegal the employment of citizens of the Russian Republic in the Soviet army on the territories of other Soviet re-publics (Müllerson 1994, 145).

After the attacks on customs and border guards at the Medininkai bor-der station in July 1991, it was again Yeltsin who asked Gorbachev to with-draw Soviet Ministry of the Interior troops from Lithuania (Vitas 1996, 75). The Russian Republic had recognized the Lithuanian sovereignty immedi-ately after the declaration of independence on 11 March 1990 and prepared similar treaties with Estonia and Latvia. These treaties remained unsigned at that time because representatives of the popular movements in both re-publics decided to refrain from declaring full independence. Yeltsin recog-nized the independence of Estonia without delay after its declaration on 20 August 1991 (Müllerson 1994, 120).[3]

An analogous decree on Latvia followed immediately. The foreign min-ister of the Russian Republic, Andrei Kozyrev, had already in May 1991 ac-cepted the accreditation documents of Estonia's representative to the Russian Soviet Federated Socialist Republic and by so doing had "made the new [Estonian-Russian] relationship 'official' " (Dunlop 1993, 53). Although Yeltsin's proceeding sought mainly to weaken the Soviet center, it surely helped the Baltic republics become sovereign states again. This support, al-though legally doubtful, might have served as a starting point for the devel-opment of good neighborly relations between Estonia, Latvia, and Lithuania, on the one hand, and the Russian Federation, on the other. The smooth process of the Russian troop withdrawal from the Baltic states or the

---

stitution, contradicts the decision of the RSFSR Extraordinary Congress of People's Deputies, accepted on December 11, 1990, and is therefore unlawful."

3. In the decree, Yeltsin "(a) recognized the independence of Estonia, (b) required the For-eign Ministry of Russia to start negotiations with Estonia on the establishment of diplomatic re-lations, (c) called on President Gorbachev to recognize the independence of Estonia, and (d) called on the world community of states to recognize the independence of Estonia" (Müllerson 1994, 120). The Resolution on the National Independence of Estonia is reprinted in Taagepera 1993, 201–2.

restraint displayed by the Russian government in its relations with Estonia, Latvia, and Lithuania might have served the same purpose.[4] But they did not. Why they did not is discussed in this book.

This examination is not meant to minimize what has been achieved in the Baltic Sea region in terms of peaceful change. In fact, the 1990s were characterized by the absence of armed conflict, but it must be acknowledged that peaceful change has not yet been translated into expectations of peaceful change. Seeming to reflect profound threat perceptions, the security policies in Estonia, Latvia, and Lithuania during the 1990s revolved around the buildup of armed forces and integration in a military alliance (i.e., NATO)—an alliance, however, that was not too keen on inviting the Baltic states to become members as long as their relations with Russia seemed to be tense. At the same time, it should be noted that the military expenditures of the Baltic states throughout the 1990s were actually quite modest, when the apparent threat perceptions are taken into consideration, and remained below 2 percent of gross domestic product (GDP). Furthermore, the Baltic states' military policy was inconsistent. For example, in the year 2000 the London-based International Institute for Strategic Studies (IISS) reported on a large reduction in Lithuania's naval manpower (2000, 98) that contradicted both the government's numerous declarations that it would increase the military capabilities and Commander in Chief Jonas Kronkaitis's statement in 1999 that "we will settle on the present size of the fleet" (Kronkaitis 1999, 118).[5]

The question of "what went right in the Baltics" (Jutila 2002, 3) is indeed asked only infrequently. It is addressed in this book in terms of thought pat-

---

4. The Russian government "did not attempt to use its military presence to subvert or even to influence the internal political affairs of the Baltic states" (Garthoff 1994, 787). Indeed, in Baltic-Russian relations from 1991 to 1994, "they [the Russians] did not actually themselves seek out or invent disputes as part of a deliberate planned campaign to worsen relations and create insecurity" (Lieven 1996, 176).

5. The figure 560 personnel (including 280 conscripts) in the Lithuanian navy is 50 percent of the manpower given for 1998 and 1999 (1,320 personnel, 670 conscripts). The Lithuanian government, responding to a draft copy of the IISS's *The Military Balance,* has provided the figure for 2000. It is thus the official figure, however unlikely it appears. The figure is difficult to verify or falsify because there are few other open sources on Lithuania (Joanna Kidd, Defense Analyst [Navy], IISS, London, personal communication, 16 November 2000). My inquiries to the Lithuanian Ministry of National Defense remain unanswered.

terns dedicated to peaceful change. Another question asked only infrequently is whether security policy (and rhetoric) in the Baltic states aims primarily to construct security or something else. This question is difficult to answer in part because it is difficult to say what "security" is. Yet, as I try to show, security policy often has more to do with the construction of the state than with the construction of security. The construction of the state, in turn, may result in security policies that are suboptimal with respect to building a security community. Nowhere else in the Baltic Sea region can this be shown as clearly as in the Baltic states, which throughout the 1990s combined the construction of the nation-state with the construction of security, but often prioritized the former over the latter. The analysis of the Baltic states' security policies can thus help relativize some of the "illusions" identified by Fred Halliday as ingredients of the cultural turn in theorizing international relations, especially the claims that "the state is, as ever, declining in importance" and that it is increasingly being replaced by social movements and nonstate actors (2000, 55). When it comes to security, the empirical evidence of such a turn is at best mixed. Security conceptions along postmodern lines still face strong resistance among decision makers. Furthermore, the dissolution of both the Soviet Union and Yugoslavia did not result in the replacement of the nation-state by something profoundly different, but rather in several new or renewed nation-states, thus actually strengthening the nation-state as an institution. Among the new or renewed nation-states, the Baltic states figure prominently not only because of their apparent success in transforming themselves (Clemens 2001), but also because of both the attention that observers within and outside of the Baltic states have devoted to their security and integration policies and the key function assigned to them for the development of NATO-Russia relations.

However, these and other puzzles are frequently obscured by a literature that throughout the 1990s discussed Baltic security issues predominantly in the light of Estonia's, Latvia's, and Lithuania's aspirations to NATO membership and thus helped to naturalize thinking about security in terms of military security, including the buildup of armed forces and the membership of the Baltic states in the Western alliance. There can in fact often be observed a basic correspondence between, on the one hand, the conceptions of security underlying most scholarly writings on the Baltic states (quite regardless of the respective writers' epistemological orientations) and, on the other hand, the conceptions of security underlying the decision-making process in Estonia, Latvia, and Lithuania. Those authors who deviate from a "security-driven" discourse find themselves in a marginal position indeed (Browning and Joenniemi 2004, 241). Accordingly, many writings follow the official narrative and can hardly be considered works of critique. Whatever the merits of such a correspondence are, it hardly contributes to a critical investigation of the security conceptions and policies of the Baltic states and hardly breaks with institutionalized forms of the production of knowledge on security either theoretically or practically. Most prominently, it does not call into question the state-security-military nexus that prevails in the thinking about security in the Baltic states (and elsewhere) and without which the security policies of Estonia, Latvia, and Lithuania in the 1990s cannot be adequately grasped. As a consequence, a controversial debate of alternative paths to Baltic security has virtually been absent, thus bolstering a one-dimensional understanding of security that is theoretically untenable.

For example, whether the Western states should learn from the Baltic states' nonmilitary path to reindependence and can tailor their security policies according to their eastern neighbors' experiences has not even been asked. On the official level, both sides early on seem to have agreed on the reading of eastern Europe's transition as a process of one-sided adaptation rather than as a process of mutual social learning. This agreement was facilitated both by the powerful slogan of a "return to the West" or to "Europe" in eastern Europe—which repeated and confirmed western Europe's arrogant post–World War II ignorance of half the continent by identifying Western unification with a European Community (Mazower 1999, xiv)—and by the prevalent Western reading of the end of the Cold War "as a Western triumph rather than as a result of the Soviet decision to end the military confrontation" (Dalby 1997, 12).

The politicomilitary process of the 1990s is frequently depicted and retrospectively rationalized as a linear one from the restitution of Baltic independence to Baltic membership in NATO. One of the results of this depiction is the continuation of thinking about security in terms of military security. Another result is the unimaginative expansion of a military alliance (NATO) even though a centralized institution's capability to deal successfully with post–Cold War Europe's stratified security patterns has been doubted from the beginning (Albrecht 1995a, 75).

In Estonia, Latvia, and Lithuania, the peaceful retreat of the Russian military and the success of nonviolent mass movements were not translated into a corresponding understanding of security and security policies. What followed independence was a growth in the military potential, but also, and probably more important, a militarization of the way security was conceived: security was reimagined from civilian concepts to military concepts, initially in terms of the buildup of national armed forces and subsequently in terms of NATO membership. Even those policymakers who before independence had supported civilian security conceptions after independence favored the opposite. Having national armed forces was often depicted as "natural" and "simply necessary," allegedly resulting from "geopolitical" or other ostensible "necessities." Besides, not having to argue for one's own views on security brought with it the convenience of assigning to others the task of justifying their "deviating" views while one's own views were declared "logical" and "obvious." Western advisers did not disagree. Why should they? After all, in the Western states security was increasingly being advertised broadly as having political, economic, environmental, and societal dimensions, but military means could hardly be said to have lost their outstanding importance to security policy concepts even in the West.

On the political level, thus, agreement seems to prevail as to what security is. A more thorough investigation, however, reveals substantial differences. This discrepancy is not surprising because security policy is to a large extent the result of subjective processes of perception, notion, and observation as well as of subjective ideas of security. Security policy reflects knowledge attained by observation or information. It thus is a mental reflection: security is as it is thought and spoken of. Security policy mirrors change and continuity in perceptions and misperceptions, beliefs and disbeliefs, learning and unlearning. Security policies are influenced by changing perceptions of one's own and others' interests just as they are shaped by changing perceptions of one's own and others' capabilities and the meaning assigned to these capabilities.

It is a truism that what are thought to be security issues vary over time and across actors (Katzenstein 1996, 10). Whatever issues policymakers consider to be security issues, how they respond to them reflects historical and political judgments (R. Walker 1997, 67). Judgments, by definition, follow from estimations and opinions. To say that policymakers have to respond to what they define as security challenges in only one way ignores the fact that every political decision is a decision between different alternatives. In other words, security policy, like all social action, is an action that might be otherwise. Accordingly, Barry Buzan argues that "we are still far from consensus on what can and cannot legitimately be designated as a security issue or a referent object" (Buzan 2000a, 11). Elsewhere, Buzan even defines "the nature of security" as one that "defies pursuit of an agreed general definition" (1991, 16). By a consideration of security as an "essentially contested concept," it is seen as one of those concepts that "necessarily generate unsolvable debates about their meaning and application" (1991, 7). Yet this claim itself can be contested. Simon Dalby, for example, argues that *security* is a "contested term, one with multiple meanings, some of which are not at all necessarily logically linked to conventional understandings" (1997, 6), whereas David Baldwin (1997) does not think that security is a contested concept, but rather one characterized by the lack of contestation. Both positions may serve as an invitation to interrogate different points of view with respect to their plausibility.

It is worth remembering that during the Cold War the prevailing understanding of security was rather stable. Security was equated with military security, so that international relations and the related disciplines were concerned mainly with the use of force or the prevention of the use of force in interstate affairs, and the referent object of security and security policy was the nation-state. This logic collapsed to some extent with the end of the bloc rivalry, but without having yet resulted in an equally forceful post–Cold War logic. It may indeed be argued that a new post–Cold War security logic—equaling its predecessor in dominance, imperviousness, and insensibility to alternatives—is neither attainable nor desirable. In fact, the intellectual cohesion from which the disciplines allegedly benefited during the Cold War and that some authors (Walt 1991) would like to reestablish may precisely be seen as a restricting moment that prevented a constructive negotiation of different points of view on security.

Although the end of the Cold War can be seen as an invitation to a critical negotiation of the ideas and interests explicitly or implicitly underlying the use of the word *security*, and although many scholars have accepted this invitation since the mid-1990s and thus challenged the conventional scholarly wisdom from a constructivist or other nonrealist approach, this opportunity has not been grasped fully in the Baltic states. In other words, a conventional understanding and analysis of security, one that declares the survival of the nation-state (still seen as the basic unit of the international

system) superior to all other purposes, has persevered. In this rather traditional understanding, national security, evaluated in terms of the survival of
the state, outweighs in importance all other possible referent objects. Yet although adherents of realism are inclined to conceive of survival as being
permanently threatened in an anarchic environment characterized by the
absence of central rule, nation-states only exceptionally cease to exist. In history, states have indeed exhibited an astonishing capability to survive.[6] Neorealists or structural realists tell us that nation-states survive because they
balance one another—either each on its own or together with others in military alliances. Accordingly, "the only institutions which neorealists deem
worthy of serious consideration are traditional alliances" (Ruggie 1998, 7). If
this is so, then the security conceptions of the Baltic states followed and are
still following a realist script to a large extent.

The Baltic world, then, still resembles a realist world. In Estonia, Latvia,
and Lithuania, security is predominantly regarded in terms of military security and consequently in terms of the buildup of national armed forces and
membership in a military alliance. Looking back at the 1990s, we may very
well say that the main domestic policy issue in the Baltic states (especially in
Estonia and Latvia) throughout the decade was the citizenship legislation
and the integration (or not) of Russian speakers. The Baltic states' aspirations for guarantees of military assistance through membership in NATO
can be seen as the main international issue for them. The latter was also the
single most serious conflict in the field of military security in the Baltic Sea
region. With respect to security narrowly defined in military terms, regional
cooperation involving all Baltic Sea littoral states was the exception rather
than the rule. Cooperation among the three Baltic states aimed at both increasing national military capabilities and approaching NATO, but not at regional security in itself. In the 1990s, some decision makers even depicted
regional security as an oxymoron because they retained a view of the indivisibility of European security (Ministry of National Defense of the Republic
of Lithuania 1999, pt. I, IV, 3). Furthermore, the potentialities both as to re-

6. Kenneth Waltz, in his influential *Theory of International Politics*, prognosticated for the
Soviet Union a life span of at least one hundred years (1979, 95). Ironically, the Soviet Union in
the meantime joined the small club of states that "have met involuntary ends," whereas three of
the states declared dead by Waltz have experienced resurrection, Estonia, Latvia, and Lithuania
(1979, 137).

gional approaches to security and as to security community building in the Baltic Sea region have hardly been exhausted, and there has even been a lack of political interest in both issues. Being aware of the promising beginnings in the period from the mid-1980s to the early 1990s and the peaceful development of the 1990s, one might wonder why this is so.

In this book, I aim to answer this question by analyzing the security policies of Estonia, Latvia, and Lithuania and the underlying conceptions of security there. I analyze and contextualize patterns of argumentation referring to security, policy manifestations, and representations. Furthermore, I link the analysis to the prospects for security community building in the Baltic Sea region, which serves as a positive reference point. The concept of a security community thus is the analytical frame within which I treat the security policies of the Baltic states.

From the absence of an agreed general definition of *security* follows that whoever wants to approach security policy in a descriptive *and* critical manner has to establish a norm against which the actual security policy of a specific security policy establishment can be measured and evaluated. Constructively tackling the tensions between norm and actual policy may help to improve social relations. I consider this norm to be "peaceful change" and the development of a security community in the Baltic Sea region. It reflects the character of this book as a contribution to peace research: peace research cannot be neutral with respect to questions of peaceful or violent change. However, I do not offer an application of the original concept of a security community, introduced by Karl Deutsch and associates in the 1950s, to the Baltic Sea region (see Vesa and Möller 2003). Rather, the concept of a security community in both its original formulation and its social constructivist adaptations serves as a way to escape from the axioms and closures that throughout the 1990s governed the Baltic states' security policies. The idea of a security community thus serves as an inspiration to think about international politics in the Baltic Sea region in terms of integration and peaceful change and to break with the habit of "conceptualiz[ing] managed security by working within the limits of some ontologically privileged anarchy and thus imagining security as accomplished only through alliances, balances of power, hegemonies and the like" (Barnett and Adler 1998, 436).

In an analysis of security policy, it does not make much sense to adhere to what Barry Buzan, Ole Wæver, and Jaap de Wilde call a "rationalist universalism": as long as policymakers follow their own rationalism as to what security is, any "rationalist universalism will easily be 'right' on its own terms, but it will be of very little help in political analysis" (1998, 31). Because security policy follows from particular political decision makers' particular understandings of what "security" is, this book deals with these understandings and the policies that result from them. Indeed, as the title of the book indicates, "thinking peaceful change" is not the same as "thinking about peaceful change." The title is meant to indicate that in the process of mentally reflecting on security, we form and construct both security and security policy. Ole Wæver, influenced by language theory, tells us that security can be seen as "a *speech act*. In this usage, security is not of interest as a sign that refers to something more real; the utterance *itself* is the act" (1995, 55, emphasis in original). Yet speech is an expression of preceding mental reflections rather than being exclusively a reflection of wider external factors or structure. Security policy thought of in terms of a security community will be qualitatively different from security policy that follows from mental reflections on deterrence or a balance of power, for example, and it will also be spoken of differently. Likewise, security policy following from an understanding of security as derivative of the construction of the nation-state may clash with the objective of a security community if the construction of the nation-state unfolds primarily through delimitation from other states. In this case, the others tend to be treated as negative reference points against which the nation-state is being constructed rather than as potential candidates for security community building.

Security communities are characterized by peaceful change and the resolution of conflicts among groups of people without resort to violence. However, peaceful change alone does not make a security community. In order to qualify as a security community, groups of people have to *expect* and *believe in* peaceful change. They do so when a set of values perceived as compatible is translated into a sense of community. Again, the issue is not one of community, but one of a *sense* of community, a *we-feeling*. Security communities thus are to a large extent *mental constructions* underpinned by individual and collective experience as well as by shared knowledge. Security communities are also *normative* constructions in the sense that their supporters declare a specific kind of intergroup relations—namely, one based on dependable expectations of peaceful change—normatively superior to and more desirable than other kinds of intergroup relations. They share this normative commitment with peace research, which, as stated earlier, cannot be indifferent to the question of war and peace. At the same time, security communities are historically and empirically traceable institutions: peaceful change—underresearched in the discipline of international relations (Patomäki and Wæver 1995; see, however, Holsti 1991a)—does occur; expectations of peaceful change do exist and govern intergroup relations; and expectations of violent change have in some cases successfully been replaced by expectations of peaceful change.

Approaching international relations in terms of security communities means in the main to think and subsequently to speak of and construct them on the basis of expectations of peaceful change. In this case, the realists' favorite—membership in a military alliance—may no longer be prioritized because military alliances have not proven to be the best path to peaceful change, as shown throughout history. Effective security policies thought and conceptualized in terms of and aiming at security communities may thus differ from security policies constructed around worst-case scenarios, which have habitually made actors seek refuge in military alliances, thus often initiating the threats to security, the perception of which made them enter the alliance in the first place. Alliance membership may give them an often false sense of security, but others outside the membership may become nervous. This is the security dilemma: some policies aimed at security may be perceived by others as threatening, and thus they trigger counterreactions—a process of action and reaction resulting in a suboptimal outcome for actor and re-actor (Jervis 1978). Whether worst-case or best-case scenarios are followed, the objective may arguably be the same: peace, security, or at least noninvolvement in war. However, dependent on the respective ideas of security, the factual policies with which to gain peace and security may differ from one another considerably, just as may others' perceptions of these policies.

Before giving a brief overview of the book, I should clarify three terms and categories used here. First, I frequently use the term *the Baltic states* as a shorthand for Estonia, Latvia, and Lithuania. This application may be debatable in the sense that particularities and important differences between the three states may disappear behind this collective noun. The term *the Baltic states* is, however, no expression of convenience. As has been argued elsewhere, it is justified given the similarities among the three states with respect to their basic foreign and security policy orientations (Möller and Wellmann 2001, 81). Analyzing the three states independently of one another is not possible without tiresome repetitions. Furthermore, in all three states the construction of security has been unfolding in parallel and occasionally in competition with the construction of the nation-state, thus resulting in similar tensions. I offer differentiations whenever necessary.

Second, the use of the term *Baltic Sea region* requires some qualifications. That "the Baltic Sea region" has been built by policymakers, practitioners, and scholars since the late 1980s neither presupposes a unanimous understanding of what it actually is nor shows agreement on the part of different region builders as to the aims of their activities. As Wæver has rightly put it, "the Baltic 'identity' was created by the fact that various actors wanted very *different* things, but all found the 'Baltic Sea Region' a useful slogan to further these aims" (1997, 305, emphasis in original). According to Lassi Heininen's concept of regional dynamics, originally meant to describe the Euro-Arctic region, it may thus be argued that the Baltic Sea region is a region "where different multifunctional interests of different actors meet each other, also including competition and conflicts" (1999, 373) and by doing so constitute the region in the first place.

It follows from an understanding of regions as social constructions rather than as given geographical or otherwise natural entities that region building and the assigning of political meaning to a specific region by means of particular criteria are political acts (Neumann 1992, 5–31). Mapping regions indeed cannot be a politically neutral act. The issue is always one of contested interests, of inclusion and exclusion, self and other, us and them. To define regions in a specific way involves both stating who "belongs" to the region and who does not. Thus, as David Harvey writes, "the very act of naming geographical entities implies a power over them, most particularly over the way in which places, their inhabitants and their social functions get represented" (1997, 257; see also Möller and Pehkonen 2003). This is important to remember because "we live in societies and cultures where individuals are spoken *for*, much more than they speak in their own name" (Couldry 2000, 58, emphasis in original). Furthermore, the prospects of security community building in the Baltic Sea region are affected differently by different understandings of what the region is. In this book, the *Baltic Sea region* is understood in terms of *northeastern Europe*, and both terms are used interchangeably.[7] This understanding includes Estonia, Latvia, and Lithuania as well as Denmark, Finland, Norway, Sweden, the northwestern parts of the Russian Federation (including the Kaliningrad region), (northern) Germany, and (northern) Poland. The issue here is not one of presenting a timeless, general, and decontextualized understanding of the region. Instead, the context is security and security community building. The substantiation of the claim that this particular mapping of the Baltic Sea region is the appropriate one in the context of security community building follows later.

Third, readers in search of theoretical and methodological purity may be disappointed in this book. Just as the original writings on security communities avoided a self-departmentalization in one of the dominant lines of theorizing international relations at that time, my study makes use of peace research and international relations, history, and sociological conflict theory as well as, perhaps less obviously, cultural studies, human geography, social memory studies, and social psychology. Furthermore, I do not spend much time on discussing *the* method of studying international relations "since different methods have their advantages in different circumstances and can, in any case, be combined" (Couldry 2000, 67). Likewise, as to theory, I take a fairly pragmatic and nondogmatic approach. Although my discussion involves some sympathy with the cultural turn in international studies, it is not limited to social constructivist or other culturalist approaches. At the same time, the book's emphasis on the norm of peaceful change and its capacity to regulate and guide international relations clearly points in a constructivist direction. However, even realist approaches, although sidelined throughout the book, remain important because they govern to some extent official understanding of security in the Baltic states and therefore the actual security policies, which are often presented in constructivist language.

Chapter 2 introduces and contextualizes the original concept of a secu-

7. I do not discuss whether this way of seeing the Baltic Sea region is the historically most appropriate one. Among historians, this view is shared by Kirby 1995, Klinge 1997, and Troebst 1999a, 1999b. According to Kirby, Russia "has wielded a powerful influence [in the Baltic Sea region] for the past two hundred years, and will in all likelihood continue to do so" (1995, 9). According to Troebst, northeastern Europe, including the northwestern parts of Russia, shows so many internal similarities that it appears to be a distinct historical space that can be separated from neighboring spaces (1999b, 11–12). These similarities have outlasted the vicissitudes of political, economic, and social history from late antiquity to the present time.

rity community. Chapter 3 discusses some recent adaptations of the original concept in light of social constructivism and other approaches before zooming in on Alexander Wendt's work on different types of anarchy. Wendt's notion of friendship in international politics is especially relevant in the present context and is reencountered in chapters 9 and 10. Furthermore, chapter 3 introduces a new category—a Deutschian kind of anarchy—aiming both to grasp international relations in the Baltic Sea area adequately and to come to terms analytically with Baltic-Russian relations. This theme is resumed in chapter 12. Chapter 4 explores the conceptions of the state prevalent in Estonia, Latvia, and Lithuania and problematizes them with respect to both the negotiations on the Treaty on Conventional Armed Forces in Europe (CFE Treaty) and the issue of the Russian troop withdrawal. Chapter 5 presents a historical analysis of the events in the late 1980s that resulted in the reestablishment of the independence of Estonia, Latvia, and Lithuania in 1991. It interprets the policies of both the Soviet government and the Baltic popular movements in terms of peaceful change, and it elaborates on military, economic, and political measures undertaken by the Soviet government in order to prevent the Baltic republics from becoming sovereign states again.

Chapter 6 provides an introduction to security and security studies, sketching the development of the discipline from the Cold War to the recent culturalist turn. It shows that even after the end of the Cold War some policymakers continued to adhere to rather simplistic readings of security. It stresses the ostensible lack of alternatives and the patterns of argumentation referring to—and naturalizing—geography as representational modes, the main aim of which seems to be to render a constructive and critical renegotiation of the concept of security impossible. In chapter 7, I analyze the process of the Russian troop withdrawal. After long and complicated talks, the withdrawal was completed in Lithuania in August 1993 and in Latvia and Estonia one year later, paving the way for the development of national security documents in the three states, as discussed in the next chapter. Chapter 7 not only resumes the narrative thread interrupted at the end of chapter 5, but also shows that the withdrawal of the Russian troops unfolded completely peacefully; not one single shot was fired. In addition to the peaceful independence of the Baltic states and the Russian Republic's support of the Baltic popular movements, this outcome is relevant to long-term perspectives of security community building in the Baltic Sea region.

Chapter 8 then explores the writing of the national security documents as privileged textual representations of the Baltic states' security conceptions. It also provides the textual background against which the following chapters—especially chapters 10, 11, and 13—should be read.

Chapters 9 and 10 are dedicated to U.S. foreign and security policies in northeastern Europe. Both examine the substance of the "friendship" that allegedly prevails in U.S.-Baltic relations. The argument is presented in two stages. A historical examination (chapter 9) tries—and fails—to substantiate the claim that the U.S. nonrecognition of the Soviet takeover of the Baltic states was accompanied by active and substantial U.S. support for Baltic independence, justifying the designation of the United States as a "friend." Then that chapter analyzes the first Bush administration's policy toward the Baltic states in the late 1980s and, again, looks for clues with which to justify the designation "friend." In chapter 10, a policy analysis shows the development of the Clinton administration's policy toward northeastern Europe. Emphasizing the U.S. policy's ideational element and downplaying its material element, the reading is based mainly on policy manifestations and shows the policy's evolution from a rather conventional starting point to a sophisticated approach to both northeastern Europe and security. From the point of view of the Baltic states' governments, this approach may indeed have been too sophisticated, not only because it diverted attention away from the Baltic states and toward northwestern Russia, but also because it broke with the habit of equating security with military security and of declaring security the exclusive business of the state. Seen from the perspective of security community building, however, the U.S. policy is one of the more promising features of international relations in northeastern Europe. Finally, implications for "friendship" between the United States and the Baltic states are discussed.

Because one of the keys to Baltic security is often said to be found in Moscow, chapter 11 deals with Baltic-Russian relations. The focus is on the representations of Russia among Baltic decision makers rather than on the de facto Russian policies or on Russian rhetoric toward the Baltic states. This approach reflects what Anthony Burke has called "the urgent need to interrogate the images of self and other that animate (in)secure identities" (2002, 7). My theoretical point of departure in this chapter is the concept of a Deutschian kind of anarchy, subdivided into a preparatory phase and a mature phase, with the subject positions "nonfriend" and "prefriend," introduced in chapter 3, offered as a further development of Wendt's conception of anarchy. By suggesting a Deutschian kind of anarchy, I attempt to grasp analytically the characteristics of international relations in the Baltic Sea region—namely, empirical absence of armed conflict simultaneous with a lack of dependable expectations of peaceful change. In this chapter, I argue that the representations of Russia have changed from stressing predictable malevolence to emphasizing unpredictability. Both characterizations were until recently equated with a threat to Baltic security and thus have inhibited the crossing of the threshold between the preparatory phase and the mature phase of the Deutschian anarchy. I discuss whether these representations can sufficiently be explained in terms of security and suggest that seeing Russia as a negative reference point has had an important function in the construction of nation-states and collective identity in the Baltic states. This function has profound implications for security community building. The cultivation of Russia as the Other, which by virtue of its otherness helps strengthen the self, renders impossible the development of a we-feeling or a sense of community encompassing the Baltic states and Russia.

I continue this topic in chapter 12 by discussing selected representations of the past and their relevance to current security policies. By drawing on social memory studies, I tentatively argue that representations of the past through the lenses of collective memory tend to perpetuate past enmities rather then assign them to history. Images of past enmities then may determine current policies and make difficult the evolution of a we-feeling and a sense of community among former enemies.

Chapter 13 discusses paths to security chosen de facto by the decision makers in the Baltic states. It deals with the buildup of armed forces and the Baltic strategy toward NATO. The main argument of the chapter's first section is that throughout the 1990s the armed formations were given an important role with respect to constructing national identity and imagining historical continuity. Rather than discussing the armed forces' operational capabilities and quantitative developments, I focus on patterns of argumentation justifying the change from following the nonviolent path to independence to advocating military strategies to defend it. I discuss this transformation in view of the conventional state-military nexus stressed throughout the book. The emphasis on military security led to the establishment of both national armed forces and a plethora of armed formations, in particular voluntary territorial defense forces and paramilitary organizations, control over which has proven difficult. These armed formations modeled themselves on their interwar predecessors and saw themselves as the guardians of independence and the "true" national values. This self-identification was supported by the roles officially assigned to the armed formations so as to strengthen collective subjectivity. The chapter then explores the evolution of the Baltic states' strategy toward NATO. Like elsewhere, the technical-operational aspects of the Baltic preaccession strategy are not discussed. Rather, emphasis is placed on the ambivalence inherent in the Baltic governments' position toward NATO, the evolution of their position over time, and the tensions between Baltic representations of NATO and NATO's self-representations resulting from, among other things, different views of its military and political functions as well as from different interpretations of "Europe." Some speculations are included about the relevance of Baltic NATO membership to security community building in northeastern Europe. Chapter 14 summarizes the book's main points and shows how they all relate to security community building.

2

Karl W. Deutsch and Security Communities

A SECURITY COMMUNITY is a group of people which has become "integrated."

By INTEGRATION we mean the attainment, within a territory, of a "sense of community" and of institutions and practices strong enough and widespread enough to assure, for a "long" time, dependable expectations of "peaceful change" among its population.

By SENSE OF COMMUNITY we mean a belief on the part of individuals in a group that they have come to agreement on at least this one point: that common social problems must and can be resolved by processes of "peaceful change."

By PEACEFUL CHANGE we mean the resolution of social problems, normally by institutionalized procedures, without resort to large-scale physical force. (Deutsch et al. 1957, 5)

LIKE MANY OF THE EUROPEAN SCHOLARS who came from the social sciences or international law, such as Hans J. Morgenthau and John Herz, and who had to emigrate because of the coming to power in 1933 of the National Socialist regime in Germany and its consequences, Karl W. Deutsch found his academic home in the United States in the discipline of international relations. Like Morgenthau and Herz, Deutsch became a prominent figure in the U.S. academic community far beyond the narrow space of the study and teaching of international relations. Like all of his coemigrants, Deutsch, who had to leave Europe in 1938, was strongly influenced by the experience of emigration and World War II. One of the main driving forces behind his work—and behind the work of many of his fellow emigrants—was the conviction that future interstate wars ought to be made impossible. In retrospect, it is perhaps surprising how much the consequences drawn by the European scholars in exile from their own stories differed from one another. This difference shows that regardless of the number of emigrants, emigration is first and foremost an individual experience, and the similarities among emigrants are often confined to rather technical matters such as emigration permits, transit papers, or working permits.

Hans J. Morgenthau, having developed a rather pessimistic anthropology, insisted on reducing international politics to the national interest and a struggle for power. His claim that politics should be separated from ethics implicitly contained a moral component that is often neglected in writings on classical realism and on Morgenthau as its dominant figure (Olson and Groom 1991, 111–12).[1] For him, the path to peace was through the maximization of the nation-state's power—an approach understood, however, not in nihilistic terms as an end in itself, but rather as a means to an end: namely, to counterbalance the power of other nation-states and by so doing to keep the peace.

John Herz explored among other things the social and psychological constellations of domestic and foreign policies and emphasized what he called the "security and power dilemma" as one of the forces driving international affairs. He argued that, resulting from the lack of an overarching government in the international system and the system's corresponding character as one of self-help, nation-states strive to build up and maintain capabilities to defend themselves if attacked. Because there will never be real assurance that military capabilities, even when built up primarily for self-defense, will actually be used only for defensive purposes, other states are likely to regard these measures as potentially aggressive and directed against themselves (Herz 1950). Robert Jervis summarizes Herz's findings by saying that "some policies aimed at security will threaten others" (1976, 76) or will be perceived by others as threatening and are likely to trigger competitive armament—a process of action and reaction that has historically led to (at least regional) arms races and a suboptimal outcome for both actors and re-actors.

As a result of the security dilemma and the insecurity inherent in the international system, the level of armaments will increase, but not necessarily the level of security. Herz argued against an anthropological conception of a

---

1. Morgenthau was critical of "those who deny that moral principles are applicable to international politics," stating that "all human actions in some way are subject to moral judgment. We cannot act but morally because we are men" (1984, 341).

"power urge" or a social-Darwinist one of a "struggle for survival," but he nevertheless realized that "competition for security, and hence for power, is a basic situation which is unique with men and their social groups." He suggested that in order to avoid a violent escalation of interstate conflicts, the security and power dilemma has to be recognized as a basic trait of a "race whose members are conscious of the threat of mortal dangers and death" (1951, 14). Jervis adds that at the least the security dilemma's existence must not be ignored because it "will operate much more strongly if statesmen do not understand it, and do not see that their arms—sought only to secure the status quo—may alarm others and that others may arm, not because they are contemplating aggression, but because they fear attack from the first state" (1978, 181).

In contrast to Morgenthau, who remained skeptical of modern empirical and quantitative methods and who instead relied on the classical European history of political ideas even after he emigrated, Karl W. Deutsch combined broad theoretical interests and comprehensive historical knowledge with statistical and behaviorist methods—that is, methods "dedicated to the observation and explanation of why human beings behave as they do in a wide variety of social situations" (R. Snyder 1954, vi). Like Morgenthau and Herz, Deutsch explored "possible ways in which men some day might abolish war" (Deutsch et al. 1957, 3) and identified what he termed *security communities* as a path to overcoming the security dilemma and keeping the peace.

Deutsch and his associates' intellectual effort can hardly be overestimated given both the McCarthy campaign in the United States and the nuclear hysteria of the mid-1950s. The inquiries associated with the name Senator Joseph McCarthy exerted a strong pressure to conform on everyone appearing before the public. Not joining in the officially demanded anti-Soviet song could at least ruin one's career. Exploring ways to peaceful change that did not in principle exclude the possibility of transcending the borders of the military alliances was also completely at odds with the questions explored by the vast majority of Deutsch's colleagues at that time. As the editor of his 1954 study put it, "To some readers of the following pages it may seem whimsical and paradoxical to discuss international community in an atmosphere dominated by fear, hostility, and distrust" (R. Snyder 1954, v).

Indeed, such a discussion might have made a fairly exotic impression on those colleagues who were preoccupied with stabilizing the bipolar international system on the basis of U.S. nuclear superiority and a NATO strategy of massive nuclear retaliation. These colleagues bothered little about theorizing the possibilities of major peaceful international changes and instead made themselves at home with bipolarity. They argued for its superiority over competitive distributions of power for preserving the peace among the superpowers. The nonescalation of the superpower rivalry was, however, accompanied by what has aptly been called the "Third World War" (Marshall 1999), a two-dimensional moniker indicating that this third world war took place in the so-called Third World at the same time that the superpowers were keeping a peculiar peace based on mutually assured destruction and nuclear deterrence. For this reason, Western strategists were inclined rhetorically to minimize this third world war as a "cold" war, which, in a way, amounts to a mockery of its victims. Pleading for bipolarity on the basis of nuclear deterrence was also politically risky (Sagan 1993). Cold War realism and the resulting realpolitik may thus be likened to "cynicism" (Herz 1951, 247). Cold War realists helped stabilize the Cold War system that they subsequently analyzed "scientifically" and, by doing so, helped legitimize. Their work has correctly been described as one of "theorizing the status quo whose continuation, possibly even stabilization, remains the model" (Albrecht 1986, 33).[2]

Against this cynical stabilization of both the lamentable status quo of Cold War politics and the ontological assumptions underlying most realist writings at that time, Deutsch and his associates presented a counterreading of international relations based on what they called security communities. Although the concept of a security community is about peaceful change at both the domestic level and the international level, I focus on the latter. By so doing, I work in accordance with the application of the concept in the origi-

---

2. See also Ken Booth's assessment that "realists captured the semantic high ground when they tied to their philosophical flagstaff a designation ('Realism') which connotes common sense, concreteness and objectivity. But the semantic flag should not cloak the fact that Realism is not necessarily realistic. Realism is a set of ideas which have hardened into an ideology. . . . Like other ideologies, Realism has sought to legitimize and naturalize the status quo; it portrays as natural and immutable what are believed to be the prevailing characteristics in the system. Thus realists tell us that this is the best of all possible worlds. It must be so since their static view of the world does not allow that what is possible can expand" (1991a, 17).

nal writings and in most recent constructivist adaptations. It has to be noted, however, that Deutsch was not exclusively concerned with the international dimension of security community building and interstate relations. The famous chain of definitions that opens this chapter, for example, does not even mention the nation-state and defines *integration* as the attainment of a sense of community among "a population within a territory." This population does not necessarily have to be a nation, and the territory need not be a nation-state. A security community is "a group of people which has become 'integrated.' " The definition refers to "individuals in a group" who must believe that they have come to agreement that social problems can and must be solved peacefully. It is therefore only one side of the coin to represent Deutsch's explanandum as "the occurrence of groups of (two or more) *states* between whom the thought of war has become, if not impossible, at least displaced by 'dependable expectations of stable peace' " (Wiberg 2000, 289, emphasis added). Likewise, Ole Wæver's interpretation and implicit criticism that Deutsch conceived of security communities "as made up of simply 'states' " (1998, 105) confuses the factual applications of the concept of a security community with its theoretical fundamentals.

A *political community* qualifies as a security community only if it is effectively integrated. Integration depends on "sufficiently widespread compliance habits and other favorable circumstances" (Deutsch 1954, 41). Deutsch clearly states that "political amalgamation need not lead to a security community" and that "even a well-established political community with many amalgamated institutions may fail to function as a security community" (1954, 35, 41). It is therefore incorrect to say that he assumed that all political communities "will generate the assurance of nonviolent dispute settlement" (Adler and Barnett 1998a, 33). If this were the case, theorizing *security* communities would be redundant. In order to qualify as a security community, two things in particular have to be absent from domestic social interaction: preparations for civil war and preparations for secession (Deutsch 1954, 34–35). Only two situations in the Baltic states in the 1990s are occasionally referred to in the literature as autonomous movements: one in the southeastern part of Lithuania when in 1991 parts of the Slavic population of the Vilnius region sought to form an autonomous region within Lithuania (Burant and Zubek 1993, 383–84); the other in the northeastern part of Estonia when in 1993 a referendum on local autonomy was initiated. In the latter case, widespread dissatisfaction with economic developments and the Estonian Parliament's adoption of a law on the status of aliens coincided with a local power struggle, but neither large-scale violence nor a war of secession loomed behind the temporary increase in tensions (D. Smith 2002). Neither situation resulted in autonomy or secession. Thus, neglecting the domestic dimension of a security community in this book and focusing on its international dimension appear to be justified.

Why did Deutsch focus on interstate war and international security communities rather than on intrastate war and domestic integration? Here, it is useful to refer not only to his personal experiences, but also to two features of the international system at the time that he wrote. The first feature is the decline in the nation-state's controllability of international events, as observed by Deutsch: "While most nation-states are in fair control of events occurring within their respective national territories, all in substance have lost control over the external international events upon which their peace and eventually their survival may depend. . . . For much of the world, international relations appear to be characterized by increasing importance and decreasing control" (1954, 25).

The second feature concerns the degree of great-power rivalries and in particular the development of U.S. and NATO strategy at the time that studies on political community were being prepared and written, making global war—general nuclear war—"not just a theoretical possibility" (Trachtenberg 1999, 3).[3] General nuclear war, assumed to be possible at the end of the 1950s, would indeed have meant total devastation. The new NATO strategy, based on the NATO document MC 48 approved by the North Atlantic Council on 17 December 1954, lowered the threshold for the use of nuclear weapons considerably. It unequivocally emphasized nuclear retaliation, tactical as well as strategic—"no strategy up to that point, and indeed no NATO strategy since, placed such a heavy and unequivocal emphasis on rapid and massive nuclear escalation" (159)—and it rejected the possibility of a Soviet conventional attack on western Europe (158–59). Rather, U.S. and NATO officials assumed that if the Soviet Union were to launch a military attack on western Europe, this attack would from the beginning involve nuclear weapons. This assessment led NATO planners to include in MC 48 an ele-

---

3. What follows relies heavily on Marc Trachtenberg's important book on the making of the European settlement after World War II (1999). This work is cited parenthetically by page number in this paragraph and the next.

ment of nuclear retaliation, even preemption in response to a purely conventional Soviet attack, however unlikely such an attack was considered to
be. The bulk of the evidence examined by Marc Trachtenberg suggests that
MC 48 was preemptive also in the stronger sense that a full NATO nuclear
strike might follow the judgment that war was unavoidable, "yet before the
enemy had actually begun military operations" (160). As a consequence, retaliation meant destroying the "enemy force—the force to be used for the *initial* attack, and not just whatever was being held in reserve for follow-up
strikes—while it was still on the ground" (163, emphasis in original). In the
final analysis, then, "massive retaliation . . . really meant massive preemption" (162).

In addition, "throughout the 1950s, NATO did not have a real procedure
for determining when and how force would be used" (168). At the same
time, President Dwight D. Eisenhower's thinking was based on the assumption that the time factor would be crucial—that is, that in the event of war
the national survival of the United States would depend on swift decisions,
made in possibly hours if not minutes. Under these circumstances, the
United States could not afford to be bothered with long negotiations with allies or congressmen or both. From May 1957 onward, "certain commanders—especially the SAC [U.S. Strategic Air Command] commander and the
U.S. commander in Europe—were authorized to make the crucial decisions
on their own" (171) and to circumvent the political authorities if need be—in
other words, if "crucial decisions" had to be made so rapidly that contacting
higher political authorities was deemed impossible. All these factors—a
strategy officially based on massive retaliation, but unofficially on massive
nuclear preemption; the delegation of the authority to wage nuclear war to
the SAC commander and the Supreme Allied Commander, Europe; the deliberate undermining of civil authority and control; and "a full-fledged policy of nuclear sharing" (178) implicit in MC 48's emphasis on rapid action in
combination with weakly developed control patterns—helped significantly
lower the threshold of the use of nuclear weapons at that time.

As an emigrant, Deutsch was possibly more than others aware of the
fact that there would be no escape from a nuclear war by means of emigration. He and his team observed that men have historically often "wanted a
political community that would not merely keep the peace among its members but that would also be capable of acting as a unit in other ways and for
other purposes" (Deutsch et al. 1957, 31). The new age of nuclear weapons,

however, made it impossible to accept the risk of civil or interstate war in order to ensure other gains. It rather demanded "that these risks and gains must be reevaluated." Consequently, the writings on political communities focused on "the keeping of the peace among the participating units" and the realization of peaceful changes in interstate relations (Deutsch et al. 1957, 31).

These changes can be achieved by two forms of security communities, both of which are characterized by dependable expectations of peaceful change. A merging of formerly independent units in one single unit with a common government—what Deutsch called *amalgamation*—is more difficult to attain and preserve than a *pluralistic* security community, that is a community that "retains the legal independence of separate governments" (Deutsch et al. 1957, 6). Furthermore, the amalgamated type has not been found to be more stable than its pluralistic counterpart. In any case, an amalgamated security community is not on the cards for the Baltic Sea region. Even the most ambitious integration project in the region, the enlargement of the European Union (EU), will not include Russia in the foreseeable future. The Council of the Baltic Sea States (CBSS), however, does include all political actors in the region, but because of its organizational principle and mandate it cannot be considered an amalgamated political community. It is reassuring, then, that Deutsch and his associates considered pluralism "the major and most general policy goal to be sought" (1957, 163).

Deutsch, instead of exploring the causes of war, "asked the classical question: what are the conditions of peace?" (Holsti 1991a, 8–9). This question had not been realism's classical question. The ingenuity of Deutsch's approach lies precisely in this modification and in asking "vital questions which previously were left unasked (at least in the context of international relations)" (Bull 1995, 194). As Kalevi Holsti has observed, "Karl Deutsch took the classical problematic of realism and changed it slightly to generate a new research program. Instead of asking, as had most of his predecessors, [W]hat are the causes of war and the conditions of peace[?] (thereby focusing on the antecedents of war and a hypothesized set of non-existing conditions for peace), he asked, [W]hat are the sources of peace *in fact?*" (1991a, 245, emphasis in original). Accordingly, Deutsch saw the creation of the ability both "to establish controls sufficiently effective to preserve the peace" and "to insure [sic] the adequacy and timeliness of political decisions to prevent the outbreak of large-scale violence under any legal or ideological pretext whatsoever" as one of the major tasks of future political communities (1954, 28).

At the international level, a security community is characterized by "stability of expectations of continuing peaceful adjustment" (Deutsch 1954, 40). As pointed out already in chapter 1, however, peaceful change alone does not make a security community. To a large extent, security community building is a cognitive project. It deals with beliefs that may be defined as "propositions that policy makers hold to be true, even if they cannot be verified" (Holsti 1995, 273) and with expectations resulting from social practices, historical experiences, and memories of experiences (Möller 2004). The essence of international security communities is the belief in and the expectation of peaceful resolution of interstate conflicts among populations and policymakers, as well as the renunciation not only of resort to large-scale physical force but also of significant preparations for it: "The absence of . . . advance preparations for large-scale violence between any two territories or groups of people prevents any immediate outbreak of effective war between them, and it serves for this reason as the test for the existence or nonexistence of a security community among the groups concerned" (Deutsch 1954, 34).

The relation between members of a security community and nonmembers is not explored in the original works (creating considerable puzzles that I address later). The original concept of a security community thus is an essentially introverted concept describing the relations within the particular community. Belonging to a security community need not prevent its members from maintaining troops for general disposition and from preparing for large-scale physical force in general, for example against nonmembers. As a consequence, even within a security community the issue of the use of force is usually one of probability, not possibility. However, social groups within a security community take for granted that conflicts will be resolved without physical force. Indeed, within a security community, "war is no longer considered a legitimate way of settling disputes" but is replaced by "negotiation, arbitration, or the courts" (Wendt 1999, 300). The establishment of a security community, then, requires the crossing of a threshold beyond which war between the political units concerned neither appears possible (any longer) nor is being prepared for. The crossing of this threshold is called *integration*. One of the practical consequences of this feature of security community building is that it "allows a period of testing to make sure that we really have a security community before national self-defense is tapered off too far. Countries need not disarm suddenly, or even merge their armed forces, until integration is assured. As the prospect of war becomes gradually more remote, they can withdraw their troops from their common borders, unit by unit, while still retaining troops for more general disposition" (Deutsch et al. 1957, 163).

Once integration is achieved, security communities are capable of overcoming the security dilemma and of proving that the realist claim of this dilemma's omnipresence in international politics is exaggerated: fellow members of the security community will no longer perceive policies aimed at security as threatening. But how can integration be achieved? Deutsch and his associates name only two or possibly three conditions found to be "very important" for a pluralistic security community:

> the first of these [is] the compatibility of major values relevant to political decision-making. The second [is] the capacity of the participating political units or governments to respond to each other's needs, messages, and actions quickly, adequately, and without resort to violence. . . . A third essential condition for a pluralistic security community may be mutual predictability of behavior. . . . Compared with these three major conditions for a pluralistic security community, the remaining nine conditions seem to be less important. (Deutsch et al. 1957, 66–67)[4]

A compatibility of major values results in a sense of community that is indispensable for integration and much more important than the signing of declarations and "verbal assent to some or many explicit propositions." Deutsch and his colleagues pointed out that "[t]he kind of sense of community that is relevant for integration . . . [turns] out to be rather a matter of mutual sympathy and loyalties; of 'we-feeling,' trust, and mutual consideration; of partial identification in terms of self-images and interests; of mutually successful predictions of behavior, and of cooperative action in accordance with it" (1957, 36). A security community, then, is the result of diverse processes of social interaction between states and societies that in

---

4. For the remaining nine conditions, see pages 46–59. In a later part of their study (pp. 123–33), the authors emphasize only two conditions that are "essential" for both types of security communities: "compatibility of major values" and "mutual responsiveness."

themselves need not have anything to do with a search for security, but nevertheless result in a sense of community and a sense of security.

The reading of international affairs presented by Deutsch and his team anticipated many ingredients of the current social-constructivist wave in international relations theorizing. It successfully challenged the realist view of the world in several respects. First, by establishing that durable expectations of peaceful change between states and societies are indeed possible in an anarchic environment, Deutsch called into question core realist assumptions such as the security dilemma.[5] Second, he and his colleagues emphasized the construction of security communities around core areas of strength in terms of political, administrative, economic, and educational advancement (Deutsch et al. 1957, 37–39). Within security communities, nation-states, contrary to the balance-of-power theory, seem to be inclined to associate themselves with centers of strength rather than to balance them out. Third, by focusing on beliefs rather than on material factors, Deutsch rejected realism's mechanical understanding of states' behavior as following from objectively measurable material, especially military, capabilities. Without ignoring structural factors, a societal approach to the study of security, such as the one applied by Deutsch, "puts more emphasis on the political participation and social transformations and, in that sense, the social construction of security" (Väyrynen 2000, 109). Fourth, rather than treating national security as a "symbol [that] suggests protection through power" (Wolfers 1962, 149), Deutsch understood security in terms of community building. Finally, he argued that peace within a security community does not require a Leviathan, but rather a pluralistic framework involving state and nonstate actors (Olson and Groom 1991, 143). In sum, then, the "basic realist assumptions about power, military capabilities and security in inter-state relations do not hold for states within a pluralistic security community" (Starr 1992,

---

5. The notion of the omnipresence of the security dilemma has recently also been challenged from social-constructivist and critical points of view: "Claims about a security dilemma tend to trade on images of ahistorical determination, of structural necessities to which states can only respond as they must. As expressions of the claims of the modern state, however, modern security discourses rest on historical and political judgments" (R. Walker 1997, 66–67). The security dilemma is affected not only by the distribution of power and the anarchical structure of the international system, as realists hold, but also by "collective beliefs that vary independently of power" (Kowert and Legro 1996, 460).

211). Deutsch and his associates indeed produced a "corpus of theoretical and empirical work that stands as a monument of creativity that has successfully challenged major elements of the realist tradition" (Holsti 1991a, 245).

Before addressing selected theoretical developments as well as adaptations of and competitors to the original concept of a security community, I need to discuss three problematic aspects of the original concept. First, the chain of definitions quoted at the beginning of the chapter is not without problems. For example, what does "*long*" mean? The use of quotation marks around the term indicates both the desirability of some temporal qualification and the difficulty of establishing a nonarbitrary measure. The concept of a security community shares this problem with other approaches that focus on qualitative changes in world politics, such as the democratic peace proposition. As one proponent of the proposition puts it vaguely, in order for a government to qualify as a representative government "some period must have elapsed during which democratic processes and institutions could be established" (Russett 1993, 16). Reflecting this temporal problem, integration is said to be "a matter of fact, not of time" (Deutsch et al. 1957, 6). Once people have become integrated, it matters little how long the process of security community building took. Crossing the security community threshold may indeed take decades or even generations, and this threshold may be crossed and recrossed several times. A linear development toward a security community is but one of several possible paths.

With respect to the Baltic Sea region, the span of roughly fifteen years that has passed since the end of the bloc antagonism is in historical terms the blink of an eye. Any assessment, therefore, can only be a tentative one. It is important, however, that the peaceful changes of the 1990s in the region were not interrupted by regression to more violent forms of dealing with conflicts (save for some violent situations in Lithuania and Latvia in 1991). However, without subscribing to the argument used by realism in the last resort to set bounds to alternative interpretations—that international relations are always uncertain—a note of warning may be in order: security community building may fail even after a promising beginning. Likewise, a security community "may not prove stable in the event" (Deutsch 1954, 40).

Second, the term *peaceful change* also requires some qualification. In the original texts, it was defined as the absence of "large-scale physical force," and a security community was seen as one "in which there is real assurance that the members of that community will not fight each other physically" (Deutsch et al. 1957, 5). For the sake of precision, I equate peaceful international change with the absence of armed conflict and define *armed conflict*, following the Conflict Data Project, as "a contested incompatibility which concerns government and/or territory where the use of armed force between two parties, of which at least one is the government of a state, results in at least 25 battle-related deaths" per year (Wallensteen and Sollenberg 2000, 648).[6] In light of these definitions, it seems to be justified to represent international relations in the Baltic Sea region throughout the 1990s as a peaceful change, without deriding the victims of the attacks by Soviet Ministry of the Interior troops on civilians and border guards in Lithuania and Latvia in 1991.

Third, if "the entire world were integrated as a security community, wars would be automatically eliminated" (Deutsch et al. 1957, 5), or, in other words, "each nation's security must be assured through the existence of a community embracing all nations" (Deutsch 1954, 33). The realization of these ambitious aims may indeed be impossible and mere visionary perfection, but the original studies show that the elimination of wars on the regional level and regional security communities are possible: in the North Atlantic area, "subareas of integration" have been said to exist (Deutsch et al. 1957, 9). The application of the theoretical framework to one particular area has resulted in an introverted concept that focuses on the elimination of both war and expectations of war within a clearly defined political community. This approach does not address the intricate question, alluded to earlier, of the relations between members of a security community and nonmembers. Likewise, it ignores the relations between different security communities. It is a slightly irritating ingredient of the original concept that in their relations with nonmembers, members of a security community may very well construct their security *against* rather than *with* others, thus relying on a practice that is transcended within and banned from the security community. The identity of a security community may even to some extent be constructed through the otherization of nonmembers. Clearly, such a proce-

6. *Wars* require at least 1,000 battle-related deaths per year. *Intermediate armed conflicts* are defined by "at least 25 battle-related deaths per year and an accumulated total of at least 1,000 deaths, but fewer than 1,000 per year." The Conflict Data Project is at the Department of Peace and Conflict Research, Uppsala University.

dure would not be an adequate manifestation of a concept that is dedicated to keeping the peace.

Security communities are characterized by dependable expectations of peaceful change among the members of the community, but *not* between the members and nonmembers. Although within the security community the expectation of peaceful change has the upper hand, the relations between the members and nonmembers may well be based on traditional forms of dealing with conflicts, including the use and the threat of use of force. Indeed, the absence of "advance preparations for large-scale violence *between any two territories or groups of people* prevents any immediate outbreak of effective war *between them,* and it serves for this reason as the test for the existence or nonexistence of a security community *among the groups concerned*" (Deutsch 1954, 34, emphasis added). The relations between members and nonmembers may, however, be similar to conventional relations among nation-states. They may, for example, be influenced by the whole variety of misperceptions and mistrust characteristic of international politics in its realpolitik shape (see Jervis 1976). They may also be based on the logic of the security dilemma that is overcome within the security community. Furthermore, save for the dedication to peaceful change within a security community, different security communities need not be based on the same set of values. The values underlying different security communities may not be mutually supportive or may even be mutually exclusive or be perceived as mutually exclusive. The values upon which a particular mutual-response community is based and that facilitate nonviolent conflict resolution within this community may be perceived as threatening by the members of other security communities.

As a consequence, the relations between security communities may very well resemble the conventional relations between nation-states, including misperceptions, security dilemmas, mistrust, self-help, and both the use and the threat of the use of force. This is true especially because, as shown earlier, security communities require the renunciation of preparations for the use of force among the members, but *not* of preparations for the use of force between the members and nonmembers. The security community pattern explicitly permits the members of the community to maintain troops for general disposition. The concept of a security community, therefore, is no "contemporary version of Kant's pacific union" (Sørensen 1992, 406) because Kant's eternal peace among republics has as one of its preconditions the abolishment of standing armies.[7] For the same reason, it is equally un-convincing to represent NATO as a "Kantian security community" (Herman 1996, 306).

Troops within a security community may be perceived by nonmembers (some of whom may be members of other security communities) as at least potentially threatening. Even if the maintenance of military means is in-tended exclusively for defensive purposes,[8] the basic problem remains be-cause usually the issue is not one of intentions, but one of capabilities and perceptions. Furthermore, "offensive and defensive security motives often lead to the same behavior" (Jervis 1982, 359). The security community may thus in the perception of nonmembers be an *in*security community against which military means must be sustained or built up in case something goes wrong. If nonmembers perceive a security community as a potential source of threat, the result will in all likelihood be preparations for a situation in which the perceived potential threat becomes or is perceived as an actual threat—or is constructed as an actual threat for domestic or identity or other purposes. Because of a lack of trust between members and nonmembers of a security community, the members for their part will then also prepare for the worst case, and these preparations may include military means. The re-lations between security communities limited in geographical scope and number of members, especially if they border on one another, and between members of a security community and nonmembers, especially if they are geographically close to one another, are thus likely to be an extrapolation of traditional interstate relations rather than their replacement by relations in-herently more peaceful.[9] The peaceful relations between the members of a

7. For Kant, see the third preliminary article in *Zum Ewigen Frieden: Ein philosophischer Ent-wurf:* "Stehende Heere *(miles perpetuus)* sollen mit der Zeit ganz aufhören" (1984, 5).

8. For a discussion on the difficulties of distinguishing between offensive and defensive conventional weapons, see Jervis 1978, 199–206. Jervis accepts only fortifications and passive resistance as credible means for exclusively defensive purposes: fortifications because they have no ability to penetrate the enemy's land and passive resistance because "it is very hard for large numbers of people to cross the border and stage a sit-in on another's territory" (204). The necessity to require civilian support renders guerilla warfare, too, a kind of defensive warfare that does not seem to be easily exportable.

9. The assumptions begun with "especially" follow from the findings reviewed by Geller and Singer that "at the level of the dyad, the distance between states is inversely related to war-fare. Proximate dyads are more likely to engage in war than are nonproximate states. The re-

security community may thus be thwarted by the lack of (expectations of) peaceful change between members and nonmembers as well as between members of different security communities.

Like all spatially limited security communities, then, a Baltic Sea security community therefore requires at least three features in order to avoid a situation in which the peace potentiality within the community is accompanied or even thwarted by conventional outside relations. First, the community's fringes must be understood in terms of frontiers rather than boundaries because, in geographical terms, "a frontier is a zone of contact, an area," but "a boundary is a definite line of separation" (Paasi 1996, 25). Second, in order to avoid conventional security dynamics between members and nonmembers, one ingredient of the original writings must be problematized—namely, the maintenance of troops for general disposition. It is exactly these troops that nonmembers may see as threatening. Disarmament, starting with withdrawing troops from common borders, is the ultimate proof of the security community members' sincerity as to their commitment to peaceful change (Deutsch et al. 1957, 163).[10] It should also be given some credit for contributing to expectations of peaceful change between members and nonmembers, which ultimately would mean the enlargement of the security community.

Third, another ingredient of the original writings must be defended against social-constructivist adaptation: the compatibility rather than commonality of the values relevant to political decision making (discussed further in the next chapter). In essence, this procedure requires a less introverted understanding of regional security communities. It implies defining the community's security in a way that also takes into consideration nonmembers' security interests. It means thinking of security as an intersubjective process rather than as something performed by states or clusters of states in selfish isolation. It does not mean abandoning one's own security interests. Rather, it means amalgamating one's own interests with

---

sults on contiguity are even more compelling. War within dyads with a land or (narrow) water border is much more likely than between noncontiguous states" (1998, 78). Despite diverse trends toward deterritorialization (Adler and Barnett 1998a, 33), contiguity remains an important issue when it comes to security community building.

10. For the relevance of the demilitarization of borders to security community building, see also Shore 1998.

others' interests. It means defining security with others rather than independent of or against others.

Thus, with respect to the Baltic Sea region, it does not make much sense to divide the region into different groups of states and to explore the position of the particular groups of states on the continuum between security communities and noncommunities (Wallensteen et al. 1994, 38; Engberg et al. 2002, 478). For example, whether Estonia, Latvia, and Lithuania form a security community is not really relevant in the present context.[11] Likewise, treating Russia as an actor external to the region and arguing on the basis of this exclusionary practice that the member states of the Baltic region "have enough confidence that war between them is unlikely, so that they might qualify as potential security communities" (Väyrynen 2000, 126, 128) do not carry the analysis much further.

Security communities in a narrowly defined region (one, for example, without Russia) may be analytically feasible, intellectually conceivable, and perhaps, from particular political actors' point of view, even desirable. Of course, subregional security communities may be a value in themselves (as long as they do not threaten others and are not perceived as threatening by others). Consider, for example, the possibility of an emerging Lithuanian-Polish security community and compare this community in the making with the troublesome bilateral relations in the twentieth century and earlier. As late as 1991, then Lithuanian minister of defense Audrius Butkevicius reportedly referred to Poland as the greatest threat for Lithuania (Miniotaite 1998, 168), but since then both states have developed a relationship that they officially refer to as a strategic partnership. Moreover, each subregional security community may become a building block for a regional community. An all-Baltic security community does not, however, automatically follow from the existence of a handful of subregional security communities. Pluralistic security communities in parts of the region are unlikely to produce expectations of peaceful change in the whole region as long as the relationship between these security communities and nonmembers remains unsettled and as long as the renunciation of preparations for the use of force among the members of the security communities is not accompanied by a similar renunciation of military preparations toward those whom they still consider

11. For a pessimistic assessment, see Mouritzen 2001, 304; for a critique of this assessment, see Möller 2003a, 320.

nonmembers and vice versa. Subregional security communities in the Baltic Sea region may even be the condition of impossibility for a security community in northeastern Europe because they may be seen as *in*security communities by those states and societies excluded from them and because they may even construct their identity and communityness against nonmembers. Therefore, I emphasize security community building in the whole Baltic Sea region or northeastern Europe.

3

Adaptations

SINCE THE END OF THE COLD WAR, the security community pattern has gone through an intellectual revival. This is not to say that the concept was completely ignored in scholarly writings on international affairs during the Cold War.[1] Yet the Cold War, besides its military and political elements, was also a cultural conflict in the sense that the interpretative modes established in think tanks and university departments helped establish borders, intellectual and otherwise. On the basis of rigid politico-intellectual barriers, scholars doing security studies and international studies established a self-perpetuating production of a particular kind of knowledge that privileged a particular reading of the world—statist, militarized, masculinized, and ethnocentric—that was difficult to penetrate, both from within and from without (Booth 1997, 111). Although the potential regarding peaceful change inherent in the concept of a security community was acknowledged in some scholarly writings (Holsti 1985, 33), it remained largely unexplored. Even then, however, security communities transcending the military blocs were possible logically and theoretically, as were security communities encompassing members of both blocs. All the same, the concept of a security community was largely an unthought of and politically even an undesired idea.

The end of the Cold War[2] no doubt facilitated the thinking about inter-

1. See, for example, the reviews by Haas 1958, Thompson 1958, Hoffmann 1959, and Mason 1959.

2. For the sake of convenience, I follow a conventional representational mode here according to which the Cold War has come to an end with the dissolution of the Soviet Union and with the end of the military, political, economic, and ideological conflict between "the West," in particular the United States, and "the East," in particular the Soviet Union. For the argument that the mechanism of otherizing that characterized the Cold War still prevails in foreign and security policy, see Campbell 1998 and chapter 11 in this book.

national affairs in other than realist terms and helped bring to the forefront those approaches that during the bloc antagonism had been suppressed by the hegemony of realism. Thus, although it may be an exaggeration to represent the end of the Cold War as the condition of possibility for unfolding the potentiality inherent in the security community pattern, there has surely been an intellectual and political renascence of the concept, although it is still excluded from many surveys of the concept of security and therefore also banished from the collective memory of the discipline of international relations.[3] The concept has been applied to regional patterns that were not covered in the original writings[4] and has been theoretically adapted and further developed in the light of social constructivism and other approaches.[5]

As Unto Vesa has pointed out, Deutsch's original approach seems to be the *least* ambitious among a series of related studies in the 1950s focusing on the European unification process. Security communities in their pluralistic version do not require two or more individual units to form a new unit, nor do they demand the construction of a supranational organization to replace nation-states. At the same time, the concept of a security community is certainly the *most* ambitious one because integration, as defined by Deutsch, is about the emergence of dependable expectations of peaceful change rather than about institution building. It is thus "of universal concern and relevance." To explore the prospects of integration in Deutsch's sense is especially important in the current European setting, which, according to Vesa, is characterized by "both integration and disintegration processes," in both meanings (1993a, 160, 164).

Many post–Cold War policymakers have identified rhetorically with Deutsch's conceptual terminology—in particular the attractive combination of *security* and *community,* with the latter often being equated with common norms and values. Some scholars even envision the emergence of "a transnational community of *Deutschian* policy-makers" to replace their realism-inspired predecessors (Adler and Barnett 1998b, 4, emphasis in original). Others are more critical of what they regard as the frequent inclination

3. For example, Haftendorn 1991, Walt 1991, Kolodziej 1992, and Baldwin 1995 do not refer to Karl Deutsch and security communities.

4. With respect to the Baltic Sea region, see Wallensteen et al. 1994; Väyrynen 1998, 167–70; Ozolina 1999, 15–16; Vesa and Möller 2003.

5. Adler and Barnett 1998a, 1998b; Barnett and Adler 1998; Mouritzen 2001.

in the political discourse to lose the distinct elements of a security community, and they warn against confusing security community as an analytical concept with security community as a political project (Vesa and Möller 2003, 17–18). Behind the rhetorical smokescreen of community-language, realism-inspired policies—rhetorically legitimized in the language of community—may very well continue, unnoticed by many observers. What is required, then, is "for the scholar to get behind the social mask of the subject" without ignoring the possibility that the front-stage behavior may very well influence the back-stage behavior, "in ways that are neither intended nor fully understood by the actors themselves" (Sagan 1993, 257).

Furthermore, academic interpretations of the original concept are not always helpful. I have already noted the tendency to equate security communities with interstate relations. Ole Wæver, moreover, equates Deutsch's security communities with what he calls "non-war communities" (1998, 71, 104) and disregards Deutsch's distinction between no-war communities and security communities.[6] In this type of conflation, terminological and conceptual clarity is rendered difficult. According to Emanuel Adler and Michael Barnett, the "distinctive feature of a [pluralistic] security community is that a stable peace is tied to the existence of a transnational community" (1998a, 30–31). "Stable peace" and, as in Deutsch's original work, "expectations of stability" are surely two different things. By linking a transnational community with stable peace, Adler and Barnett thus go beyond Deutsch, who insisted on the possibility that security communities, although based on expectations of stability, "may not prove stable in the event" (Deutsch 1954, 40). Håkan Wiberg designates as one of Deutsch's basic assumptions the idea that security communities "evolve from military alliances based on common threats" (2004, 210), although Deutsch stated that military alliances in and by themselves "[do] not seem to be very helpful" to integration. Indeed, they seem to be "relatively poor pathways" toward integration, and foreign military threats are often found to be helpful to but not essential for integration (Deutsch et al. 1957, 202). Barry Buzan thus correctly states that many scholars refer to security communities "without looking too far be-

6. Deutsch refers to the League of Nations as "a no-war community. . . . In contrast to a security community, however, the possibility of war is still expected and to some extent preparations are made for it. Sanctions may include continuing defensive preparations for self-help by members" (1954, 41).

yond the basic definitions" (2000b, 154). However, Buzan himself is not even strict about the basic definitions: by equating regional integration with an amalgamated security community (2000a, 3), he ignores that integration may also result from pluralism.

What, then, accounts for groups of people moving toward a security community? Important preconditions for security community building are often said to be found in the economic sphere, following a logic according to which "the evolution of mutual, binding economic ties promotes stable peace" (Väyrynen 2000, 108). Deutsch and his team, however, emphasized that strong economic ties among the units to be integrated were helpful, but not essential (1957, 157). With respect to pluralistic security communities in particular, they saw "no real evidence that the expectation of economic gain . . . is widespread in the sense of mass feeling" (1957, 144). Thus, the increase in trade relations among the Baltic Sea littoral states, the growth in foreign investments, and the diversification of trade patterns during the 1990s are not likely to be automatically translated into a sense of community. Conversely, the economic competition between Estonia, Latvia, and Lithuania with respect to, for example, transit to and from Russia does not make the development of a security community impossible. Neither does Russia's economic weakness in the 1990s nor the asymmetry in the patterns of trade and foreign investments that can still be observed (Polkowski 2000). In any case, although economic transactions may be helpful to security community building, their importance should not be overemphasized.

The writings of the 1990s on security communities introduced many fresh ideas. However, no clear picture has emerged from them. It is important to remember that, regardless of the end of the Cold War, the threat of nuclear annihilation has not disappeared entirely. Thus, the insights motivating the original study on security communities as sketched in the previous chapter are still to some extent valid. It is also important to note that security communities do not necessarily have to be the result of processes of social interaction, in the course of which groups of people consciously and deliberately aim at security. This path to a security community may very well be the exception. Thus, the idea that groups of people are consciously moving toward a security community is somewhat misleading. Rather, security communities are the outcome of an increase in social interaction, the aim of which is either security or something else, and it is a task for empirical analysis to find out whether a given process of security community building is security driven. The EU serves as an example of the first option (conscious movement toward a security community), whereas the security community encompassing the Nordic states arguably exemplifies the second option (Browning and Joenniemi 2004, 241). Both processes have resulted in a sense of community and therefore security. Security communities may be initiated by both state actors and nonstate actors, with, again, the EU exemplifying the first option and the Nordic states illustrating the second option (Wiberg 2000; Mouritzen 2001).

Security communities do not require common formal institutions, but formal institutions may support the process of security community building to an extent unforeseen by Deutsch and his colleagues (Vesa and Möller 2003, 161, 168). They may provide integration capabilities and thus help state and other actors come to terms with the integration load resulting from an increase in social interaction (Deutsch 1954, 43). However, common institutions may also follow the emergence of a security community rather than the other way round (Wallensteen et al. 1994, 35). Initial steps toward a security community may not be immediately apparent as steps toward a security community and may be identified as such only with the benefit of hindsight. Formal and multilateral security organizations have proven to be weak in or entirely missing from the Nordic security community (Wæver 1998, 74), but they are now claimed to be relevant to security community building by dint of their capability to "reflect a belief that security is interdependent and should be overseen by a collective body" (Adler and Barnett 1998a, 52). In Deutsch's writings, military alliances were not found to be particularly suitable vehicles for security community building, and especially excessive military commitments were found to have disintegrating effects (Deutsch et al. 1957, 190–91). However, in these writings it was already noted that a military alliance such as NATO may evolve into a security community by transforming itself into a primarily political, economic, and social organization (Deutsch et al. 1957, 203). Alexander Wendt has resumed this idea by arguing that NATO as a military alliance that survived the disappearance of its chief enemy without itself disintegrating should be considered a "collective security system with an expectation of permanence" (1999, 302). This interpretation does not automatically make NATO a security community, however, as other authors are inclined to argue (Risse-Kappen 1996).

From a social-constructivist and path-dependent point of view, Adler and Barnett have suggested a three-stage model of security community building that wisely abstains from claiming to be *the* theory of security community building. By arguing that a security community may be the result of the initially implicit but subsequently deliberate actions of policymakers who indeed have the desire to develop such a community, the authors limit their conceptualization to target-oriented strategies by policymakers. They thus sideline the idea that security communities may equally well be the inadvertent, incidental result of policies aiming at very different objectives and that they may occur casually in connection with or resulting from something else. As these authors acknowledge, their model presents only one particular conceptualization. It is therefore not necessary to sketch their theory in detail here. Neither does it make much sense to criticize them for particular ingredients of their model—such as the preference assigned to the role of governments rather than to the role of civil society in the initial phase (Adler and Barnett 1998a, 50), which follows a rather conventional understanding of security communities as resulting from state actors' activities.

Their claim, however, that among the many possible trigger mechanisms for the initiation of what slowly develops into a security community there is a "mutual security threat" (1998a, 50) is irritating. In the historical case studies underlying the original writings, the effects of foreign military threats were usually transitory. They often resulted in military alliances, and military alliances may or may not evolve into security communities. In fact, "more permanent unions derived their main support from other factors," and although foreign military threats "at times served a useful purpose," they sometimes had "exactly the opposite effect: they induced a state of fear or at least of intense preoccupation among the political elites of the privileged political unit and rendered them less able or less willing to pay attention to the needs of weaker or less privileged units, or to make concessions to them" (Deutsch et al. 1957, 45). It may thus be suggested here (and is discussed at some length in chapter 10) that the "disappearance of the security argument" (Joenniemi 1998, 69) may be as important to the evolution of a security community as the existence of the security argument; it may possibly be even more important.

In the 1990s, the academic discussion of the concept of a pluralistic security community also revolved around the question of the concept's proximity to the democratic peace proposition. For example, Harvey Starr argues that the democratic peace is a more narrow construction than a pluralistic security community (1997, 153).[7] This is so because, as Adler and Barnett (1998b, 13) emphasize, Deutsch's security community pattern is not linked with or dependent on a specific type of state organizing principle. Thus, security communities can also be possible outcomes of the social interaction between states and societies that do not follow those liberal-democratic forms of political organization celebrated in the literature on the democratic peace, but that are nevertheless in accordance with one another regarding the peaceful resolution of conflicts.

Moreover, the democratic peace proposition does not require the renunciation of preparations for the use of force, but merely the empirical absence of war among democratic polities. Bruce Russett, however, suggests that *"to use or threaten to use force is not usually normatively acceptable behavior in disputes between democracies,* even in the form of symbolic, ritualized bargaining behavior" (1993, 42, emphasis in original), thus moving security communities and liberal democracies closer to each other. Democratic peace thus seems to be a possible path to a pluralistic security community, the more so because a security community is usually not, as Starr proposes (1997, 155), characterized by the absence of the military option, but rather by the expectations, shared by the members of the security community, that the available military means will not be used against them. That liberal democracies may form a security community does not, however, mean that democratic peace is the only path along which a pluralistic security community may emerge. Indeed, security communities are a difficult case for theory building. There seem to be as many ways to a security community as there are security communities. The possibilities to model future security communities after existing ones appear to be limited.

Adler and Barnett, while reviewing the original writings on security communities, claim that "states within a pluralistic security community possess a compatibility of core values" (1998b, 7). In developing a new framework for the study of security communities, the authors then place emphasis on a "shared structure of meanings and identity," which renders expectations of peaceful change possible. Digging into sociological and

7. The democratic peace proposition basically claims that *"there is a virtual absence of war among dyads of democratic polities . . .* and is meant to explain the lack of war between democracies" (Starr 1997, 153, emphasis in original).

philosophical writings on communities, they identify as a characteristic of a community that its members "have shared identities, values, and meanings." According to Adler and Barnett, "there is no *a priori* reason why [the qualities of a community] should be limited to the territorial state." A pluralistic security community, defined by the authors as "a transnational region comprised of sovereign states whose people maintain dependable expectations of peaceful change," is thus linked to the existence of structures of shared values among its members (1998a, 30–32). They declare again later that "compatibility of core values and a collective identity are necessary for the development of security communities" (58).

This claim deviates from the original concept, which emphasizes only compatible values and mutual responsiveness. Given the social-constructivist nature of Adler and Barnett's research, it is not surprising that they focus on community and identity, but by doing so they change the concept of a security community considerably and limit possible paths to a security community to those paths resulting in shared norms and values across the board. In the original concept, the members of a security community need not have values in common other than the belief in peaceful change; they have to identify with one another only in terms of security. This identification may be achieved through community building and shared identities, but community building is not the only path to a security community. A collective identity among the members of a security community is but one of several possible manifestations of such a community.

As early as 1994, Peter Wallensteen and associates, while exploring the prospects for a security community in the Baltic Sea region, to some extent anticipated this shift of emphasis. They concluded that uncertainties remain regardless of a growth in commonality of values and responsiveness. In particular, they saw individual routes to military security among the Baltic Sea states, developments in the Russian Federation, and weakly developed institutions as obstacles to security community building. By linking "emerging" security communities in the Baltic Sea region with, among other things, the "expectation that democratic conditions are durable" (Wallensteen et al. 1994, 39), and by equating common values with new democratic and market institutions, the authors linked security communities closely to the democratic peace proposition. Here, the issue is neither one of compatible values nor one of common values, but rather one of common liberal-democratic values and institutions.

This linkage disregards the idea that expectations of peaceful change are not necessarily dependent on the organization of states and societies along liberal-democratic lines. Indeed, "non-democracies are not destined to external aggression or internal war; they too can live in peace and be members of a security community" (Väyrynen 2000, 118).[8] Security communities thus can emerge among states and societies that do not share liberal democracy in its western European style as an organizing principle, but that nevertheless agree on one fundamental thing—namely, that mutual conflicts must and can be resolved without recourse to large-scale physical force. Apart from peaceful change, the issue within security communities is one of compatible rather than common values. Compatible values means values that are tolerant of one another and capable of coexisting, values that are not mutually exclusive, values that are mutually tolerated by their carriers. Even values that are mutually exclusive may not be a problem as long as their carriers do not exhibit a missionary attitude. The concept of a security community, thus, is equally interested in identity and difference (Deutsch et al. 1957, 123–26; Möller 2003a, 317–21). Furthermore, emphasizing compatibility of values rather than commonality renders possible the regarding of different sets of values as equally valuable. That a security community allows and appreciates difference is politically important: a security community exhibiting a tolerant attitude toward difference in its internal relations can be expected to be perceived by nonmembers as less threatening than a security community that internally displays intolerance by, for example, "nudg[ing] and occasionally coerc[ing] others to maintain a collective stance" (Adler and Barnett 1998a, 39).

Furthermore, it has been argued that EU and NATO norms may exert "disciplining effects" on EU and NATO candidates (Mouritzen 2001, 305). This may be so.[9] The result, however, can be a relationship based on values

8. Väyrynen continues by saying that the probability of external war and internal instability is higher for nondemocracies than for democracies.

9. In political reality, there can often be observed a "pattern of deftly miming Western rhetoric and playing by the so-called Western rules while producing political effects different from those envisioned by Western benefactors." As a consequence, there is a "growing body of research that approaches post-socialist transformations not as a wholesale adoption of norms but as a selective appropriation of narratives to advance specific political strategies of specific political groups in post-socialist countries" (Kuus 2002, 303).

that are internalized only superficially, adhered to either to avoid punishment of exclusion or to bolster self-interest, but neither because the "actor fully accepts [a norm's] claims on himself" nor because the "actors identify with others' expectations, relating to them as a part of themselves" (Wendt 1999, 272–73). The result is at best a weak social relationship, insufficient for the development of mutual trust and an enduring we-feeling necessary for security community building.

Furthermore, a reference to Wendt's "master variables" for collective identity as regards security, which he discusses in the context of regional integration theory, is useful here. Wendt identifies, among other things, *homogeneity* as an "efficient cause of collective identity formation" but adds two important caveats. First, "as actors become alike along some dimensions they may differentiate themselves along other, even trivial, ones in a 'narcissism of small differences.'" Second, "as actors become more alike there is less potential for a division of labor between them," thus reducing the degree of interdependence and the perception of a common fate among them, which in turn inhibits the formation of a collective identity. Wendt argues that as a consequence "there is little theoretical reason to think that a convergence of corporate and even type identities will in itself generate prosocial security policies and thus collective identity. And there is much evidence to the contrary" (1999, 353–57).[10]

At first sight, then, the original emphasis on compatibility seems to be less demanding than the focus on commonality in recent writings on security communities, but in a sense it is more ambitious. The original concept may be said to have thought of security communities as political and social spaces where multiple norms, values, and identities meet and potentially interact with one another in a positive manner. This conception necessitates the readiness to endure tensions and conflicts and requires the willingness to develop a political space that "works *with* difference and not by *reducing* difference" (Couldry 2000, 21–22, emphasis in original). If we see conflict as a "natural and inevitable part of all human social relationships" (Laue 1987, 17), and if we define it as, among other things, a "struggle over values" (Coser 1956, 8), then this reading of security communities takes seriously the

---

10. *Corporate identities* refers to intrinsic, self-organizing qualities that constitute actor individuality; *type identities* refers to shared characteristics, intrinsic to actors, with social meaning given by membership rules (Wendt 1999, 224–27).

multiplicity and complexity of values, identities, and differences within and between groups of people. Indeed, "people identify themselves with several territorial communities, stories and discourses at the same time" as well as "with a number of groups, distinguished by more or less visible social, cultural and spatial boundaries" (Paasi 1996, 46–47).

Furthermore, conflict "has stabilizing functions and becomes an integrating component of the relationship," provided that the basic assumptions upon which the relation is founded are not contradicted (Coser 1956, 80). In a security community, the basic assumptions are nonviolence and peaceful change. In fact, emphasizing compatibility of values appears to be a way to conceive of security communities in a nondominative and nonhierarchical way, which, in turn, may help mitigate the inside-outside *problematique* alluded to in the previous chapter, which is inherent in every spatially limited conception of a security community.

The capability to engage in security community building requires a capacity to deal with complexity in a non-Manichaean and nonantagonistic manner. Complexity theory, as applied to the Baltic states by Walter Clemens, emphasizes, among other things, democratic self-organization, coevolution of one actor with others, and trust as necessary conditions for cooperation for mutual gain. It "converges with liberal peace theory and a normative theory of mutual gain grounded in interdependence" and parallels the democratic peace proposition in that it holds that "a capacity to meet complex challenges arises from creative self-organization, liberated from destructive anarchy and from the dead weight of rigid hierarchy." Human beings are not prisoners of zero-sum logics, but may indeed seek to create values with others for mutual gain: " 'value-creating' assumes that the pie can be expanded, permitting each side to gain ever more" (2001, 16, 245, and 17). But "value creating" can be pursued along different paths; in the sphere of security, for example, it can lead to collective security or collective defense arrangements. In the long term, economic, political, and cultural cooperation may result in a sense of community strong enough for security community building, but security cooperation may also give preference to "value creating" by the joining of an alliance, which may alienate those units that are excluded and thus render difficult the building of a security community with them. Complexity theory posits that people have to deal constructively with complexity if they wish to survive, but how to deal with complexity, besides avoiding both rigid order and chaos, cannot be prescribed by the theory. Likewise, the theory also cannot say which security issues are more important than others or how to respond to them. It cannot say whose security should be promoted and how. Accordingly, Clemens discusses ten different possible ways to enhance Baltic security—without, however, suggesting security community building. This gap is surprising because the concept of a security community obviously is a very effective way to deal with complexity. Furthermore, Clemens's reference to the psychological concept of the use of graduated and reciprocal initiatives in tension reduction in order to "break vicious cycles of distrust" (2001, 203) may be usefully applied to security community building. However, by discussing some of the approaches as "way[s] to reduce dangers from the East" and by representing Russia's policy toward the Baltic states in the 1990s as "economic, psychological, and political warfare on the Balts" (2001, 219, 203), Clemens reveals his implicit skepticism about the perspectives for security community building in the region.

Whereas the equation of security communities with liberal democracies has considerably narrowed the original concept and the application of complexity theory to the Baltic states has ignored the concept of a security community, Adler and Barnett have modified and expanded the concept of a pluralistic security community by adding both a postsovereign element and an element that constitutes the collective security paradigm: mutual aid. They differentiate between two forms of pluralistic security communities: loosely coupled ones and tightly coupled ones. Whereas loosely coupled security communities exhibit the basic ingredient of security communities— that is, dependable expectations of peaceful change—tightly coupled ones are even more ambitious than Wendt's friendship pattern (introduced in the next paragraph). For tightly coupled security communities, we can add to the basic ingredient two more features: first, and closely connected with Wendt, "they have a 'mutual aid' society in which they construct collective system arrangements." Second, "they possess a system of rule that lies somewhere between a sovereign state and a regional, centralized, government; that is, it is something of a post-sovereign system, endowed with common supranational, transnational, and national institutions and some form of a collective security system" (Adler and Barnett 1998a, 30).

Wendt is less ambitious than Adler and Barnett but still more ambitious than the original concept of a pluralistic security community. According to him, the role structure of "friendship" in international politics—as opposed to the role structures of "rivalry" and "enmity"—is based on two independent yet equally necessary rules: "Friendship is a role structure within which states expect each other to observe two simple rules: (1) disputes will be settled without war or the threat of war (the rule of nonviolence); and (2) they will fight as a team if the security of any one is threatened by a third party (the rule of mutual aid)" (Wendt 1999, 298–99). Although the rules are derived directly from Deutsch—note in particular the emphasis on expectations—the second rule is more associated with the collective security paradigm.

Taken together, both rules are congruent neither with security communities nor with systems of collective security. A pluralistic security community is neither a system of collective security nor a military alliance—that is, a system of collective defense.[11] Within a pluralistic security community, nonviolent ways of settling disputes have replaced violent ways. Being a member of a security community, however, does not oblige the particular member to come to the assistance of its fellow members if these fellow members find themselves under attack by nonmembers or even by members, which is against the logic of a security community, but nevertheless possible. If a security community were to oblige its members to mutual assistance, then one of the classical examples of pluralistic security communities in the original text, the Swedish-Norwegian pluralistic security community, would be no security community at all given the reluctance on the part of the Swedish government to assist Norway militarily in World War II. Sweden and Norway nevertheless serve as an example of a pluralistic security community precisely because it does not require its members to assist in case of an aggression against one of its members.

A system of collective security is based on the promise of mutual assistance in case of an attack carried out both by members of the security system and by nonmembers, rather independent of the question of whether the assisting member is as threatened in a given situation as is the member to whom assistance is given. Collective security thus is based on the notion of

11. Wendt defines *friendship* as a combination of "security communities" and "systems of collective security." He differentiates these three categories from systems of collective defense—that is, military alliances—by stressing the temporally open-endedness of "friendship." An *alliance*, in contrast, is defined as "a temporary, mutually expedient arrangement within rivalry, or perhaps enmity" that does not last forever (1999, 299).

universal assistance quite regardless of whether one's own security is or is perceived to be at stake under the given circumstances. It is not, however, based on the assumption that disputes will generally be settled without recourse to violence. If such an assumption prevailed, collective security would not be necessary. Precisely because the use of violence is not ruled out, potential aggressors are deterred by means of the threat of collective punishment, while potential victims of aggression are protected by the promise of mutual assistance.

Whereas the collective security paradigm has usually been associated with a worldwide system such as the UN, Wendt suggests that collective security can also function on the regional level. He thus bridges the gap between the emphasis on regional integration in the original empirical applications of the concept of a security community, on the one hand, and collective security's original emphasis on global security, on the other. By so doing, he solves the problem that a pluralistic security community, strictly speaking, does not require a mutual aid mechanism and that a mutual aid mechanism is necessary only as long as there is no security community. Yet regional integration may very well require a commitment to mutual assistance in case of an aggression from outside the region. By regionalizing both the concept of a security community and the concept of collective security, he reconciles them with each other. Wendt sees two reasons for introducing regional systems of collective security: "one is that states may operate on an 'all for one, one for all' basis within relatively autonomous regional subsystems or security complexes, but not with outsiders. . . . The other possibility is that even when a balance of power system dominates the global level, states within each bloc might collaborate not because they perceive the other bloc as a threat to their individual security, but because they believe in a team approach to security with the members of their bloc" (1999, 301).

Wendt's role structure of friendship (defined next) is important for other reasons apart from its emphasis on regional security and its proximity to the concept of a security community. I use it in chapters 9, 10, 11, and 12 to analyze the state of the relations between the Baltic states and the United States as well as between the Baltic states and the Russian Federation. It is therefore necessary to spend some time on a systematic reading of Wendt's typology of cultures of anarchy, one of which is the Kantian anarchy based on the role structure of friendship.

Like any role structure in the international system, friendship is a property of the system. The system is an anarchy conventionally defined by its lack of centralized authority. Understanding the international system in terms of anarchy may be "old" and "trite," but it is "nevertheless sometimes useful" (Hurd 1999, 399), especially if it is decoupled from the structural-realist notion that anarchy in itself necessarily leads to a specific outcome (Waltz 1979). As Wendt claims, the absence of centralized authority in itself does not necessarily have specific effects, such as a balance of power or permanent insecurity. Neither does it preordain or necessitate a specific behavior of the units—that is, nation-states—such as searching for security in alliance systems. Thus, according to Wendt, "anarchy *as such* is an empty vessel and has no intrinsic logic; anarchies only acquire logics as a function of the structure of what we put inside them. . . . Anarchy is a nothing, and nothings cannot be structures" (1999, 249, 309, emphasis in original). To fill the "empty vessel" anarchy with meaning, Wendt, without claiming exhaustion, establishes three kinds of anarchy based on different role structures—in other words, on different cultural representations of the Other stemming from socially shared knowledge in terms of which the Self defines its own role identity. The structures are distinguished from one another in respect to "shared understandings governing organized violence" (1999, 313) and the resulting subject positions.

The shared understanding that governs organized violence in what Wendt calls a Hobbesian anarchy is the expectation that war may occur at any moment and that it is about the survival of the units; the Other's sovereign right to existence is not recognized. Accordingly, the subject position in a Hobbesian culture is the subject position of an enemy. The shared understanding that governs organized violence prevalent in a Lockean anarchy is that violence may occur, but neither at any moment nor in order to eliminate or dominate. Rather, violence may serve narrower interests and does in principle respect both the Self's and the Other's right to sovereign existence. The subject position in a Lockean anarchy thus is one of a rival. The subject position in a Kantian anarchy, finally, is the subject position of a friend. Friendship patterns here are characterized by both nonviolence and mutual assistance and are based on the principles of nonintervention and sovereignty.[12] Representations of the Other follow from the structure of the inter-

12. See pages 259–78 for the Hobbesian anarchy, pages 279–97 for the Lockean anarchy, and pages 297–308 for the Kantian anarchy. Note that in Wendt's use these labels are meant as

national system. Considering one another as friends, for example, follows from specific, Kantian characteristics of the international system that are not present in what Wendt calls Hobbesian and Lockean structures. Role structure is a property of the system, whereas role identity is a property of the agents, but is constituted by the specific features of the particular anarchy.

According to Wendt, states are to be considered friends if they represent the structure of the international system in a way that makes them base their mutual relationships on both the rule of nonviolence and the rule of mutual aid. There will be friendship only if both rules, which are independent of each other, are observed. Friendship is about national security and does not rule out conflict or competition in other issue areas. Like the concept of a security community, then, it requires the various parties' identification with one another only in terms of security. It may be added that, strictly speaking, friendship—like a security community—does not rule out conflict in the realm of national security, but guarantees treatment of conflict according to the rules. Friendship is not temporally limited, although, like any social relationship, it may end at some time or other.

Understanding interstate relations in terms of friendship may be criticized as pushing anthropomorphism to extremes, but it is no more anthropomorphic than understanding interstate relations in terms of enmity, an approach that has a long tradition in international studies and peace research. Furthermore, foreign policy representatives do refer to other states as friends, and there is no reason to treat these references a priori as mere rhetoric devoid of substance. Cultivating friendship seems to be a part of the construction of interstate relations and security. Exploring the substance of friendship designations should therefore be a part of analyzing interstate relations and the construction of security.

Now, it might be argued that understanding interstate relations in terms of friendship is naïve, idealistic, and potentially even detrimental to national security because, as structural realists are inclined to argue, the anarchic structure of the international system forces states, if they wish to survive, to realize their interests quite regardless of friendship and without much consideration for friends. Furthermore, as Wendt admits, "history

---

"metaphors or stylized representations" and do not claim close adherence to the views of Hobbes, Locke, and Kant (247).

suggests that few states remain friends for long anyway" (1999, 298), and yesterday's friends may very well be tomorrow's rivals. Yet, at the same time, in international politics interstate relations have evolved that do not seem to fit into the classical realist pattern (e.g., United States and Canada; Germany and France; the Nordic states) and that therefore have to be explained by means of different theoretical approaches. In all these cases, the lack of violence in interstate relations did not follow from a lack of opportunities for violence, but rather from a deliberate decision to deal with conflict in a peaceful manner. Finally, the friendship pattern may be criticized from the rationalist point of understanding states as "self-interested utility-maximizers" (Wendt 1999, 298). In this argument, there is a place for considerateness toward friends, but only as long as it does not run counter to self-interests. As Wendt claims, however, incorporating the interests of others into one's own interests need not mean sacrificing one's own interests. It may mean expanding them.

Friendship patterns—like cultural-behavioral patterns in general—may be internalized to different degrees. Wendt differentiates three levels of internalization that have to be treated as ideal types rarely found in pure isolation (1999, 266–78). Put simply, the first level of internalization is about force, the second about price, and the third about legitimacy. Force means that a state behaves according to a certain set of behavioral patterns because if it did not do so, punishment would follow certainly and immediately. Here, internalization is weak or even nonexistent. Behavioral patterns are observed because nonobservance would result in counterreactions. In the absence of the certainty of punishment, the behavioral patterns would in all likelihood not be observed. Price is closely associated with a rationalist understanding of political behavior in terms of a cost-benefit ratio. Behavioral patterns are observed if and only if the observance is considered to be in one's self-interest. The observance of behavioral patterns and of the values for which they stand does not follow from having assigned to them intrinsic worth, but rather from seeing them as a vehicle with which to further one's own interests. If the observance of certain behavioral patterns is deemed detrimental to self-interest, then nonobservance is likely to follow. This conception implies that on the second level of internalization the actor has a reasonable choice whether to accept certain values, whereas on the first level the choice is between only observance and punishment.

On the third level, finally, behavioral patterns are perceived as legitimate, not only as useful for furthering self-interest or out of fear of punishment. They are followed quite regardless of whether they are seen as serving one's interests in a given situation. This adherence to behavioral patterns does not mean that self-interests are sacrificed in favor of altruistic behavior and observance of general values and cultural patterns. Rather, it means that one's interests are perceived as being in good hands within the general cultural patterns even if in the given situation they may be impeded by observing general rules. On the level of legitimacy, values are accepted and respected as such: legitimacy, according to Wendt, means "that states identify with each other, seeing each other's security not just as instrumentally related to their own, but as literally being their own" (1999, 305). The model of degrees of internalization is constructed as concentric circles building upon one another rather than as separate phases secluded from and relieving one another.[13] From this understanding, it follows that even within friendship patterns that are in principle based on third-degree internalization, the reasoning in individual cases may ensue from second-degree logic in terms of instrumentality and self-interest or perhaps even from first-degree logic.

As noted earlier, security communities are based on the value of nonviolence and dependable expectations of peaceful change. These expectations are based on the compatibility of major values relevant to political decision making, resulting in a we-feeling and a security identification among the fellow members of the particular security community—that is, each person identifies his or her security with the security of the other member(s) and vice versa. Within security communities, at least the value of nonviolence in dealing with mutual conflicts is internalized to the third degree and finds its expression in the lack of preparations for and expectations of warfare against fellow members of the community. Seen in its ideal-typical form, the other's security interests are amalgamated with one's own interests and vice versa: "International interests are now part of the national interest, not just interests that states have to advance in order to advance their separate national interests; friendship is a preference over an outcome, not just a prefer-

13. The fact that "internalizing a culture involves the formation of a collective identity should not blind us to the possibility that egoistic identities may still be important. The picture here is one of 'concentric circles' of identification, in which the nature and effects of collective identity vary from case to case, not one of altruism across the board" (Wendt 1999, 338).

ence over a strategy" (Wendt 1999, 305). The rules are observed even though in individual cases nonobservance would result in a preferable outcome. They are observed because they are represented and understood as intrinsically worthy. Friendship patterns are based on the expectation of stability, but—as in the case of security communities—disillusionment may follow. Likewise, although patterns of friendship imply the expectation of endurance, they may not prove stable in the event. Thus, they are about probability. To some extent they are speculative: one expects the other(s) to behave according to the rules; one reckons on the other(s) coming to one's assistance; one even "knows" because it is socially defined as reality that the other(s) will not use violence to resolve mutual conflicts, yet there is no 100 percent certainty. Friendship patterns are even more uncertain than security communities because the we-feeling within security communities is accompanied, strengthened, or even created by a lack of preparations for intra-community warfare, thus giving it a material base.

Assuming the role position of a friend may follow from a specific, Kantian representation of the international system. However, at the foreign policy level, a different interpretation may also be possible. Others may be represented as friends because the international system and its subsystems are *not* understood as being constituted by the logic of the Kantian kind of anarchy, but rather by the logic of a Lockean kind of anarchy. Because the international system is not seen as being based on nonviolence and mutual assistance, bilateral links are looked for to provide the security that the system is ostensibly unable to deliver. In this case, friendship may be based on the same combination of rules, but these rules are likely to be weaker, more instrumental, and more untrustworthy because they are not constituted by the perceived structure of the system, but rather constructed contrary to it. Not only do they not reflect the system's structure, but they are cultivated as remedies against the system's ostensible deficiencies. Rhetorically adopting the role position of a friend and referring to others as friends thus does not necessarily reflect an understanding of the international system as being based on a Kantian kind of anarchy. It may instead mirror a much more skeptical view of the system and its degree of violence (for more on this possibility, see chapter 10).

Although Wendt's typology is useful, understanding a fourth kind of anarchy located between the Lockean and the Kantian versions is also necessary in the present context. According to the terminology I use in this book, the fourth kind of anarchy can be called a *Deutschian anarchy*. Such anarchy is, like Wendt's Lockean and Kantian kinds of anarchy, composed of sovereign states that recognize and respect one another as such. Furthermore, the Deutschian kind of anarchy is to be subdivided into two phases: first, a preparatory phase characterized empirically by the absence of violence, and, second, a mature phase defined by the absence of violence, by dependable expectations of peaceful change, and by a corresponding lack of preparations for warfare among the members of the (sub)system. To define an international (sub)system as a "mature Deutschian anarchy" thus is another label for a pluralistic security community.

Thus, rather than adding the ambitious element of mutual aid to the basic definition of a security community and then calling the resulting pattern of social interaction either "friendship" or "a tightly coupled security community," the Deutschian kind of anarchy subdivides the basic definition into two stages. The subject position in such anarchy may be labeled *nonfriend* during the first, preparatory phase and *prefriend* during the second, mature phase. Progress from the first to the second phase (and beyond to the role identity of *friend*) as indicated in the labels is possible but not guaranteed. The preparatory phase thus may "prepare" the agents for a phase that may never occur. Regression from the second to the first stage is also possible.

I treat Deutschian anarchy as a category on its own merits because of both its explanatory power with respect to the current international system and its particular saliency. I argue that, owing to the ambitiousness of Wendt's understanding of friendship, the gap in the typology between a Lockean anarchy and a Kantian anarchy is simply too wide to grasp the current international system in its entirety. The Deutschian anarchy, during both of its phases, is clearly distinct from a Lockean anarchy, not to mention from a Hobbesian anarchy, because violence does not occur any longer. It is also clearly distinct from a Kantian anarchy because it does not include the principle of mutual assistance. Hence, it follows that a Deutschian anarchy during both of its phases is more ambitious than a Lockean anarchy, but less ambitious than a Kantian anarchy. As Wendt defines different understandings that govern organized violence as the mark with which to distinguish different anarchies from one another, he shares with Deutsch the emphasis on and the respect for the importance of the violence/nonviolence relationship in international politics, which is somehow obscured in some current writings on security communities.[14] Different regional subsystems of the international system may be based on different role structures—that is, on different understandings governing organized violence.

Nonfriendship and *prefriendship* are neither just the opposite of friendship nor just other words for enmity or rivalry. They are not the opposite of friendship because an inverted version of the rule of nonviolence would mean entering the domain of enmity. In other words, although nonfriends and prefriends do not expect to assist one another, they will also not expect their mutual relations to be based on the rule of violence. If this were the case, they would be enemies, but not nonfriends or prefriends: permanent expectation of violence or even a rule of violence is reserved for patterns based on enmity. The role structure of enmity is a particularly hostile relationship of potentially unlimited use of violence based on the nonrecognition of one another's right to existence. In general, an enemy may be defined as "one that hates, and wishes or seeks to injure another" or as "one who tries or wishes to harm or attack."[15] In international politics, Wendt's definition of *enmity* is equally rigorous: "enemies are constituted by representations of the Other as an actor who (1) does not recognize the right of the Self to exist as an autonomous being, and therefore (2) will not willingly limit its violence toward the Self" (1999, 260). Enemies at any time expect to become involved in and therefore prepare for military conflict with each other.

Rivals, in contrast, do in principle accept one another's right to sovereign existence. This acceptance, however, does not prevent them from engaging in warfare for realizing narrower goals than eliminating others. The mutually conceded right to sovereign existence does not include the right to be free from violence. As a consequence, rivals prepare for military conflict because they "know" that they will be involved in warfare at some point— not permanently, but once in a while. The term *enmity* thus describes an ex-

14. Wendt writes that "the deep structure of an international system is formed by the shared understandings governing organized violence, which are a key element of its political culture" (1999, 313). Wæver, in contrast, writes that "Deutsch conceives of a security community only as a *non-war* community" (1998, 76, emphasis in original). The disturbing words here are *nonwar community* and *only*.

15. *The Shorter Oxford English Dictionary on Historical Principles* (Oxford: Clarendon Press, 1973), 1:656; *The Advanced Learner's Dictionary of Current English* (London: Oxford University Press, 1963), 327.

tremely antagonistic relationship based on the perception that the other's intention and willingness to use violence is unlimited. Rivalry is less antagonistic and war prone, but in an international system characterized by rivalry war will occur at least from time to time. In the current international system—or at least in some of its subsystems, such as the one analyzed here—bilateral relations are not regularly based on the position of either enmity or rivalry. Individual identities are increasingly being derived from representations of the structure of the system that do not follow a Lockean or Hobbesian kind of anarchy.

Although the current international system is not based on the rule of violence, it is premature to expect the system to function and the agents to define Self and Other according to the role structure of a Kantian kind of anarchy. This is in part a consequence of the ambitiousness of Wendt's definition of such an anarchy that combines pluralistic security communities with collective security. Thus, the wide and analytically open space between a Hobbesian or a Lockean anarchy, on the one hand, and a Kantian anarchy, on the other hand, requires further differentiation, for it cannot be taken for granted that anarchies based on the principle of nonviolence at the same time follow the principle of mutual assistance. Put differently, although states may not use violence to solve their mutual conflicts, this nonviolence is often still far from mutual assistance in case of need. The aspirations of many east and central European states during the 1990s to enter a military alliance and the reluctance on the part of the established NATO member states to follow the example of the WTO in the early 1990s may illustrate this claim: the members of NATO do not seem to understand the international system as a Kantian kind of anarchy in which mutual assistance is internalized, regardless of whether one is a member of a military alliance.

Moving beyond a Lockean kind of anarchy is nevertheless an important step forward toward banning the use of violence in international politics. This step is the first in justifying the claim asserted earlier that the Deutschian kind of anarchy is of particular saliency in international politics. Introducing this kind of anarchy as an intermediate stage between a Lockean anarchy and a Kantian anarchy is meant to help capture this step forward analytically. It aims at coming closer to an appropriate description of the current state of the international system or (some of) its subsystems and at enabling analytically more precise assessments of interstate relations.

The crossing of the threshold between the Lockean and the Deutschian kinds of anarchy may also be the condition of possibility for the effective involvement of nonstate and societal actors in constructing security. As shown in chapter 10, the recent involvement of societal actors in security policy has required the replacement of a strict military reading of security with a more comprehensive and civilian understanding based on network building, cross-border cooperation, people-to-people contacts, and so on. In particular in the far north, such an understanding was unimaginable during the preceding decades because of the transformation of the region into an important area of military operations and nuclear confrontation. But also in the area directly adjoining the Baltic Sea, the military logic of the superpower conflict prevented societal actors from effectively influencing the official security agenda.

As long as military thinking dominates security thinking, societal actors representing an alternative understanding of security are likely to remain ignored by the official security establishment, and they can expect their suggestions to be met with outright refusal. In both Hobbesian and Lockean kinds of anarchy, security therefore is deemed state business, and the respective experts jealously guard their monopoly over the construction of security. The state's authorities are preoccupied with security understood in terms of survival because they perceive it as threatened—war may occur at any moment or at least once in a while. Security is constructed accordingly—that is, with military means quite regardless of the security dilemma that may result as a consequence. Security thus is defined narrowly: when the survival of the state or at least major assets such as borders are at stake, a demilitarization of the security agenda is unlikely. Nonmilitary aspects of security will probably be represented as a luxury the state cannot afford at this point. As a corollary of this representation, security is represented as too important to let the society perform any active role in defining it. On the contrary, citizens' engagement is seen as potentially detrimental to the bargaining position of the state vis-à-vis its enemies or rivals and, as a consequence, as a threat to the state's security and ultimately to its survivability.

Only after crossing the threshold between the Lockean and the Deutschian kinds of anarchy may the society be granted admission to the realm of security or may the realm of security be expanded in such a manner that societal actors can become involved in it. In other words, in this crossing, although the state's security authorities are still claiming a monopoly on traditional security issues, parallel security discourses may unfold and challenge the official view. This is not to say that an alternative view on security can in principle not be developed during a Lockean kind of anarchy. The likelihood that alternative views will successfully challenge the official view is increased considerably, however, if they are accompanied by the security authorities' reevaluation of the saliency of conventional security. A precondition for downgrading traditional security and opening the door to alternative perspectives on security is the reinterpretation of the official understanding of organized violence—in other words, taking the step from a Lockean kind of anarchy to a Deutschian kind of anarchy. This reasoning is a second justification of the claim asserted earlier that the Deutschian kind of anarchy is of particular saliency in international security politics, indeed transforming them into international security relations.

The transformation of the security agenda may occur at least on the surface. Security authorities are unlikely to abandon their field of expertise (which is often connected with a high social reputation and material privileges) completely and without resistance. However, once allowed into the official view on security, alternative views may develop a self-dynamic that may run counter to the official authorities' originally rather modest intentions. This argument also in part explains why official representations are eager to include at least parts of alternative views once these views have proven publicly attractive, widely held, and legitimate. It should be noted, however, that the incorporation of alternative security thinking or at least of its language into the official agenda may also be an effective way to render it harmless by paying lip service to it while continuing to adhere to conventional security practices. Thus, entering the official security agenda guarantees neither the replacement of the conventional views by alternative views nor an enduring alteration of security practices.

4

Security and the State

EXPECTATIONS OF PEACEFUL CHANGE may follow from experiences of peaceful change. Experiences of peaceful change can be translated into trust, which can be defined as "a combination of past experience and hope for the future" (Väyrynen 2000, 115). However, the translation of experience into expectation may be time-consuming, and time in itself does not make for a security community (Wallensteen et al. 1994, 39). Experience of peaceful change must be accompanied by, for example, an increase in social interaction and communication between states and societies. Yet such an increase can also be a burden. It requires the development of what Deutsch called integration capabilities. International organizations may help develop such capabilities by, for example, providing technical assistance and making its members adhere to the same set of basic standards, such as respect of human rights, thus increasing political reliability.

Emanuel Adler and Michael Barnett assert that in tightly coupled security communities states may even transfer parts of their authority to the transnational community, which may find its institutional expression in an international organization (1998a, 36). Being accepted as a member of a transnational community may in turn increase a state's legitimacy. States that only recently gained their independence, however, may be unwilling to share authority with an organization. State building and state consolidation thus may be at odds with the tightening of a security community through the transfer of authority to an international organization comprising the members of the community. International organizations may also be helpful to security community building in other ways.

Although international organizations were underresearched in the original (Deutsch) study—probably because the system of international organizations and their spheres of responsibility were not as developed then as they are now—it seems to be a reasonable assumption that they can contribute to an increase in mutual responsiveness. In Deutsch and colleagues' terminology, they may provide for "more information about one another, more attention to that information, more joint operations, and more actual contact" (1957, 164). Information, however, may be misunderstood. It is therefore necessary to ensure not only that information is received, but also that it is adequately understood. Regular consultations among the members of an international organization within the framework of this organization, conflict settlement or mediation by the organization, and regular monitoring of the domestic and international affairs in its sphere of responsibility may help the members deal with the integration load and adequately perceive and understand messages and information, with "adequate" meaning in accordance with the source's intention.

International organizations have certainly contributed to conflict resolution in the Baltic Sea region (Birckenbach 1997; Zaagman 1999; Poleshuk 2001). The most obvious candidate for regional integration in the Baltic Sea area is the Council of the Baltic Sea States because it comprises all Baltic Sea littoral states and is tailored to the region's requirements. It is important to note that the council does not officially deal with what is conventionally considered to be security issues, especially military security issues (for exceptions, see Knudsen and Neumann 1995, 4). What appears to be a deficiency with respect to conventional security is, however, an asset with respect to security community building, which in turn has repercussions for security conventionally defined. Furthermore, although there is some potential for a negative overlap between the many organizations in the Baltic Sea region (Hubel and Gaenzle 2001, 22–23), it seems to be more adequate to refer to their activities in terms of constructive redundancy: together many organizations have achieved what one single organization would have found difficult to achieve (Vesa and Möller 2003, 114). However, the presence of international organizations in itself does not make for a security community. As the case of NATO enlargement in the Baltic Sea region indicates, the presence and enlargement of an international organization may even render security community building more difficult (see chapter 13).

Important prerequisites for security community building may be found at the domestic level. The relationship between domestic security communities, on the one hand, and international, pluralistic ones, on the other, was also underresearched in the original studies. Deutsch and his colleagues did not deal systematically with the question of whether a pluralistic security community requires integrated members—that is, members who have internalized expectations of peaceful change domestically. Their focus on the north Atlantic area, consisting of apparently integrated nation-states, was an invitation to put aside this question. The question can be phrased differently: Are domestic security communities a necessary or a sufficient condition for international security community building? Or may international integration emerge despite the nonexistence of domestic integration? It has been said that security begins at home (Booth 1991b, 349). Does security community building also begin at home?

Recent international relations scholarship has established a strong relationship between domestic and international politics. It is, for example, a part of the democratic peace proposition to argue that the peace among democracies can to some extent be explained in terms of translating the norm of nonviolence in domestic politics into nonviolence in international politics. As Bruce Russett puts it, "the *culture, perceptions, and practices* that permit compromise and the peaceful resolution of conflicts without the threat of violence within countries come to apply across national boundaries toward other democratic countries" (1993, 31, emphasis in original). Borrowing from this idea, Adler and Barnett argue that it is important to the nonviolent settlement of international conflicts "that states govern their domestic behavior in ways that are consistent with the community" (1998a, 36). Here, domestic security communities appear to be a necessary but not a sufficient condition for the emergence of an international security community. Likewise, Raimo Väyrynen argues that in order for a security community to emerge, "its members must be internally in peace, at least to such an extent that domestic instability does not spill over to other countries of the region" (2000, 118).

In a different yet related context, Clive Archer emphasizes the link between internal cohesion in the Nordic states and the development of what he calls the Nordic zone of peace (1996, 452). The concept of a zone of peace covers five elements: there should be no or little interstate war in the region; there should be no or little war between the states of the region and other states; there should be no or little armed conflict in the region in the form of civil war or armed uprisings; there should be no or few military interventions by armed forces from the region in other parts of the world (with the possible exception of internationally sanctioned actions); and there should be little or no expectation of any of these elements, thereby creating a region of "low tension." These conditions are closely associated with Deutsch's concept of a pluralistic security community. It has been argued, however, that "a zone of peace has a foundation in the relations of states; a pluralistic security community rests on the social foundations of community between individuals and societies" (Holsti 1996, 148). This social element is obscured to some extent in Archer's typology (but not in his analysis). It is more important here, however, that he links the Nordic zone of peace to, among other things, the internal cohesion of the Nordic states (and, as a result, to the cohesion in the Nordic region), thus establishing a connection between the domestic and the international realms. According to Archer, within the Nordic states a culture has developed that "place[s] an emphasis on social peace," and within the Nordic polities organizational frameworks have developed that "[favor] the institutionalization of this social peace and its externalization" (1996, 462).

Archer furthermore stresses the importance of the strategic position of the Nordic states to the evolution of the Nordic zone of peace and especially stresses that "the strategic meaning of geography can change" (1996, 459). Indeed, the meaning assigned to geography is flexible. Otherwise, the transformation of the Nordic region from a zone of almost permanent warfare to a zone of peace or a pluralistic security community would not have been possible. This interpretative flexibility supports Väyrynen's suggestion that future research on security communities should include the analysis of "the internal transformations and shifting boundaries of regional systems focusing on identities, networks, and externalities rather than [on] static territorial characteristics" (2003, 39).

The idea of a relationship between domestic and international politics enjoys popularity not only among constructivists and adherents to the democratic peace proposition. For example, recent international relations social science has discussed nation-states in terms of state strength and weakness, the assumption being that state strength is an indispensable prerequisite for peaceful domestic and international social relations. Furthermore, state strength or weakness has been coupled with the prospects for a security community. According to Barry Buzan, for example, in a regional setting "groups of weak states can make the formation of security communities extremely difficult because of the way in which they export their domestic instabilities to each other" (1991, 106). Weak states furthermore "simply define the conditions of insecurity for most of their citizens" (1991, 106); that is, they mark the opposite of dependable expectations of peaceful change and internal security communities: they "either do not have, or have failed to create, a domestic political and societal consensus of sufficient strength to eliminate the large-scale use of force as a major and continuing element in the domestic political life of the nation" (1991, 99).

According to Buzan's approach, in order to qualify as a strong state, a state has to exhibit an idea, a physical basis, and an institutional expression. The state's physical basis, derived from the understanding of power in realist writings, means population, territory, national resources, and wealth. The institutional expression comprises the entire machinery of government. Whereas these two categories are relatively tangible, the idea of the state is not. It is the most central but at the same time the most abstract component of the model. The idea of the state is located on the sociopolitical level and is considered the essence of the state. Without a deeply rooted idea of the state among the population, state institutions will find it difficult to function and survive. Without an idea (or ideas) of the state, the population cannot exhibit loyalty toward the state. The idea of the state thus is about "what binds the people into a socio-political and territorial entity" (Buzan 1991, 70). The two main sources for the idea of the state are to be found in the nation and in organizing ideologies. Buzan defines the nation as "a large group of people sharing the same cultural, and possibly the same ethnic and racial, heritage." Nations are "products of a closely shared history" (1991, 70). Organizing ideologies (or binding ideas or a state's political identity) are the expression of the basic values and norms to which state and society are committed.

The transformation of weak states into strong states seems to be a major policy objective in order to increase security, nationally and internationally (as long as one adheres to the state as the basic unit of the international system), but there is at least one problem with this assumption. Owing to its focus on national security, building strong states "may also have negative consequences for the security of many individuals and groups caught up in the process" because they are excluded from the imagined national community in a given situation. Indeed, "the state-nation process is seldom benign" (1991, 106). Yet, despite the danger of collateral damage, Buzan gives preference to national security rather than to individual and international security. Although the state may be a source of threat, it is nevertheless represented as *the* object of security.

In Buzan's assessment, weak states are more often than not a result of the process of decolonization. Likewise, Kalevi Holsti argues that the shift of war from the *inter*state level to the *intra*state level has been mainly a result of the process of decolonization and state building in the so-called Third World (1996, 77–78). In many postcolonial states, governments and populations have failed to fill the formal frame provided by the international recognition of the state with political contents that can give meaning and provide loyalty to the state, its political system, and its representatives. Instead, these states are characterized by the lack of at least one of the following traditional elements of statehood: a defined territory, a permanent population, an effective government, and the capacity to enter into treaty relations with other states. The particular practice of international recognition of the postcolonial states and the ensuing state-building process or the lack thereof have often resulted in weak states that have increasingly become both subject and object of warfare. Similar to Buzan, Holsti argues that weak states render the emergence of a security community impossible, not only because these states are internal *in*security communities, but also because their weakness transcends borders and entangles neighbors. For example, neighbors may find it tempting to try to support either the weak state's government or secessionist groups or to exploit its actual weakness for their own purposes.

What, then, are the conditions of peace within state and society? According to Holsti, the degree of a given state's strength proves to be decisive for its (in)capability to deal with conflicts in a nonviolent manner. Holsti establishes a correlation between the degree of strength a state possesses and its capability to deal with conflicts in a peaceful way. Put briefly, "strong states are an essential ingredient to peace within and between societies" (1996, 13). In one of several points of similarity between Holsti and Buzan, Holsti argues that state strength is not to be measured in a state's military or economic capabilities, but rather "in the capacity of the state to command loyalty—the right to rule—to extract the resources necessary to rule and provide services, to maintain that essential element of sovereignty, a monopoly over the legitimate use of force within defined territorial limits, and to operate within the context of a consensus-based political community" (1996, 82–83).

Consensus is seen here not with regard to every single political decision, but rather concerning the basic norms and values underlying the political system and its constitutional order. Holsti adopts Buzan's three pillars of state strength: the idea, the physical base, and the institutional expression of the state. In Holsti's description, the idea of the state "represents history, tradition, culture, nationality, and ideology. It is, in a sense, the *affective* aspect of the state, the forces, sentiments, and ideas that distinguish political communities from each other" (1996, 83, emphasis in original). Yet, whereas Buzan, without offering a quantifiable measure, identifies sociopolitical cohesion as the decisive indicator of the degree of strength a state possesses, Holsti prefers legitimacy as his key category. Legitimacy, in his view, is the sum of the existence of eight building blocks for modern statehood.[1]

With Buzan's emphasis on national security and with Holsti's focus on the classical European nation-state, both approaches serve particularly well as starting points for the discussion of security and the state in this chapter. They do so not because this discussion is about state strength or weakness in the Baltic states, but rather because they paradigmatically illustrate the basic lines of thought underlying the construction of the state in Estonia, Latvia, and Lithuania, without which the construction of security cannot be adequately understood. In the context of this book, two ingredients of the discussion on state strength are of particular importance.

First, with respect to security community building in the Baltic Sea region, the basic problem does not seem to lie in the empirical absence of state strength in Estonia, Latvia, and Lithuania. This is not the place for a detailed discussion of the internal developments in the Baltic states throughout the 1990s,[2] but it can be said that with the possible exception of a coherent national identity, the typical features of weak states or clusters of weak states cannot be established for Estonia, Latvia, and Lithuania.[3] It is true that during the 1990s Estonia and Latvia did not have a state border treaty with Russia (the Lithuanian-Russian border treaty was signed in 1997 and ratified by

1. These ingredients are: an implicit social contract; consensus on the political rules of the game; equal access to decisions and allocations; clear distinction between private gain and public service; effective sovereignty; ideological consensus/pragmatic politics; civilian control of the military; and international consensus on territorial limits and state legitimacy.

2. For the internal developments in the Baltic states, see Taagepera 1993, Lieven 1994, Dreifelds 1996, Vardys and Sedaitis 1997, and Clemens 2001.

3. According to Buzan, these features are: a high level of political violence; a conspicuous role for political police in citizens' everyday lives; major political conflict over what ideology will be used to organize the state; lack of a coherent national identity; lack of a clear and observed hierarchy of political authority; and a high degree of state control over the media (1991, 100).

the Lithuanian Seimas and the Russian Duma in 1999 and 2003, respectively; Estonia and Russia signed a border treaty in May 2005). However, although the Russian Federation tried to exploit the missing border treaties to postpone the withdrawal of troops from Estonian and Latvian territory and to argue against Baltic membership in NATO, the lack of border treaties has not proven to be a trigger of serious international tensions.

Second, I have recapitulated Buzan's and Holsti's writings at some length because they deliver the historical and ideational background on the basis of which the Baltic nation-states and subsequently Baltic security are being constructed. Their conceptualization of state strength is based on the specific features of the nation-state in the design established after the Peace of Westphalia. Indeed, Holsti acknowledges that his conception is derived almost exclusively from a single political tradition—namely, the modern, sovereignty-based European nation-state (1996, 96). This narrow approach may be problematic from an overall perspective, but with respect to the Baltic states it is not. In support of his claim that this conception is relevant even in the post–Cold War world, Holsti furthermore argues that the "fundamental rules and institutions underlying international relationships have not changed with the end of the Cold War. Indeed, they have been strengthened" (1999, 289). The Westphalian nation-state no doubt serves as the ideal type after which the Baltic nation-states were being modeled throughout the 1990s. Baltic security policies cannot be understood without bearing in mind this particular conception of the state. It is reflected in, among other things, attempts to strengthen borders rather than to make them permeable as well as in the prioritization of sovereignty and the emphasis on military means to defend it. If the state's national military capacities are considered insufficient, then the application for membership in a military alliance is a logical consequence. If, as this book shows, the construction of the nation-state in the Baltic states has unfolded with Russia as a negative reference point, then an application for membership in NATO (even if it results in a certain curtailment of sovereignty) is as logical a consequence as is the renunciation of, for example, Russian security guarantees.

In a sense, then, it does not really matter in the present context at which point on the continuum between strong and weak the Baltic states are currently located or in which direction they are moving. What matters are the basic way of thinking of state and statehood—and consequently of security—among Baltic decision makers and the transformation of these thought patterns into foreign and security policies. While referring to security, Baltic decision makers are often constructing the nation-state. With respect to international integration (in the Deutschian sense), constructing the nation-state frequently results in suboptimal security policies. It may even result in a clash between both aims—state building and security building. The Baltic states' position toward the CFE Treaty is a case in point and is discussed here although it violates the chronological narrative. At the end of this chapter, I sketch the logic underlying the demands for the Russian troop withdrawal (and elaborate on it in chapter 7) because in this case the construction of the nation-state and the construction of security went hand in hand.

Because the Soviet occupation of the Baltic states was never recognized by the majority of states—in other words, because the incorporation of the Baltic states into the Soviet Union in violation of international law was not validated by subsequent recognition—for Baltic decision makers the issue could only be one of restoring independence and terminating an illegal occupation (Müllerson 1994, 122). Because Estonia, Latvia, and Lithuania were never legally part of the Soviet Union, they could neither secede nor become Soviet successor states. Furthermore, the restoration of Baltic independence preceded the dissolution of the Soviet Union. As outlined in the next chapter, Mikhail Gorbachev in 1990 reluctantly began to consider the possibility of Baltic independence, but he thought of it in terms of secession. Baltic decision makers, in contrast, considered the 1990 Soviet Law on Secession irrelevant to the Baltic case, not only because it made secession practically difficult, but also because it did not accord with their understanding of state continuity. Seceding from the Soviet Union on the basis of Soviet laws could have been interpreted as recognition of the rightfulness of applying Soviet laws to Estonia, Latvia, and Lithuania. This interpretation, in turn, could have been understood as a retrospective acknowledgment of the legality of their incorporation into the Soviet Union. This understanding had to be avoided, and state continuity had to be emphasized because "a state can be founded only once, once and for all" (Meri 2000b).

This view—with its interlocking ingredients of termination of occupation, restoration of independence, return to Europe, and preservation of Western values—was politically important. Emphasizing state continuity was deemed indispensable as a provider of legitimacy behind the strivings for sovereignty. It also aimed to increase pressure on the Western governments to restore diplomatic relations. However, the principle of sovereignty that the Baltic independence movements wanted to be applied to their cases carries with it a fundamental problem for aspiring sovereign states: "regarding the existing states as sovereign undermines the option of new states" (Lehti 1999, 417). Accordingly, no Western government responded in the affirmative to the March 1990 declaration of Lithuania's independence. In fact, to "recognize the breakaway unit would [have been] a hostile act against the 'parent' state" (Holsti 1995, 131). At that time, no Western government wanted to perform hostile acts against the Soviet leadership, the stabilization of which was a major policy objective, especially of the U.S. administration and the German government. In other words, most Western governments were very cautious with respect to providing Estonia, Latvia, and Lithuania "a form of external legitimacy and support" (Holsti 1995, 131).[4] Stressing both the restoration of independence and the Western nonrecognition policy, Baltic decision makers aimed at exerting pressure on the Western governments and making it more difficult for them to refuse to restore diplomatic relations. That "the West, having never recognized Soviet annexation, was morally bound to recognize independence" was in fact a widely held belief among Baltic politicians at that time (Lieven 1994, 235). How, then, did the reading of state continuity influence security policy decisions at that time?

Only a short period of time after the international recognition of the independence of the Baltic states, their governments had to clarify their position on the CFE Treaty.[5] This clarification carried with it the problem of having to choose between two objectives considered equally desirable but hardly attainable at the same time. As sketched earlier, the official view of *restoring* independence ruled out an interpretation that the Baltic states were Soviet successor states. As a consequence, the idea of entering the CFE

4. The first Western government to announce its willingness to restore diplomatic relations with Lithuania was Iceland, on 12 February 1991.

5. As parts of the Soviet Union, Estonia, Latvia, and Lithuania were part of the CFE regime from treaty signature in November 1990 until their independence in late 1991. For the negotiations on the CFE Treaty, see the Stockholm International Peace Research Institute (SIPRI) yearbooks' chapters on conventional forces in Europe. See also Hartmann, Heydrich, and Meyer-Landrut 1994 and Sharp 1998 for the Baltic Sea region, as well as Bolving 2001 and Lachowski 2002 for the Baltic states.

Treaty in 1991 as Soviet successor states (or later under the Tashkent Document of eight former Soviet republics) was rejected. The Baltic governments at that time did not consider entering the CFE Treaty as states independent of both the Western and the Eastern Group of States, although the delegations of several former WTO member states demanded they do so. Afterward, doing so became almost impossible owing to the structure of the CFE Treaty (Schmidt 1998, 1–2). Denouncing the CFE Treaty as a "Cold War relic due to the three Baltic countries being non-participants," as it is said to be sometimes done by Estonian officials (Huang 2001, 30), obscures the fact that the nonparticipation of the Baltic states is the result of a deliberate political decision made in 1991, resulting in self-exclusion from the treaty.[6]

The delegations of several non-Soviet former WTO member states criticized the Baltic governments' position on the CFE Treaty. In particular, the Polish, Czechoslovakian, and Hungarian delegations would have preferred that the three Baltic states "be subject to the same arms control regime as their neighbors" (Sharp 1992, 466). The Polish request to call upon the Baltic states to enter the CFE Treaty was not included in the statement of the chairman of the Joint Consultative Group (JCG) of 18 October 1991 defining the JCG's position on the Baltic states. Subsequently, the Polish delegation in a national declaration requested the Baltic states to enter the treaty (Hartmann, Heydrich, and Meyer-Landrut 1994, 216). The Baltic delegations' decision was accepted during negotiations because the Baltic states were considered special cases, and their potential military strength was seen as insignificant. The parties to the treaty formally agreed that the now independent Baltic states be excluded from the area of application as defined in the treaty. The state parties also recognized that inspections of treaty-limited equipment (TLE) on the territory of Estonia, Latvia, and Lithuania be subject to these states' consent and participation. In order to avoid an extraordinary

---

6. Baltic nonparticipation in the treaty was also supported by the head of the U.S. delegation to the Joint Consultative Group, arguing that Baltic participation might complicate the ratification process in the Soviet Union and increase the Soviet military's resistance (Sharp 1998, 425–26). Furthermore, in the autumn of 1991, fears were widespread among Baltic decision makers that Baltic participation in the treaty might result in a prolonged Soviet presence and influence in the region (Lachowski 2002, 21).

conference, the JCG agreed upon a joint legally binding declaration in the gray zone of international law (Meyer, Müller, and Schmidt 1996, 21).

Becoming a party to the CFE Treaty, however, would not only have prevented irritation between the Baltic states and several non-Soviet former WTO member states, but also have brought with it the distribution of TLE, permitted in the Soviet Baltic Military District (MD) under the CFE Treaty, among the Russian Federation's Kaliningrad region *and* the Baltic states, thereby reducing the Russian troops stationed in Kaliningrad considerably. By refusing to become members of the CFE Treaty and by disassociating themselves from any Soviet rights and obligations, the Baltic governments themselves limited their influence on troops stationed in close vicinity. Baltic calls for a demilitarization of the Kaliningrad oblast have to be seen in this connection,[7] for all that remained of the Baltic MD was Kaliningrad, and TLE "permitted in the Baltic MD under the CFE Treaty could be concentrated there for the duration of the Treaty" (Sharp 1992, 470). From the Baltic governments' point of view, this situation was undesirable in connection with the Russian (former Soviet) troop withdrawal from the German Democratic Republic, Poland, and the Baltic states. The Baltic governments would have liked the troops to have been directed toward mainland Russia instead of the Kaliningrad oblast.

Excluding themselves from the chance to influence future treaty changes by opting for nonmembership also brought with it consequences in 1996, when the CFE Treaty was changed to what was interpreted by Baltic observers as disadvantageous for and threatening to the Baltic states. For three years, Russia was allowed to station a significant number of armored combat vehicles (ACVs), considerably exceeding the allowed number under the original treaty, in the Pskov region neighboring Estonia and Latvia. The presidents of the Baltic states noted in a joint declaration on 16 June 1996 that they would have liked "other legitimately concerned countries"—meaning, less cryptically, themselves—to have been involved in the negotiations

---

7. According to the Basics of National Security of the Republic of Lithuania of 19 December 1996 (amended as of 4 June 1998), one of the main tasks of the Lithuanian foreign policy shall be "to strive to demilitarize the Karaliaucius (Kaliningrad) Region and its development, provided that this development does not contradict the interests of Lithuania" (pt. 1, chap. 5; *Valstybes zinios* [1997], 2:2–20). Entering the CFE Treaty would have facilitated this task.

(Norkus 1998, 159).[8] They would have been involved, however, had they decided to join the CFE Treaty in 1991.

Baltic nonparticipation in the CFE Treaty was also discussed in terms of an obstacle to Baltic NATO membership. In November 2002, the Baltic states were finally invited to join NATO, but for some time during the 1990s their nonparticipation in the CFE Treaty apparently increased some established NATO members' unwillingness to invite them to join. To be sure, "NATO as such is not a signatory of the CFE Treaty. . . . Therefore, from a legal point of view, NATO's enlargement per se has no impact on the Treaty" (NATO 1995, chap. 2, B, §21). Being no signatory of the treaty does not, however, exculpate the alliance from reflecting on the consequences of its enlargement decisions for conventional arms control in Europe. In the case of the Baltic states, the combination of proposed NATO membership and their nonparticipation in the CFE Treaty seemed to result in a situation that the leadership of the Russian Federation and some NATO members could have found undesirable. NATO, in a crisis, can deploy troops on the territory of the Baltic states uncontrolled by conventional arms control regimes, "except those which stem from information and inspection provisions in accordance with the Vienna Document" (Bolving 2001, 41). This deployment might result in a substantial regional destabilization, a possibility that contradicts NATO's reiterated declaration that "the enlargement process including the associated military arrangements will threaten no-one and contribute to a developing broad European security architecture throughout the whole of

---

8. Russia was permitted to deploy in the Pskov region 600 ACVs as opposed to the previous ceiling of 180. Insufficient suitable storage facilities, however, made unlikely a significant increase in TLE in real terms in Pskov. On 27 May 1997, the NATO-Russia Founding Act reacted to the problems resulting from the CFE Treaty's relevance to nonparties to the treaty by stipulating in chapter 4 that NATO and Russia, together with the other parties to the CFE Treaty, would in future cooperate "to adapt the CFE Treaty to enhance its viability and effectiveness, taking into account Europe's changing security environment and the legitimate security interests of all OSCE participating States." See Founding Act on Mutual Relations, Cooperation, and Security Between NATO and the Russian Federation, Paris, 27 May 1997, pt. 4, Political-Military Matters. For later modifications of the treaty, see Schmidt 2000. The adapted CFE Treaty, signed on 19 November 1999, permits Russia to station in the Pskov region only 31 tanks, 139 ACVs, and 91 artillery systems as opposed to 1,800 tanks, 600 ACVs, and 2,400 artillery systems under the old CFE regime.

Europe, enhancing security and stability for all" (NATO 1995, chap. 2, C, §27).

This contradiction did not seem to increase some current NATO members' enthusiasm to invite the Baltic states to join NATO. Inviting them even though they had not joined the CFE Treaty would undermine a treaty that, according to NATO, is "the cornerstone of European security" (1995, chap. 2, B, §21). The treaty then would no longer prevent regional destabilization. As a result, the treaty's value as such could be called into question. Undermining one of the cornerstones of European security by way of inviting the Baltic states to membership was apparently seen as an unwelcome result of NATO's enlargement process, the consequences of which some NATO members seemed unwilling to accept at that time.

But even joining the treaty on the basis of the Adapted CFE Treaty's openness to all European Organization for Security and Cooperation in Europe (OSCE) members would not solve the problems entirely. To begin with, accession would not be easy because the JCG decides the terms of accession on the basis of consensus. The JCG members' right to veto, the vagueness of the procedure for accession, and the importance of the terms of accession in connection with Baltic NATO membership might result in Baltic CFE membership only "after hard political and military negotiations on the accession terms" (Bolving 2001, 52). Even after Baltic CFE accession, some problems would remain. Exceptional Temporary Deployment (ETD)[9] would include the option of considerable military reinforcement in both the Baltic states and in a Russian-Belarusian union. Both options would be destabilizing and thus unwelcome. Therefore, even joining the treaty, which Lithuania stated it might be interested in doing at the occasion of the OSCE Summit in Istanbul as early as 19 November 1999, would require "additional arms control agreements" or at least "mutual military restraint" in the Baltic Sea region (Schmidt 2000, 35). That the Baltic states were eventually invited to join NATO regardless of the CFE Treaty situation does not render irrelevant the present discussion, although NATO has indicated that it will "exercise re-

---

9. Through ETD, "a State Party's Territorial Ceiling can be exceeded temporarily . . . allowing up to 459 battle tanks, 723 ACVs, and 420 Artillery Pieces. . . . Furthermore, it should be remarked that there is no legal time limit on the Temporary Deployments, which means that the receiving States Parties can host Temporary Deployments for an unlimited period of time as long as the provisions for notification are acknowledged" (Bolving 2001, 85).

straint in making conventional deployments on the territory of the new members," including the Baltic states (Lachowski and Sjögren 2004, 721).[10]

The current state of affairs reflects the current unpopularity of both arms control (see Bohlen 2003) and thinking of European security in seemingly old-fashioned terms of a balance of power. It also reflects the existent uncertainty as to the future of the CFE Treaty after the enlargement of NATO (Lachowski and Sjögren 2004, 713). The case of the Baltic states' position toward the CFE Treaty is also indicative of the condition of tension between the construction of nation-states and the construction of security, which to some extent characterized the security policies of the Baltic states in the early 1990s. In a different situation, however, both policy objectives were mutually supportive.

From 1991 to 1993–94, no other issue kept the foreign and security policy elites of the Baltic states as busy as the negotiations on and the process of the withdrawal of 120,000 Russian (former Soviet) troops from the territories of Lithuania, Latvia, and Estonia. On 24 December 1991, the Russian government took over all the rights and obligations of the Soviet Union, including the responsibility for the former Soviet troops stationed abroad. On the same day, the Lithuanian government addressed the Russian government, demanding the complete withdrawal of the former Soviet troops from Lithuanian territory. As long as Soviet/Russian troops in considerable number were stationed on their territories, "the Baltic states [were] living in a kind of semi-independence," as Estonian scholar Aare Raid put it (1993, 46), because they themselves could de facto control their territories only in a limited manner. This problem might very well have stretched the interim period during which the declarations of independence did not yet correspond with the recognition of independence by the international community. Both the failed coup d'état in Moscow and the Baltic independence movements' claim that they were restoring independence rather than seceding from the Soviet Union no doubt contributed to the relatively swift recognition of Estonian, Latvian, and Lithuanian independence (or the reestablishment of diplomatic relations), regardless of the absence of one of the fundamental preconditions for sovereignty: control of one's own territory.

10. Some Baltic politicians were not so cautious. In June 2003, for example, the Lithuanian defense minister asked NATO to set up military bases in the Baltic states. He arguably did so in order to qualify as an equal rather than as a "second-rate member" (Lachowski and Sjögren 2004, 721).

Even afterward, as long as 120,000 foreign troops were stationed on their territories, the Baltic states could hardly be seen as *sovereign* in the word's traditional meaning. To be sure, recognition of sovereignty means "that at least formally a state has an equal status in the eyes of Others" (Wendt 1999, 237), but from Baltic political decision makers' point of view, to have an equal status only formally was hardly attractive. Thus, the desired return to Europe was to be accompanied by both the withdrawal of the foreign troops and subsequently the buildup of national armed forces in Estonia, Latvia, and Lithuania. Historically, the post-Westphalian conception of the state is inseparably connected with the ruling classes' perception that if they wish to survive, a permanently combat-ready, "stable military might exclusively obedient to them" is needed. The buildup, maintenance, and enlargement of standing armies, of "permanent, bureaucratically organized, disciplined and totally loyal military apparatuses," is a legacy of the post-Westphalian conception of statehood (Krippendorff 1985, 274). The buildup of armed forces is surely an act of domination in the sense that it is a permanent struggle over the appropriation of social and financial resources and their transformation into means of warfare. It has to be enforced against resistance, and it aims at, among other things, strengthening the particular mode of domination and conception of statehood on which a given nation-state rests (Geyer 1984, 9–23). Among other things, the post-Westphalian conception of the state is based on the principle of territoriality, which includes spatial exclusion; the sovereigns' reciprocal assurance of one another's legitimacy as well as their noninterference in one another's internal affairs; the recognition of the nation-state as a sovereign entity under international law; and the presence of national armed forces as well as the absence of foreign armed forces.[11] However, "when Estonia, together with other Baltic states, re-arrived in the international arena in 1991 under the banners of national sovereignty and state independence in pursuit of securing them, it discovered that conceptualizations such as independence and sovereignty had been called into question and many former certainties in international politics had become uncertain. . . . The most discouraging finding was, of course, that [the] whole concept of nation-state was going out of style, at least in Europe" (Järve 1996, 226–27).

11. Armed forces may be stationed on foreign territories with the consent of the host country, for example, on the basis of alliance treaties.

All the same, the political elites of the Baltic states nevertheless followed a rather conventional conception of statehood during the 1990s. This conception was at odds with many political trends in western Europe, especially in the EU, but it was a logical corollary of the reading of state continuity prevalent in the Baltic states and discussed earlier. For example, the resumption of traditions broken off in 1940 found expression in the reintroduction of the 1922 Latvian Constitution in 1992 (Dreifelds 1996, 31), the "re-establishment of most of the major historical parties after 1988" in Lithuania (Krupavicius 1998, 165), and "politicians casting themselves in the role of their childhood heroes" (Lieven 1994, 55). Relying almost exclusively on voluntary defense forces immediately before and after the establishment of independence did not result simply from the lack of armed formations. It also aspired to invent military traditions by revitalizing the pre-1940 volunteer defense organizations that had existed in all three Baltic countries. Lithuania's national voluntary defense forces symbolically represented continuity of the Lithuanian state by using replicas of the uniforms of the interwar National Guard (Jæger 1997, 26). The paramilitary units of the interwar period, however, can hardly be seen as a model for democratic armed formations. They were recruited mainly from rural-conservative circles and were hardly accessible to left-wing circles, and they served as the authoritarian presidential governments' armed branches (Kerner 1994, 31). Anatol Lieven has observed that "for lack of any other tradition the new Baltic States are modeling not just their political symbolism, but their political ideologies, parties and state institutions, on those of the period of independence between 1918 and 1940. . . . The problem is that the Europe many Baltic politicians seek to return to is not the Europe of today, but that of the 1920s and 1930s, when the Baltic was first independent" (1994, 55, 374).

The traditional conception of the nation-state also made the political elites in the Baltic states equate independence with sovereignty, sovereignty with security, and security with the military—in other words, the state-security-military nexus characteristic of the conventional nation-state. In Mare Haab's words, "security is identified with sovereignty, and the chances of defending sovereignty are seen primarily in terms of military means" (1994, 148). Building up national armed forces was as clear a consequence of this conception of statehood as demanding that foreign troops leave immediately and unconditionally. It also brought with it a pronounced emphasis on the military as a symbol of this particular type of nation-state. Consequently, "displaying the Lithuanian flag," showing that Lithuania is a normal, Western nation-state, deeply rooted in Western tradition, has been identified as one of the "very useful" tasks of the small Lithuanian navy, which, as a result of its military insignificance, has indeed primarily a symbolic function (Nekrasas 1996, 69). It is thus insufficient to discuss the Baltic states' armed forces only in terms of their ability or inability to defend the Baltic states in case of a foreign aggression, their potentiality in respect to international military cooperation, or their degree of adaptation to NATO standards. Rather, the "highly symbolic, normative nature of militaries and their weaponry" requires an understanding of the military, which adds to their operational capabilities the symbolic linkage with "sovereign status as a nation, with modernization, and with social legitimacy" (Eyre and Suchman 1996, 86–87).

Demanding the withdrawal of foreign troops not only served as a corollary of constructing a modern nation-state, but was equally considered mandatory in light of constructing security because, from the point of view of the Baltic states' political decision makers, the presence of the Soviet/Russian army amounted to an offense against national security. I resume this narrative in chapter 7.

5

Peaceful Change in the Soviet Union

SECURITY COMMUNITIES have a history (Adler and Barnett 1998a, 49). This history does not start with what can later be identified as the initial steps toward a security community, but rather much earlier. In fact, security community building cannot be adequately understood without taking into consideration the prehistory of events that, up to a certain point, made the development of a security community impossible. In many cases, history and memory will provide obstacles to security community building. Historical experience will frequently intervene in and render difficult the process of integration. It will often shape collective memory in such a way that the emergence of a sense of community among a group of people is deemed unlikely, if not impossible, by this very group.

However, the absence of peaceful change in the history of intergroup relations does not make impossible the subsequent development of a security community. People are no mere puppets on the strings of their memories, and they can effectively break with their memories by assigning them to history, relieving them from the task of governing present expectations and politics. This process is difficult and time-consuming, but it is not impossible. The Nordic security community is a case in point, having replaced, since 1814, an insecurity community involving about sixty wars within and between the Nordic states as well as between them and their neighbors in five centuries (Wiberg 2000, 291) with what has gradually evolved into a pluralistic security community or a zone of peace. Sworn enmities such as those between Germany and France can be replaced by profoundly more benevolent patterns of social interaction.

In Baltic-Russian relations, the history of an insecurity community is primarily a reflection of Baltic-Soviet relations that, however, cannot be put aside as irrelevant in the context of this book just because the Soviet Union does not exist anymore. To paraphrase Bruce MacLaury, Baltic-Russian relations are a new phenomenon, but they can only be built on the foundation of the preceding stage of relations between the Baltic republics and the Soviet Union (1994, ix). This preceding stage of relations included, from the Baltic point of view, victimization through war, occupation, collectivization, imprisonment, deportation, and the loss of independence—all or some of which were integral ingredients of the lives of many Latvians, Lithuanians, and Estonians in the twentieth century. Referring to past victimization carries with it the danger of seeing the present and future primarily through the lenses of the past, but cognitive systems are tightly connected with individual and collective experiences as well as with overall history and, as Anatol Lieven has put it, with "myth as history and history as myth" (1994, 118).

Individual experiences of the Soviet time persist to some extent in the post-Soviet time and arguably result in a specific inertia of the belief system. This belief system can be assumed to be relatively impervious to change because "what is experienced as history by one generation becomes structure for the next" (Skultans 1998, 103). This structure, in turn, makes security community building and even thinking about international relations in terms of peaceful change a complicated issue: in Baltic-Russian relations, experience does not seem to support the notion of peaceful change. Or does it?

To be sure, in the terminology applied in this book, the Soviet Union was an amalgamated political community, but it was not a security community. From the 1970s onward, it was increasingly based on "systemic, anonymous and diffuse" forms of violence (Smolar 1999–2000, 95) rather than on open, direct, and ideologically fanatical forms as in some of the preceding decades. But before glasnost and perestroika, these less obvious forms of violence served mainly to veil the character of the state, which continued to be fundamentally determined by "comprehensive and penetrative central control" (Baev 1997, 175). Before glasnost and perestroika, the Soviet Union was a political community, but not a security community. What it was during glasnost and perestroika is more difficult to say. This chapter attempts to give an answer to this question.

This puzzle is not the only one with respect to the disintegration of the Soviet Union. As Robert Herman has put it, both the "depth and the breadth" of the changes in Soviet foreign policy accomplished after the coming to power of Mikhail Gorbachev and the initiation of *novoye mishleniye* (new thinking) in foreign policy in late 1987 and early 1988, "from the unilateral steps to reduce and restructure military forces to give them a less offensive cast to the rejection of force to maintain communist domination in Eastern Europe and to prevent the dissolution of the Soviet Union itself, went far beyond what conventional international relations theories can accommodate as well as what knowledgeable area specialists imagined was possible" (1996, 272).[1]

This is not to suggest that the Soviet leadership casually accepted the decline of the Soviet Union, the independence of the Baltic states, and the subsequent dissolution of the Union of Soviet Socialist Republics (USSR). Although, as Michael MccGwyre has put it, the Baltic states could claim "an inherent right to independence" because their accession to the Soviet Union had involved military coercion and rigged referenda; although they could "serve as the major economic interface between Western Europe and the Soviet Union"; and although strategic imperatives for maintaining Soviet rule in the Baltic region had declined in importance, Gorbachev seems to have been prevented from a more flexible approach toward the Baltic republics by the possibility that their secession "would start the disintegration of the Soviet Union" (1991, 353–54). Indeed, as was observed at that time, "even very small countries like Lithuania may play a very important role since its independence can be achieved only by fundamental change of the Soviet 'internal empire' " (Ágh 1991, 86). Furthermore, with respect to the possibility that the independent Baltic states would serve as an economic interface between the West and Russia, it should be remembered that the Soviet leadership intended to use the Baltic republics as a showroom for its economic restructuring. Here, rather than in the immense territory of the whole Soviet Union, the success of glasnost and perestroika and with it the reformability of the USSR could be demonstrated to the West, or so it was hoped. The Soviet leadership indeed undertook a variety of measures aimed at reversing the process of disintegration and in particular at stopping Baltic independ-

---

1. See also MccGwyre 1991, 174–344. Basic texts on Soviet new thinking in international relations can be found in Hirsch 1989, 403–533. That these texts were indeed translated into policy was a paradoxical result of the structure of the Soviet state. Although it "hindered the adoption of new ideas, it also insured [*sic*] consolidation of those ideas that were adopted by the leadership. . . . In particular, once Gorbachev had adopted the new political thinking, its diffusion and dissemination were more readily accomplished (than in a more pluralist setting) because of the extraordinary agenda-setting powers and control over media resources held by the top elites" (Checkel 1997, xi, 89).

ence aspirations. In the final analysis, however, these measures were half-hearted, ambiguous, and spontaneous rather than deliberately planned and, at least in the beginning, far from intransigent.

In this chapter, I interpret the policies of the central Soviet leadership toward the Baltic republics in terms of peaceful change. The Soviet leadership had arrived at a point at which the use of large-scale physical force against the Baltic republics was ruled out. It seems to have been replaced by the conviction that the Baltic problem had to be settled by peaceful means. That this one-sided conviction was not synchronized with the expectations of the Baltic republics' political representatives and that the Soviet leadership did not entirely refrain from using force testify to the conclusion that the Soviet Union could not be considered a security community at that point. This disagreement, however, is not surprising because perestroika, glasnost, and *novoye mishleniye* in foreign policy had been launched only in late 1987 and early 1988. A time span of less than four years seems hardly to be sufficient for the emergence of mutual trust and expectations of peaceful resolution of intrastate conflicts—especially in light of almost fifty years based on the opposite and a less restrictive use of force by the Soviet leadership in other parts of the USSR at that time.

The renunciation of the use of large-scale physical force against the Baltic republics may however be seen as a first and serious step by the Soviet leadership to transform the Soviet Union into a mutual-response union that was to be based on negotiation, arbitration, and the courts rather than on arbitrary, unjust, and often violent means of dealing with conflicts. Soviet-Baltic relations may thus be seen in terms of a security community in the making, characterized at that time by a lack of mutuality regarding expectations of peaceful change. The community character certainly needed more time to unfold and come to fruition. Interpreting the policies of the Soviet leadership toward the Baltic republics in light of peaceful change and changes in Soviet thought patterns may be more appropriate than focusing on Western material superiority and the nonviolent character of the Baltic popular movements. Their nonviolence was certainly a necessary condition, but hardly a sufficient one for their success. To explain this success exclusively in terms of the movements' nonviolence misses the other side of the coin: the nonviolent means used by the Soviet leadership.

Soviet measures against the Baltic republics included military, economic, and political means, each of which I address in turn. Although I emphasize the Soviet policy toward the Baltic republics in this chapter, I also include many aspects of the Baltic popular movements' policies in the analysis. My interpretation shows that even a process as politically contentious as the restoration of the Baltic states' independence could unfold largely peacefully. Every single confrontation between the Soviet authorities and the adherents to the Soviet Union, on the one hand, and the popular movements, on the other—as the chapter shows, there were confrontations in abundance—could have resulted in a violent escalation of the conflict. Essentially, none of them did. Like the process of the Russian troop withdrawal from the Baltic states, discussed in chapter 7, the nonviolent restoration of Baltic independence may have been seen—and still may be seen—as one of the formative events for a security community in the Baltic Sea region.

Military Measures

The most prominent example of the use of military force in order to intimidate the Baltic popular movements was the occupation by force of the television and radio center in Vilnius by Soviet Ministry of the Interior troops on 14 January 1991, which cost fifteen civilians their lives. The intervention by "Black Beret" troops[2] was preceded by the seizure of the main printing plant in Riga by Omon troops on 2 January, the occupation of several Lithuanian Communist Party buildings, and the brief occupation of the Vilnius television center on 9 January. It was followed by similar actions against the Latvian Ministry of the Interior in Riga on 19 January, which left five civilians dead. The official justification for the use of means of violence was the establishment of compliance with the Soviet Constitution and in particular with the Soviet Law on Conscription.

The Soviet leadership's struggle with the political representatives of the Baltic republics over compliance with the Soviet conscription law in general and over conscripts in particular had a long previous history that need not be narrated at this point. Suffice it to say that as early as March 1990 Soviet

2. The "Black Berets" were subordinate to the Soviet Ministry of the Interior and consisted of special assignment troops (Spetsnatz) and special-purpose militia squads (Omon) (Sharp 1991, 460).

paratroopers had occupied Communist Party buildings and the Higher School of Marxism-Leninism in Vilnius in order to capture Lithuanian army deserters. Moreover, the period from winter 1990–91 to August 1991 was overshadowed by both repeated interventions of Soviet paratroopers to capture Baltic deserters from the Soviet army and a series of violent Omon attacks against newly established Baltic border guards, culminating on 31 July 1991 in a brutal attack on the Lithuanian border post in Medininkai in which seven border guards were killed. What is more important in the present context than the mere brutality and senselessness of the attacks is that the Omon activities in Lithuania and Latvia in January 1991 were not an all-out, deliberate military intervention to stop the Baltic independence process. Rather, they were a "provocation" (Jakowlew 2003, 595), reflecting "considerable confusion" on the part of the soldiers and resembling "a conspiracy in which too few people had been consulted, too few were fully committed, and too many were afraid of being left holding the buck" (Lieven 1994, 202). Moreover, the weeks following January 1991 "were essentially a prolonged anti-climax. Even the bloody attack by Omon on the Interior Ministry in Riga . . . was more like the lashing of a reptile's severed tail than part of a co-ordinated plan for the reconquest of the Baltic States" (Lieven 1994, 254).

One can thus hardly say that the Soviet leadership made use of all the military means at its disposal to stop the Baltic states from breaking away from the Soviet Union. In January 1991, the army was not employed, and the troops from the Interior Ministry and their leaders were responsible for most of the attacks. In August 1991, only one military branch, the paratroopers, was involved, even though the commanders of the Baltic MD actively supported the coup (Meyer 1991–92, 30).

The Soviet president's role in these events remains somewhat dubious. According to Raymund Garthoff, relying on the then press assistant to the president, Gorbachev "had been fed a steady stream of false and misleading information on events and public opinion by his conservative security advisers. . . . Gorbachev was not even informed of the events" in Vilnius on 14 January 1991. The Soviet president, however, clearly failed to respond appropriately. He "not only did not go to Vilnius, he did not disassociate himself from the situation and make clear the unacceptability of vigilante salvation committees or military commanders acting without proper authority." Only on 23 January did Gorbachev express his "most sincere condolences" (1994, 452–53).[3] This utterance was certainly much too late and much too weak a response to the January events. Gorbachev was politically responsible anyway. According to Garthoff, however, the Soviet leader apparently did not initiate a deliberate military strategy with which to stop Baltic independence, regardless of diverse verbal statements threatening, at least implicitly, the use of force. Of course, these statements could have been understood by Soviet hardliners as at least an indirect call for and a justification of the use of force. Indeed, the violent operations in January 1991 could be seen to be in accordance with a decree issued on 1 December 1990 in which Gorbachev authorized "to take whatever measures were necessary to ensure that the military power of the union continued to function" (Jonson 1991, 132).

If we compare the combined capabilities of the Soviet Union with the capabilities of Estonia, Latvia, and Lithuania at that time, it is obvious that the Soviet leadership had much more military means in its power than it actually made use of. At that time, the Soviet Union had stationed two motor rifle divisions and one airborne division in Lithuania, one tank division in Latvia, and one motor rifle division in Estonia. There were three air force bases and one naval base in Lithuania; five air force bases and two naval bases in Latvia; and five air force bases and two naval bases in Estonia. In addition, both the Leningrad MD, including the Seventy-sixth Airborne Division in Pskov, and the Kaliningrad oblast, which hosted two motor rifle divisions and one tank division, were nearby (Baev 1996, 163). In line with the CFE Treaty, the Soviet Union in February 1991 declared for the Baltic republics 506 battle tanks, 1,892 ACVs, 33 attack helicopters, and 382 combat aircraft, among other things.[4]

The quantity of Soviet troops stationed in the Baltic republics at that time underscores these figures. Although loyalty to the central Soviet government and the Ministry of Defense could not be taken for granted, the number of Soviet troops was overwhelming. Furthermore, troops could easily have been brought in from other Soviet republics or from the non-Baltic part of the Baltic MD.[5] Although estimations as to the number of Soviet

3. The so-called National Salvation Committees, aimed at the restoration of Soviet rule, were founded in December 1990 in the three republics. Their membership was kept secret.

4. *Arms Control Today* 1992, 44.

5. Established after World War II and abolished in November 1991, the Baltic MD comprised the Soviet republics of Estonia, Latvia, and Lithuania and the Russian Republic's Kalin-

armed forces in the Baltic republics are difficult, it has been suggested that in 1984 there were 122,480 Soviet soldiers with 147,480 family members in the Estonian Soviet Socialist Republic (SSR) (Cunningham 1997, 54). According to an émigré organization's assessment, there were about 125,000 Soviet troops with 25,000 dependents in Estonia in the late 1980s (Clemens 1991, 99–100). Thus, "before independence in 1991, one person out of every ten living in Estonia was a Soviet soldier" (Bonn International Center for Conversion 1996, 187).

Although the withdrawal of Soviet forces from eastern Europe and the demobilization of Soviet troops were considerable in 1989, the western MDs seem to have been excluded from Soviet force reductions at that time. Given the importance attached to these forces by Western military planners, it has been speculated that if the Baltic MD had been included, the Soviet General Staff "would surely have seized the opportunity to demonstrate the decreased threat to Western Europe" (Sharp 1990, 471). Soviet force reductions at that time thus seem to have been motivated primarily by Soviet security interests rather than by political considerations in regard to East-West confidence building. According to data provided by a leading journal of military affairs, in the spring of 1991 there were all in all 135,000 to 140,000 troops under the Baltic command, among them 75,000 army troops as well as 53,000 air force personnel with 340 aircrafts. On the whole, there were six Soviet army divisions in the Baltic republics, including one airborne division and one training division.[6]

The Baltic republics, for their part, were virtually empty-handed at that time. They had at their disposal only limited numbers of militia, parliamentary guards, border guards, and voluntary forces, which were indisputably dedicated to defend the republics, but would have to do so almost without any arms (with the possible exception of hunting rifles). It is necessary to dwell on this issue for a moment to underscore the vast discrepancies between Soviet and Baltic capabilities to exert organized violence at that time and to show that the dynamics of action were on the side of the Baltic popular movements, which *acted* while the Soviet central authorities only *reacted*. The Baltic popular movements indeed began establishing parallel military

---

ingrad oblast. With only few exceptions, figures on military personnel and equipment in the Baltic MD do not differentiate between the Baltic Soviet republics and the Kaliningrad oblast.

6. *Österreichische Militärische Zeitschrift* 1991b.

formations on the territories of the Baltic republics and by so doing challenged the Soviet central authorities. This procedure would have been considered unacceptable by any central government of any state and probably would have triggered serious counterreactions. The Soviet leadership's reaction, however, was reserved.

The Baltic popular movements had emphasized the necessity to establish their own—that is, non-Soviet—defense formations at an early stage. They repeated this claim subsequently at various occasions. As early as 1–2 June 1988, the desirability of territorial military formations and military training in the Latvian language was adopted in a resolution at the occasion of the Latvian Writers' Union expanded plenum "Pressing Problems of Soviet Latvian Culture on the Eve of the 19th CPSU Conference" (Muiznieks 1993, 192–93). Public discussion on the issue of serving in Latvian armed forces continued in a Latvian Popular Front official document of 18 December 1988 entitled "Attitudes to the Armed Forces of the USSR." The Second Congress of the Latvian Popular Front in 1989 devoted an entire chapter of its program to (Soviet) demilitarization and anticipated many issues that were bound to reappear in the Latvian-Russian negotiations on Russian troop withdrawal following August 1991—for example,

> to establish a commission in order to control the armed forces and their relation to Latvian laws; to stop flows of retired Soviet officers to Latvia; to stop the creation of new military bases in Latvia; to take control over territories belonging to the Soviet army which had not been used for military purposes; the defense ministry of the USSR should pay rent for used Latvian soil; to work out principles for alternative service; to stop military training at schools and universities; and to remove military educational establishments from Latvia. (Ozolina 1998, 138–39)

In December 1990, Latvia was the first Baltic republic to pass a law on state borders after having started to control its borders as early as October by means of customs and militia posts. The events of January 1991 led to the haphazard formation of voluntary territorial defense forces and the inauguration on 24 January 1991 of the Department of Public Safety (Jundzis 1995, 553). In Estonia, the Kaitseliit (National Defense League) was reestablished as early as 17 February 1990 (Haab 1995, 41). Soviet legislation on the draft was suspended in March (Allison 1993, 6). On 12 April, the Estonian Parliament passed a law according to which Estonian citizens were prohibited from serving in the Soviet armed forces.[7] In September 1990, the establishment of border guards began with the Economic Border Defense Service and the subsequent ratification of the Law on the Economic Order on 22 October 1990. The establishment of the border guard was completed on 1 November 1990.[8]

Subsequent to the declaration of independence on 11 March 1990, the chairman of the Lithuanian Supreme Council, Vytautas Landsbergis, indicated that he "was prepared to grant the Soviets the right to maintain military installations in the country" (Krickus 1993, 173). The declaration of independence was, however, accompanied by a resolution that the Soviet military service law was no longer binding in Lithuania. Three days later Landsbergis declared the Soviet conscription law null and void. In April 1990, the Department of National Defense was founded, which in October closed all Soviet military commissariats.[9] Territorial defense forces were being created in which civil resistance and psychological defense were to be taught. According to Audrius Butkevicius, since 23 May 1990 general director of the Department of National Defense, a professional border guard should take care of Lithuania's economic protection and exports. The volunteers of the object and property protection forces were to support police and border guards. The only armed formations would be the two hundred troops for special tasks, which would have to protect the Parliament and governmental buildings in times of crisis (Kerner, Stopinski, and Weiland 1993, 30–33; Kerner et al. 1998, 125–27).

In late 1990, the Lithuanian Supreme Council brought about the legal foundation of the realization of this defense concept and passed four laws on universal conscription, territorial defense, border guards, and voluntary service. On 17 January 1991, the council officially founded the Voluntary Service of National Defense, which on 15 February was declared responsible for protecting the Parliament, a task formerly assigned to the National Guard, a paramilitary volunteer auxiliary founded in 1990 especially for object protection (Vitas 1996, 75). However weak and inchoate the Lithuanian armed formations were at that time—and to call them "armed" formations

7. *Österreichische Militärische Zeitschrift* 1990.

8. *Estonia Today* 1998.

9. *Österreichische Militärische Zeitschrift* 1991a.

is actually an exaggeration—the recruitment of volunteer forces and border guards nevertheless served as the justification for the Soviet central authorities to send in paratroopers in order to compel compliance with Soviet legislation and to realize conscription into the Soviet armed forces.[10]

At that time, the realization of Baltic military plans was still in its initial stages. Military formations subordinate not to the Soviet central authorities but to the democratically legitimized political representatives of the Baltic republics existed only in the minds of ambitious politicians, hardly in reality. Many observers have stated that the Baltic armed forces were to be built up from scratch, especially after the initial plans had failed to incorporate into the Baltic armed forces parts of the Soviet military equipment stationed in their territory. In October 1991, for example, some Baltic officials demanded that equipment of the Soviet army be left behind for use by their new national armed forces. This demand met with resistance not only by Soviet/Russian officials, but also by some representatives of several of the former non-Soviet WTO states because it would have meant considerable and uncontrolled armament of the Baltic states. In addition, such practice could have served as a precedent for Soviet successor states such as Ukraine (Sharp 1992, 466).[11]

Yet more important in the present context than the necessity to build up armed forces from scratch—once the Baltic governments had decided to build up their own national armed forces in the first place—is that they started to do so before the international recognition of the Baltic states' independence in early autumn 1991. According to Mária Huber, in early 1990 Butkevicius had already recruited 1,200 Lithuanian officers from the Soviet

10. For example, as early as 22 March 1990 Gorbachev had urged Landsbergis to stop the recruitment for volunteer forces and to comply with Soviet legislation. Even earlier, immediately after the declaration of independence, Soviet paratroopers seized Lithuanian deserters from the Soviet army. On 25 March 1990, the Soviet military occupied the central press building in Vilnius and five days later the republic's prosecutor office. Gorbachev's warning that he would send Soviet troops to Vilnius to arrest deserters and to ensure Lithuanian compliance with Soviet demands was repeated by Marshall Achromejev on 23 April 1990 in a BBC interview, as well as by Gorbachev and Achromejev in November 1990 (*Österreichische Militärische Zeitschrift* 1990; Vardys and Sedaitis 1997, 163–64).

11. Some Baltic officials also demanded Soviet army equipment be left behind in compensation for the Baltic states' military equipment confiscated by and incorporated into the Soviet armed forces in 1940 (Petersen 1992, 38).

army and 2,100 border guards. Thus, the Baltic authorities began to build up military formations in opposition to the Soviet armed forces, which saw themselves as the sole legitimate armed forces in Estonia, Latvia, and Lithuania at that time.

In consideration of this extraordinarily explosive situation, the limited use of Soviet force against the Baltic peoples, however tragic every single case of actual use undoubtedly was, indicates an appreciation of the nonuse of force on the part of the Soviet government at that time, which was completely at variance with the traditional Soviet understanding of the use of force.[12] It might be argued that the Soviet leadership refrained from the use of violence not because of its dedication to peaceful change, but rather because it feared punishment, especially U.S. military countermeasures. This argument, however, ignores the U.S. government's policy priorities at that time. Neither is it in accord with U.S. policy toward the Baltic independence movements in the late 1980s and early 1990s, and it is not supported by evidence regarding U.S.-Soviet-Baltic relations (discussed in detail in chapter 9).

Economic Measures

It was primarily in the economic realm where the Soviet leadership signaled its willingness to make concessions. This willingness reflected both the emphasis on economic self-determination prevalent in Baltic popular movements' initial agendas and the Soviet central government's intention to assign to the Baltic republics the role of an economic laboratory. It was neither accidental nor hypocritical that Gorbachev sent his greetings to the founding congress of Lithuania's popular movement Sajudis, the Lithuanian Reconstruction Movement, on 22 October 1988, stating that "Sajudis was a positive factor in perestroika" (Vardys 1991, 16). Even earlier, in the summer of 1988, the central authorities in Moscow protected the Sajudis Initia-

---

12. The basic decision on the nonuse of force seems to have been made in the very early stages of perestroika. In 1985, Gorbachev is said to have seen the need to end the Cold War and the Soviet war in Afghanistan and to decentralize the economy. The members of the Politburo reportedly agreed that all political steps had to unfold evolutionarily and without any kind of violence (Jakowlew 2003, 454).

tive Group from "possible repression by the Lithuanian Communist Party" (K. Girnius 1991, 57).

From the point of view of the Baltic popular fronts, the economic realm was the most appropriate sphere to measure its own freedom of action. After all, by launching perestroika, the Soviet government had acknowledged the urgency of economic reforms. It could therefore be expected to react relatively open-mindedly to initiatives by the republics. In contrast, political demands could be assumed to be met with stronger resistance. Relatively moderate political demands by the popular fronts in their incipiency were certainly in part conditioned by "tactical limitations" (G. Smith 1996, 159) in order to avoid too strong reactions by the Soviet government. However, the emphasis on economic self-determination should not be reduced to tactical considerations: it was temporarily an important objective in itself. The leaders of Estonia, Latvia, Lithuania, Georgia, Moldova, Ukraine, and Russia "were quite certain that getting rid of the Union structures would not only make them 'free,' but would also dramatically reduce their economic burden—thus opening the way to prosperity" (Baev 1997, 177).

Accordingly, the quest for economic self-determination was an important driving force behind the emergence of Baltic popular movements. Initially, they hoped to realize it through economic autonomy, later through economic sovereignty, and finally through complete economic independence. One impetus behind these aspirations was the historical evidence of economic near similarity between Sweden and Finland, on the one hand, and Latvia and Estonia, on the other, in the 1930s. From this fact followed the assumption that the Baltic states would have been better off economically and that the living standards would have developed in a way similar to those of Finland and Sweden if the Baltic states had not been incorporated into the Soviet Union. This issue was not only an economic issue, but also one of self-esteem because collective self-images always depend in part on relationships to others (Wendt 1999, 236). The comparison with the Nordic states in the self-assessment of economic and social performance helps explain why in the Soviet Union—in contrast to nineteenth-century western Europe—ethnic mobilization was strongest in those parts where the most successful ethnic groups in terms of educational, occupational, and often political attainment were located. For these groups, the evaluation of their social and economic situation seems to have primarily been an issue of comparison with western Europe rather than with the other Soviet republics (Roeder 1991, 197).

Another impetus behind the quest for economic self-determination was the decline in growth rates of income and industrial production in the Baltic Soviet republics between 1980 and 1985 as compared with the previous five-year period, a decline that reflected industrial maturity and hidden inflation (Götz-Coenenburg 1990; Misiunas and Taagepera 1993, 288–89). Therefore, the economists of the popular movements saw economic self-determination as a necessary condition for recovering lost economic ground and for bridging the gap between Baltic and western European living standards. Demanding economic self-determination indeed preceded claiming political independence. The Estonian Supreme Soviet, for example, declared its economic sovereignty as early as 28 November 1988. In contrast, the sovereignty declaration of 16 November 1988 was not a declaration *of* sovereignty, but a declaration " 'about' a sovereignty already seen to exist in the Soviet constitution. . . . It did not imply that the Estonian SSR had decided to leave the USSR." According to this declaration, "the future status of the republic within the Soviet Union should be determined by a Union Treaty" (Taagepera 1993, 145–46). In mid-1988, the Supreme Council of the Lithuanian SSR had already adopted a resolution titled "On Increasing Economic Independence of the Lithuanian SSR," based on the document "The Conception of Economic Independence of the Lithuanian SSR" prepared by a group of reform economists in the first half of 1988. This conception only vaguely envisaged the possibility of direct trade relations between Lithuania and states not belonging to the Soviet Union. Moreover, it "did not trespass on the main requirement of 'perestroika'—to preserve the USSR. The conception restrict[ed] itself to the economic independence within the USSR and [did] not refer to the perspective of reestablishing the independence of Lithuania" (Simenas 1997, 29).

Even Sajudis at its founding congress on 22 October 1988 confirmed that its objectives were basically in accordance with perestroika—that is, that "sovereignty would be executed within a Leninist Soviet federation" (Vardys 1991, 16). Economic self-determination was initially not seen primarily or exclusively as a way to subsequent political independence. Indeed, "starting with demands for increased 'sovereignty,' linguistic separatism, and various forms of economic autonomy, nationalist forces steadily extended their influence and the scope of other demands, until these coalesced into a call for full political independence" (MccGwyer 1991, 351). National independence thus entered the popular movements' agenda as a result of an evolutionary process. The movements had other aims as well at that point: cultural self-determination, ecological preservation, liberation from the Soviet army, and so on. However, whereas at least some of these aims, such as the protection of the environment, served as " 'intermediary'-problems for reaching the political goals" (Narusk 1997, 13), as proxies with which to camouflage more far-reaching political aims, or as catalysts that were soon to lose importance and be replaced by others, economic self-determination was an end in itself, at least for some time.

In particular, the so-called Four-Man Proposal, often associated with the name of Edgar Savisaar and published in the media in September 1987, can be mentioned in this context. Making use of new thinking among Soviet economic reformers, the plan demanded, among other things, the subordination to the Estonian SSR of all economic subjects located in the territory of the Estonian SSR as well as expanding factory-level autonomy to Estonia as a whole. The plan's direct results were negligible: even some reformers in Moscow were skeptical of economic autonomy, and the central bureaucrats obstructed the implementation of the agreement of April 1988 to turn over seven economic sectors to Estonia. Nonetheless, the plan at least initiated an informed debate about the possibilities and limits of economic self-determination (Taagepera 1993, 128–30).

In Lithuania, Sajudis called for moral, spiritual, and political independence in its three declarations of November 1988, February 1989, and May 1989, respectively. But the notion of economic self-determination certainly played a major role in the situation. Indeed, Sajudis "made radical demands concerning private ownership and management" (Vardys 1991, 12) that went far beyond what Gorbachev initially had intended in launching perestroika. Meetings to discuss, among other things, economic independence were held as early as summer 1988. Even the Lithuanian Communist Party—at that time still in bondage to Moscow—started to think about models for future economic self-management. In Latvia, the Writers' Union demanded republican self-financing and local control over natural resources as early as June 1988. At the same time, the plenum demanded political rights for Latvia as a sovereign member of the Soviet Federation (Muiznieks

1993, 192–93). Demands for economic self-determination thus had the precedence of claiming political independence. Afterward, both objectives were amalgamated and bound up inseparably with one another.

It was precisely in the economic realm that the Soviet leadership most clearly applied a carrot-and-stick policy toward the Baltic republics, but even here it did so only half-heartedly and experienced very soon its own vulnerability. As the most notable "carrot," Gorbachev in July 1989 announced that economic accountability or self-financing *(khozraschet)* would be effective in the Baltic republics as of 1 January 1990,[13] as well as limited foreign trade rights and some minor form of fiscal authority. From the Baltic popular movements' point of view, the attraction of economic accountability seems to have been rather limited from the outset. It certainly offered much less than the experts of the popular fronts were hoping for. However, it contained at least one important ingredient: economic accountability defined "each national territory as a single and coherent entity and [gave] its administrators a frame of reference for making decisions" (A. Trapans 1991, 98). It thus paved the way for "an unprecedented measure of economic self-determination" during what has been called, with respect to Latvia, the second or sovereignty phase of transition, 4 May 1990 to 21 August 1991 (Dreifelds 1996, 17).

What is more—and this certainly did not coincide with its initiators' intentions—economic accountability supported the secession of the Baltic republics from the Soviet Union and their subsequent development, for it emphasized national, or republicwide, territory and assigned responsibilities to the republics' authorities. Republicwide economic autonomy facilitated the breakup of the Soviet Union along the borders of the republics and provided the republics' authorities with the necessary administrative skills to start taking care of themselves. Thus, there are similarities to what Rogers

13. "An entity functioning according to the principles of *khozraschet* is not subsidized; it is expected to be financially self-supporting and enjoys a measure of financial autonomy. *Khozraschet* institutions are supposed to practice 'businesslike management' and are given more normative types of assignments. The role of various financial levers and credit (and hence the banking system) is slated to increase substantially, although the banking changes will be gradual. Soviet reform documents generally project an image of autonomous, self-financing socialist enterprises, freed from the petty tutelage of ministries and involved in markets and competition" (A. Trapans 1991, 89).

Brubaker calls the Soviet "institutionalized multinationality"—the definition in national terms of the "component *parts* of the [Soviet] state and the citizenry" rather than of "the state or citizenry as a *whole*." The Soviet Union could disappear in a surprisingly orderly fashion not least "because the successor units already existed as internal quasi-nationstates," including at least some degree of economic authority and administrative skills (1996, 29, 41, emphasis in original).[14]

Gorbachev certainly considered obligingness in economic issues a means with which to make the refusal of political concessions palatable to the Baltic popular movements. The Baltic popular fronts, however, either saw economic concessions increasingly as a path to political concessions or regarded the degree of economic autonomy assigned to the republics by Gorbachev as insufficient. Their conviction grew that the degree of economic self-determination they desired was to be realized only outside of the Soviet Union. In both ways, the result was the same, namely a radicalization of political demands. Gorbachev, when it became clear that his strategy had failed, radicalized himself—presumably under pressure from conservative hardliners in Moscow—and reached for the "stick." Reacting to the Lithuanian declaration of independence, Gorbachev, on 18 April 1990, among other things initiated the cut off of approximately 80 percent of the oil and natural-gas deliveries to Lithuania and closed the republic's borders to prevent supplies reaching it from abroad.[15] After the Lithuanian government announced on 29 June a moratorium of one hundred days on the declaration of independence in reaction to the embargo, Gorbachev soon lifted the embargo.

Although the embargo carried with it some negative consequences for the all-Soviet economy, Gorbachev did not necessarily have to give in when the Lithuanian government declared the moratorium on the declaration of independence. The Soviet government was certainly in a much stronger economic position than the extraordinarily weak Lithuanian republic,

---

14. According to Brubaker, "no other state has gone so far in sponsoring, codifying, institutionalizing, even (in some cases) inventing nationhood and nationality on the sub-state level, while at the same time doing nothing to institutionalize them on the level of the state as a whole" (1996, 29).

15. Ironically, electrical production by the disliked Ignalina Nuclear Power Plant, the low safety standards of which had triggered the environmental protest movement in Lithuania, now helped compensate for the interrupted oil and gas deliveries (Senn 1996, 177).

which was heavily dependent upon intra-Soviet trade.[16] Moreover, on 20 April 1990 the White House played down the economic sanctions as "another unfortunate step" (Vardys and Sedaitis 1997, 170), and four days later U.S. president George Bush, "to the surprise of many . . . rejected U.S. economic sanctions against the Soviet Union as a response to the Soviet economic measures against Lithuania" (Garthoff 1994, 424), thereby ignoring a nonbinding resolution by the U.S. Congress (Vardys and Sedaitis 1997, 171) and indirectly encouraging the Soviet leadership to continue its economic pressure on the Baltic republic.

The moratorium was to last one hundred days and not more. It did not at all change the objectives of the Lithuanian popular movement, which at that time could aptly be called an independence movement. The Lithuanian Parliament had made it abundantly clear that legal actions to restore Lithuanian independence were to be resumed after the moratorium and in case negotiations were broken off (Vardys and Sedaitis 1997, 172).[17] Independence remained the aim. Nevertheless, Gorbachev lifted the embargo for the most part in exchange for only modest concessions by the Lithuanian government. Furthermore, the Soviet leadership did not exert full-fledged economic pressure against Latvia and Estonia in the shape of, for example, an embargo against all three Baltic republics, thereby cutting them off entirely from inter-republic economic exchange and from energy supplies upon which they were heavily dependent.

---

16. "Energy supplies and some food products were rationed. Tens of thousands of workers who stood idle in factories that could not operate were taken care of by payments of 'vacation' salaries that drained the republic's budget. . . . Car and truck transportation stopped, public transportation thinned, and people were forced to bicycle or walk. The blockade cost the republic tens of millions of rubles in damages and lost production. According to Lithuanian data, 35,000 industrial and construction workers lost their jobs, 435 enterprises could operate only part time, 6 large industrial enterprises were completely closed down, and the republic lost 415.5 million rubles worth of production and 125 million rubles for the state budget up to July 1" (Vardys and Sedaitis 1997, 166).

17. During these one hundred days, official bilateral negotiations between the USSR and the Republic of Lithuania were to begin. These negotiations never took place because, among other things, the Soviet authorities were not willing to negotiate with representatives of the *Republic* of Lithuania, but only with those of the *Soviet* Republic of Lithuania. Negotiating with the representatives of the Republic of Lithuania would have meant an indirect recognition of the Lithuanian declaration of independence on the part of the Soviet authorities (Simenas 1997, 38).

Political Measures

The political disintegration of the Soviet Union can essentially be understood in terms of the decline and collapse of a single-party system—that is,
a political system based on the leading role of the Communist Party of
the Soviet Union (CPSU) (Gill 1994). It is thus hardly surprising that the political measures undertaken by the central Soviet authorities were weak,
unimaginative, and in general off the point. Politically, the Soviet leadership
practically let everything happen and declared disintegration invalid
retroactively, but did almost nothing to enforce the invalidation. This weakness followed from the unexpected challenge posed by the popular movements in the Baltic republics and the growing unrest within the Communist
parties of the republics. It also reflected the beginning of the CPSU confidence crisis at the all-union level. In 1989, CPSU membership started to decline for the first time since 1954. The loss of party members at that time was
not dramatic at the all-union level. The sharp decline in the number of CPSU
candidates from 701,730 in 1985 to 372,104 in 1990 was nevertheless an indicator of the beginning and rapidly increasing confidence crisis (Gill 1994,
101).

   In the Baltic republics, the erosion of the Communist Party was accelerated by the emerging popular movements' challenge to the monopoly of
rule claimed by the CPSU. As to the Communist Party of Lithuania, it is important to note that from the 1960s onward Lithuanians had entered the
Communist Party and started their march through the institutions—which
was not inconsistent with Soviet nationalities policies—aiming to Lithuani-
anize the outcome of the political decision-making process, which certainly
was inconsistent with the ideas of the all-union authorities (Roeder 1991,
203–10; Rikmann 1999, 63).[18] Now, for the first time in Soviet Baltic history,
they saw the possibility to influence political decision making outside of
and increasingly in competition with the Communist Party. The public support was impressive. Sajudis, for example, won in thirty-six of forty-two
electoral districts in the elections to the Congress of People's Deputies on 26
March 1989 (Vardys and Sedaitis 1997, 144). Furthermore, on 7 December

---

18. In 1959, ethnic Lithuanians constituted 55.7 percent of the members of the Lithuanian
Communist Party; in 1965, 61.5 percent; in 1970, 67.1 percent; in 1975, 68.5 percent; in 1980, 69.4
percent; in 1986, 70.4 percent (Misiunas and Taagepera 1993, 281, 360).

1989, the Lithuanian Supreme Soviet amended the republic's Constitution and by a vote of 237 to 1 eliminated Article 6, which guaranteed the Communist Party's leading and guiding force (Gill 1994, 99; Senn 1996, 176; Vardys and Sedaitis 1997, 152), with the Latvian Supreme Soviet to follow with a similar decision on 11 January 1990 (Gill 1994, 210).[19]

In view of the weak reactions of the central authorities in Moscow to the challenge posed by the political developments in the republics, I analyze first the CPSU's relationship to the Communist parties in the republics and then its relationship to the popular movements. The focus on Lithuania is justified by some important developments in Lithuania during this period: the defeat of the CPSU at the elections to the Congress of the People's Deputies in March 1990 and Sajudis's corresponding victory; the disassociation of the Communist Party of Lithuania from the CPSU in December 1989; Sajudis's victory at the elections to the Lithuanian Supreme Soviet in February and March 1990; and the swift radicalization of the independence claims (moving from moral to spiritual to full political independence) from November 1988 to March 1990. These developments called the central authorities' attention increasingly to the developments in Lithuania. The situation in Lithuania thus exemplifies the central authorities' helplessness and lack of planning during this period in respect to the question of how to deal with the Baltic republics.[20]

Soviet institutionalized multinationality (Brubaker 1996) or ethnofederalism (Roeder 1991) was based on the idea of indigenizing the republics' civilian *nomenclatura*.[21] Indigenization *(korenizatsija)* meant assigning to representatives of the titular nations of the republics most of the decisive positions within the Communist power edifice (Communist Party, administration, higher educational institutions, Academies of Science, and so on). As a

19. Even *Pravda* and subsequently Gorbachev declared that Article 6 "might not be essential." A reevaluation of the article, however, should follow from "a careful consideration of the Constitution" without political pressure (Gill 1994, 99).

20. For Latvia, see J. Trapans 1991c, 36–41; Muiznieks 1993, 196–201; Karklins 1994, 77–80, 90–101; G. Smith 1996, 155–62. For Estonia, see Taagepera 1993, 170–77; Kionka and Vetik 1996, 137–40; Lauristin and Vihalemm 1997, 84–91.

21. The following discussion relies on Roeder 1991. Following Skak, *nomenclatura* refers to "a system for the recruitment of politically reliable persons—trusted party members—to positions of importance within state and society, such as enterprise managers, leaders of institutions and public servants (e.g. diplomats)" (1996, 92).

result, "since the mid-1950s, most leadership positions and high-ranking posts in practically all the republics belonged to members of the local nationalities" (Zaslavsky 1993, 37). From the all-union point of view—and, of course, the all-union point of view did not necessarily correspond with the reasons that made individual members of the titular nations enter the party—*korenizatsija* aimed to monopolize public expression, increase loyalty to the center, and channel and control potential protest by preventing its public expression in all but officially approved ways or by banning and punishing it altogether. Not least, the indigenous elites were responsible for preventing counterelites from arising and institutionalizing. Loyalties with the indigenous elites were cemented by a substantial lengthening of the party elites' term of office, with the first secretary of the Lithuanian Communist Party, Antanas Snieckus, as an unofficial record holder. He led the party from "underground days in 1936" to his death in 1974 (Misiunas and Taagepera 1993, 205).

For some decades, this system of using "nationalism as an additional instrument of control (and a safety valve)" (Venclova 1991, 48) was fairly successful. Its success, however, depended on the all-union authorities' capability to deliver material rewards, benefits, status, and the prospects of upward mobility within the power structure for the republican elites. An ever-growing number of rising experts and specialists of the titular nations had to be integrated into the Communist power edifice, which, as a result, became increasingly saturated—the more so because purging these ethnic cadres had become an increasingly unpopular way to deal with the problem since the 1960s and was becoming counterproductive in terms of efficiency from the point of view of the authorities in Moscow considering the cadres' lengthening term of office and the strengthening of their power base at home.

Korenizatsija was designed, in Mette Skak's words, to "create national elites who could be entrusted with the implementation of sovietization targets," but it was also "instrumental in strengthening feelings of ethnicity and national pride" (1996, 84–85). This ambiguity was not a severe problem as long as the Soviet economy provided the financial and budgetary means to integrate the rising indigenous elites into the power structure by constantly expanding the structure. The post-Stalinist decentralization had already offered a variety of possibilities for the proliferation of the power apparatus. Yet during the 1970s and 1980s, a growing "disparity between the increasing demands for rewards and mobility opportunities and the diminished capability to meet those demands" became apparent: "Ethnic cadres were forced to intensify their pressure on Moscow to gain additional resources, and consequently, competition for the same scarce resources grew among ethnic communities. But declining growth rates left Moscow with even fewer resources to respond to rising demands" (Roeder 1991, 213–14).

Besides increasing competition among the elites of the titular nations, a further result of declining growth rates on the all-union level was a growing redistribution of resources among the republics to the detriment of the more advanced ones. This redistribution pattern led to a relative stagnation of the elites' social, material, and status position in the more advanced republics and further infringed on their capabilities to integrate rising experts into the satiated power structure. As a corollary of this development, the Communist cadres of the republics sought to enlarge the resources under their control by, among other things, demanding economic autonomy. This demand was not an exclusive feature of the popular movements. Indeed, "in all three Baltic republics calls for 'regional economic accountability,' 'territorial cost accounting,' and 'self-financing' . . . supported the attempts on the part of republican cadres to wrest control of industries from the centralized ministries" (Roeder 1991, 219–20).

The Communist cadres' strategy was a risky business for at least three reasons. First, all-union authorities' counterreactions could not in principle be ruled out. In parallel with the republic elites' increasing inability to realize the central authorities' interests and their own aspirations to disassociate themselves from the CPSU, the central authorities might have started to look for personal alternatives in the republics. Second, strategies to defend the privileged positions of the titular elites could have been met with violent reactions on the part of the nontitular nations in the republics. Third, potential counterelites within the republics could have benefited from the cadres' strategy, and forces could have been unleashed that would subsequently be difficult to control.

From January to November 1989, nearly 9,000 party members left the Lithuanian Communist Party. This exodus may or may not have been a reflection of the party elite's growing incapability to provide resources and mobility opportunities in material and social terms (Brill Olcott 1990, 36). The pressure on the party elites to react in one way or another increased in any case. The party's Seventeenth Plenum argued in favor of greater independence from the CPSU as early as February 1989. At the occasion of the Eighteenth Plenum, held only four months later on 24 June 1989, many speakers called for the creation of an autonomous party (K. Girnius 1991, 60). Until October 1990, however, a complete schism could be avoided, albeit at the price of the coexistence of two mutually exclusive positions within the party: the Communist Party of Lithuania "was to be an independent party with its own program and statutes, but also to be an integral part of a renewed CPSU" (Gill 1994, 92). Even the Baltic republics' Communist leaders now considered autonomy in the socioeconomic realm as offered by the Soviet leadership an insufficient incentive to back off the popular movements and to avoid being swallowed by them.

As a consequence, the Lithuanian Communist Party made steps to distance itself from the CPSU by, among other things, halting the construction of the Ignalina Nuclear Power Plant and reversing the Russification of education (Vardys 1991, 15). On 20 December 1989, at the Twentieth Lithuanian Communist Party Congress, 855 delegates split off from and declared themselves independent of the CPSU and soon named the new party organization the Lithuanian Democratic Labor Party (LDDP) (Misiunas and Taagepera 1993, 324). One day later 160 delegates declared their adherence to the CPSU. Gorbachev's reaction was weak. Of course, he insisted on the party's unity and denied the republics' parties the right to cut loose from the CPSU. At the same time, he had no more to offer the Communist parties of the republics than a greater degree of autonomy over the resolution of local issues and a greater degree of codetermination on national issues (Gill 1994, 93, 206). Gorbachev's attempt to bring the Lithuanian republic's Communist Party back on track during a personal visit following the party's "declaration of independence" profoundly failed.

"The emergence of Sajudis was initially made possible by Mikhail Gorbachev's *glasnost* and *perestroika*" (Vardys 1991, 11). Sajudis was an umbrella organization without formal membership. Almost one-half of the founding members of the Initiative Group were indeed members of the Communist Party (Lieven 1994, 226). Indeed, Sajudis and its Estonian and Latvian counterparts were "not . . . as cohesive and single-minded in their period of emergence as accounts found in the West sometimes imply" (J. Trapans 1991b, 5). As shown earlier, Gorbachev's position toward Sajudis had initially been neither reluctant nor confrontational. As long as Sajudis operated within the framework of the Soviet understanding of ethnicity in terms of instrumentality—that is, focusing on social and economic interests—Gorbachev did not in principle disagree with the popular movement. Yet Sajudis's turn toward a more "primordial" understanding of ethnicity—one focusing on identity, culture, and symbols—found the Soviet leadership profoundly perplexed because this understanding was completely at odds with the principles of Soviet ethnofederalism. As a consequence, Gorbachev did not find an appropriate answer when Landsbergis "agreed that all instrumental issues were negotiable but not the primordial ones—that is, the symbolic declaration of independence" (Roeder 1991, 232). As long as Sajudis operated within the limits of perestroika, and as long as perestroika had not been questioned in principle by hardliners in Moscow, Gorbachev welcomed and (mis)understood Sajudis as a tool with which to test the reformability of the Soviet economy.

As noted earlier, in the summer of 1988 the central authorities in Moscow helped protect the Sajudis Initiative Group from possible repression by the conservative Lithuanian branch of the Communist Party. Although Sajudis initially "was not inimical to the Party but simply wished to be independent from it," the Lithuanian Communist Party's leadership encountered Sajudis with suspicion and was concerned with defending "Party prerogatives in the face of Sajudis's growing power" (Vardys 1991, 13, 15).[22] The conservatism of Lithuania's Communist Party branch became evident at the occasion of the nineteenth Communist Party conference held in Moscow in late June 1988. The delegates chosen by the Lithuanian Communist Party represented the old *nomenclatura*'s way of thinking and were appointed by the leadership, although the central authorities' intentions had been to show the progressivity and reform-mindedness of the Communist Party. The procedure applied by the Lithuanian Communist Party—appointing rather than electing delegates—showed exactly the opposite and

22. Brill Olcott adds that "First Secretary Songaila immediately attacked the [Sajudis Initiative] group as 'antisocialist,' ensuring its isolation from officially sanctioned reform groups" (1990, 33). This isolation required a great deal of tactical maneuvering on the part of Sajudis. Landsbergis's 1989 emphasis on Sajudis's "close relationship with the republic's Party authorities," upon which "the ability of his movement to conduct its activities" depended, seems to be a case in point (Roeder 1991, 211). For a more positive assessment of the Communist Party cadres in the Baltic republics, see Roeder 1991, 211.

thus contributed to its alienation from Sajudis (Vardys and Sedaitis 1997, 101–2).

Regarding Sajudis as a positive factor in perestroika, Gorbachev for his part had even sent greetings to Sajudis's founding congress in October 1988. How narrow his economic latitude was became apparent only four weeks later when he refused economic proposals by the Baltic popular movements as unacceptable. Moreover, he faced the dilemma that economic concessions to Sajudis's demands might be understood as an invitation to further political demands. This would be especially true if the economic concessions came too late and were only half-hearted, modest, and not unequivocally supported by the central CPSU authorities in Moscow, where Gorbachev encountered increasing resistance. The refusal of concessions, however, would almost inevitably lead to a radicalization of the popular movements and to an understanding that agreement with Gorbachev was impossible anyway. After all, Sajudis was neither easily mollified by concessions nor intimidated by a lack thereof. Moreover, with the benefit of hindsight, it is apparent that Gorbachev's reaction to Sajudis was based on a profound underestimation of the movement's momentum, its degree of radicalization within a relatively short period of time, and its support by the population.

Of course, Sajudis's 20 November 1988 declaration of moral independence, its 16 February 1989 declaration of spiritual independence, and the Lithuanian Supreme Soviet's 18 May 1989 declaration of republican sovereignty did not help ease the tensions between Sajudis and the central authorities in Moscow. Neither did Sajudis's victory at the elections to the Congress of People's Deputies in March 1989 or the victory of candidates supported by Sajudis at the multiparty elections to the Lithuanian Supreme Soviet on 24 February 1990 and in follow-up elections two weeks later. Furthermore, "when Estonia had declared its republican sovereignty in November 1988, the Politburo had threatened it with repercussions, but these had never come" (Vardys 1991, 20). The relinquishment of sanctions on the part of the Politburo contributed both to its image as a paper tiger, preoccupied with itself and with the Soviet economy's negative growth rates from 1986 to 1990, and to the radicalization of the popular movements, which led to the declaration of full independence by the Lithuanian Supreme Soviet on 11 March 1990.

Two months earlier, Gorbachev, while visiting Vilnius, had "reluctantly conceded the theoretical possibility of secession, though he argued for liberalization within a continuing union" (Garthoff 1994, 419). Two days before the independence declaration was made, he introduced new terms for Lithuanian independence, which hardly increased his popularity. For example, he offered independence in exchange for full compensation for all Soviet postwar investments in Lithuania, amounting to $33 billion; the restoration of Lithuania's 1939 borders; and the secession of Klaipeda to the USSR (Brill Olcott 1990, 39–40; Vardys and Sedaitis 1997, 155).

History always includes the alternative possibilities that did not materialize. To say that the Soviet policy in the late 1980s and early 1990s was dedicated to the idea of peaceful change thus carries with it the necessity to make some retrospective speculative remarks about what the Soviet leadership could have done at that point had its policy been based on different ideas. Three possible reactions have already been pointed out. First, the central authorities could have made use of military means on a large scale. They had at their disposal almost all the means they needed to destroy the opposition movement, including secret service, police, public prosecution, jurisdiction, imprisonment, and the military (Kis 1999–2000, 21). Second, the Soviet central authorities could have imposed an enduring economic and energy embargo on the three Baltic republics. Third, the party authorities could have prevented Sajudis (and its counterparts in Estonia and Latvia) from running their own candidates for the 26 March 1989 elections to the Congress of People's Deputies and for the 1990 elections in the republics. The Soviet government could also have cancelled the elections altogether in order to avoid the delegitimization of the Communist Party that was to be expected in democratic elections. It did none of these things. Indeed, the Soviet authorities did not interfere in the 24 February 1990 elections in Lithuania even though Gorbachev "knew that *Sajudis* would demand full independence after his visit there in January 1990" (Gvosdev 1995, 21). Thus, "the Lithuanian position [on independence] was never seriously challenged by Gorbachev" (Park 1991, 259).

Nor did the Soviet authorities prevent the referenda on independence held in Lithuania on 9 February 1991 and in Estonia and Latvia on 3 March 1991. They could easily have hindered them by, for example, positioning a handful of soldiers in front of each polling station. Once the referenda were held, and once they had resulted in strong support for independence, however, the legitimacy of the strivings for independence and their dynamics were considerably enhanced. After all, the Soviet Constitution—even after the changes adopted by the USSR Supreme Soviet on 1 December 1989—included in Article 72 the right of each republic to secede from the USSR, specified in the 1990 Law on Secession (Clemens 1991, 244–45).[23]

In Estonia, "the higher the percentage of non-Estonians, the more there were 'no' votes" (Raitviir 1996, 381) in the referendum on independence.[24] Nevertheless, independence was supported in the referenda by large portions of those parts of the population in Latvia and Estonia who were later temporarily completely excluded from the political community as noncitizens or aliens. In Estonia, approximately 30 percent of the part of the population that is here colloquially referred to as "non-Estonians" voted for independence.[25] The share of naysayers in the referendum was much less than the share of "non-Estonians" in the population (Taagepera 1993, 194).[26]

23. According to the 1977 Soviet Constitution, the so-called titular nations of the republics had the right to secede from the USSR, but "there was no procedure provided for the implementation of this right" (Müllerson 1994, 75). The 1990 Law on Secession required, among other things, a two-thirds approval in a republicwide referendum without prior campaigning and a five-year waiting period to sort out disparate economic claims. During this period, "Soviet law would remain in force and a call by 10 percent of the population would force a new referendum." Afterward, the Soviet Congress of People's Deputies had to ratify the republic's secession. Finally, "the seceding republic must also pay for the resettlement for all those who wish to remain within the USSR" (Brill Olcott 1990, 43). This Law on Secession, in the final analysis, provided for the condition of near impossibility for secession. As a consequence, Baltic politicians argued that what was at stake was the end of an illegal occupation and therefore the restoration of independence rather than secession from the USSR. Such an interpretation made the Soviet Law on Secession appear irrelevant to the Baltic case.

24. In predominantly non-Estonian areas, the option of sovereignty within a confederation was offered and favored by, for example, 65 percent, 79 percent, and almost 90 percent in Kohtla-Järve, Narva, and Sillamäe, respectively (Kaplan 1993, 216).

25. It makes a substantial difference whether one refers to the Russian-speaking part of the population in the Baltic states as non-Latvians, non-Lithuanians, non-Estonians, Russian speakers, the russophone population, Baltic Russians, occupants, colonizers, settler community, noncitizens, or aliens. As a result of a particular linguistic designation, a particular part of the population may be treated "normally" as anybody else, or it may not be. Referring to a particular part of the population in a specific way may also facilitate its inclusion into or exclusion from the political community.

26. In percentage of votes, the yes/no ratio was 77.8/21.4 percent in Estonia; 73.7/24.7 percent in Latvia, and 90.5/5.5 percent in Lithuania, with 82.9 percent of all eligible voters participating in Estonia, 87.6 in Latvia, and 84.7 in Lithuania (Taagepera 1993, 194; Karklins 1994, 101–2; Raitviir 1996, 348–49; Vardys and Sedaitis 1997, 183). In Estonia, the question was: "Do

In Latvia, the nationality of the voters was not registered at the polling station, but by way of calculation it was concluded that at least one-third of "non-Latvians" supported Latvian independence in the referendum (Karklins 1994, 101–3; Dreifelds 1996, 78; Latvian Human Rights Committee 1999, 9). Most important, by supporting independence, the "non-Latvians" and "non-Estonians" delegitimized the central Soviet government's claim to speak and act on behalf of the "Baltic Russians." After the referenda, the question was indeed not whether but rather when and on which conditions the Baltic republics were to become sovereign states again. From this point of view, it did not make much sense for the Soviet government to declare the referenda invalid retroactively and to organize its own referenda—the die was cast.

The Soviet government could also have prevented its own delegitimization by not agreeing to the establishment of a commission to explore the Molotov-Ribbentrop Pact, as it had done regarding similar commissions during the preceding decades. It could furthermore have influenced the composition of the commission by vetoing any of the nominees suggested by the Baltic popular movements or could have influenced the commission's findings. It did neither. Once the commission was established, and once it had come to the conclusion that the Baltic states had indeed been incorporated into the Soviet Union illegally[27] on the basis of the two notorious secret protocols attached to the 23 August 1939 German-Soviet Non-

---

you want the restoration of the state sovereignty and independence of the Republic of Estonia?" (See also note 24.) Former Soviet servicemen had no right to vote, but their family members did. In Latvia, the question was: "Are you for a democratic and independent Republic of Latvia?" All permanent residents of Latvia could participate in the referendum. Consequently, Soviet soldiers on active duty were barred from voting because they were not registered as permanent residents. In Lithuania, the question was: "Do you favor [the idea] that the Lithuanian state should be an independent democratic republic?" As a percentage of all eligible voters, the "yes" figures corresponded to 64.5, 64.6, and 76.4 percent in Estonia, Latvia, and Lithuania, respectively.

27. The commission came to the following conclusions: "1) The Nazi-Soviet Non-Aggression Treaty had a secret protocol; 2) The secret protocol was a deviation from Leninist principles of Soviet foreign policy. Moreover, the delineation of spheres of interest between Germany and the Soviet Union contradicted the sovereignty and independence of other countries, with which the USSR had signed treaties; and 3) Stalin and Molotov conducted clandestine negotiations with Germany on the secret protocols. The protocols therefore did not reflect

Aggression Treaty and the 28 September 1939 German-Soviet Border and Friendship Treaty—a conclusion anticipated at the occasion of the extended plenum of the Latvian Writers' Union in June 1988 and confirmed by both the Supreme Soviet in Moscow in summer 1989 and the Congress of People's Deputies in late 1989—a relegitimization of Soviet domination of whatever kind over the Baltic states without consent of the Baltic peoples was virtually impossible.[28] Here, glasnost seems to have been the condition of impossibility for perestroika.

Peaceful Change in the Soviet Union

Simon Dalby writes that the "script of the end of the Cold War, as a Western triumph rather than as a result of the Soviet decision to end the military confrontation, adds to the ideological support for maintaining the institutions of the Cold War and modeling future policies on this apparently successful formation" (1997, 12). Furthermore, it underestimates the substantial changes introduced in Soviet foreign policy after Gorbachov's coming to power and the initiation of *novoye mishleniye* in foreign policy in late 1987 and early 1988.[29]

These changes were characterized by at least four elements. First, in superpower relations, the idea of common security took the place of unilateral or alliance-based paths to security. Second, the idea of a constant and irreconcilable struggle based on class values between socialism and capitalism

the will of the Soviet people, which does not bear responsibility for them" (U.S. Commission on Security and Cooperation in Europe 1990, 74).

28. For the commission's exploration of the Molotov-Ribbentrop Pact, see Muiznieks 1993, 192–93; Lieven 1994, 222; G. Smith 1996, 160; Vardys and Sedaitis 1997, 150. For personal accounts of two of the chief protagonists in this exploration, see Vulfsons 1998, 79–86; and Jakowlew 2003, 496–501, 767–87. The Estonian government newspaper *Rahva Haal* published the secret protocols on 10–11 August 1989. It was the first Communist Party organ to do so. The Sajudis newsletter *Sajudzio zinios* had published the protocols already on 5 August.

29. Gorbachev spoke of "new political thinking" for the first time publicly in a speech given before the British Parliament in September 1984. Reflecting his responsibility within the Soviet *nomenclatura* and his initial interest in reforming the Soviet economy, it took some time until new thinking entered foreign, security, and defense policy. Gorbachev was quick to learn that the success of the economic reform was in large part dependent on "lifting the shadow of a further Soviet military buildup in the 1990s and beyond" (Meyer 1988, 128).

was abandoned for the benefit of universal human values. Third, it was recognized that "resort to force or threats of force is neither an efficacious nor a legitimate way to resolve interstate conflicts" (Herman 1996, 271–72). This recognition included a ban on military intrusions of major powers, including the Soviet Union itself, "in the internal affairs of lesser brethren" (Mcc-Gwyer 1991, 186). Finally, Gorbachev's reform group had the courage to see that Soviet policy practices helped sustain the Cold War, to reassess Western intentions, and to change Soviet policy. Indeed, "someone else in his place might have found a more aggressive solution to a decline in power" (Wendt 1995, 80).

The interpretation of the Baltic states' independence as the triumph of three small, victimized nations following a David versus Goliath scheme of fighting courageously, nonviolently, and efficaciously against the apparently overwhelming opponent has been carefully conserved as a part of the myth of the Baltic states' foundation (Bildt 1994, 75; Roberts 1994, 191–94; Jundzis 1995, 550; Clemens 2001, 39–54).[30] And, of course, this script is not wrong. It is, however, only one part of the full story and neglects the dedication to peaceful change on the part of the Soviet government, at least in its relations with the Baltic republics. If the Soviet leadership had decided to crush the Baltic popular movements, it would obviously have had in its power the necessary means to do so. Apparently, the leadership decided to forgo the use of large-scale force while dealing with the Baltic issue: "It was becoming clear to a growing number of reformers that the USSR could not take its place among the world's 'civilized nations' if it insisted on violating at home and abroad the norms of behavior governing Western democratic regimes' treatment of their own citizens and relations between member states of that community" (Herman 1996, 296).

However important the Western states' acceptance of the Soviet Union as a "normal" nation-state was—not least as a condition for the possibility of the "common European house," Gorbachev's design for a security system with which to replace the existing military alliances[31]—even more impor-

30. However, according to Huber, Landsbergis started to advocate armed resistance in January 1991 (2002, 169).

31. Although the common European house, "based as it was on the [Conference on Security and Cooperation in Europe] model, would have included the United States and Canada" (Fierke 1997, 244), some of Gorbachev's statements revealed an "anti-American thrust" (Mcc-

tant was the fact that, in the words of Soviet foreign minister Eduard She-vardnadze, "if we were to use force then it would be the end of perestroika. ... If force is used, it will mean that the enemies of perestroika have triumphed" (cited in Herman 1996, 308).[32] The insight that "security is a political problem which requires political solutions" (Light 1988, 309) was accompanied by the conviction that the survivability of perestroika was dependent on a ban on the use of force. Thus, Shevardnadze, at the occasion of the Plenary Session of the USSR Supreme Soviet on 23 October 1989, described perestroika's impact on Soviet foreign policy as follows: "We can talk about a nonviolent world and the triumph of the force of right over the right of force, relying on the rule-of-law nature of the Soviet state and its adherence to the rejection of the use of force within the country" (Shevard-nadze 1989). Likewise, in 1989 Shevardnadze told U.S. secretary of state James Baker that Moscow would not use force against the border republics (Clemens 2001, 42).

The introduction of a policy of nonviolence and the dedication to peaceful change thus appear to have been integral ingredients of the "Soviet foreign policy revolution" (Herman 1996) launched in the train of Gorbachev's coming to power, upon which the success of perestroika in part hinged. It is thus insufficient to speculate that the Soviet authorities were prevented from imposing "central rule by military means [because] such actions would seem to have gone against [Gorbachev's] own instincts" (Lieven 1994, 230) or his need for Western support. Paraphrasing Herman, resort to force or threats of force was considered neither an efficacious nor a legitimate way to resolve intrastate conflicts.[33] Indeed, that the Soviet military

---

Gwyer 1991, 213) by not suggesting that the United States had a place in this design. This omission, in turn, made many Western observers understand the initiative as an updated version of the traditional Soviet objective of disembarking the United States from the European continent.

32. Eduard Shevardnadze's role can hardly be overestimated. According to Ekedahl and Goodman, "Shevardnadze believed that members of his generation had acquired a '1956 complex' for the rest of their lives—rejecting the use of force as a political instrument. Commitment to the nonuse of force became one of his most important contributions to the end of communism and the Cold War, permitting the virtually non-violent demise of the Soviet empire and the bloodless dissolution of the Soviet Union itself" (2001, 16). Shevardnadze resigned in December 1990, three weeks before the violent intervention in Lithuania.

33. Herman argues that "a parallel argument about Soviet identity and rejection of force in Eastern Europe could be made concerning the dissolution of the USSR itself" (1996, 308).

"supported antigovernment forces [in the Baltic republics] by, for example, dropping Interfront leaflets from military helicopters" (Karklins 1994, 121) supports this interpretation by way of its insignificance, harmlessness, and nonviolence.[34]

In the language applied in this book, the Soviet leadership at that point seems to have crossed the threshold beyond which it no longer considered the use of large-scale force in the Baltic Sea region a legitimate way of settling interstate and intrastate conflicts. Instead, political and economic measures were applied—albeit ambiguously, half-heartedly, and inconsistently, as shown earlier. The use of only minor forms of force in January and in the summer of 1991 supports rather than disproves this interpretation because of its undecidedness, ambivalence, and lack of consistency and central command. It was a far cry from an all-out approach and does not seem to have been initiated by the Soviet political leadership (although Gorbachev's role in authorizing it remains something of an enigma). Thus, if the Soviet leadership is being blamed for the use of force in 1991, it should also be given some credit for the use of only minor forms of force.

That the Baltic peoples and especially their political representatives did not subscribe to this point of view, however, is hardly surprising. Security community building needs time, and the short period of time since the introduction of perestroika, glasnost, and *novoye mishleniye* was clearly insufficient to create a mutual-response community. The dissolution of the Soviet Union on 24 December 1991 brought an end to this analytically challenging test case of security community building.

34. "Founded in October 1988, Interfront was dedicated to the defense of socialism and the unity of the USSR and emphasized the leading role of the CPSU" (Karklins 1994, 80). For Estonia, see Ilves 1991.

6

The Construction of Security in the Baltic States

TO SAY THAT SOMETHING IS CONSTRUCTED means in this book that it is a product of human activity. In the social world, this assessment almost amounts to a truism because "all social phenomena are *constructions* provided historically through human activity" (Berger and Luckmann 1967, 123, emphasis in original). Constructing security thus means that security and security policies are also products of human activity. They are also, like all human actions, an expression of power relations within society and do not follow from cosmic or natural laws, divine will, or geography. Rather, they are the result of human decisions and choices between different alternatives, reflecting interests, and identities. As briefly noted in chapter 4, security policy is a conflict over the appropriation of resources, and the buildup of armed forces is an act of domination that has to be realized against resistance. Security policy, therefore, should not be taken for granted. Claiming to represent the "national interest" while designing and implementing security policy is often under suspicion of ideology, with *ideology* basically meaning that "a particular definition of reality comes to be attached to a concrete power interest" (Berger and Luckmann 1967, 141). Security policy contributes to the construction and imagination of a particular internal order that benefits the members of the security establishment.

With respect to the Baltic states, for example, it has been argued that the security policy during the 1990s was a reflection of the self-interests of the newly created armed forces as an organized interest group rather than an expression of an abstract "national interest" (Heinemann-Grüder 2002). This assessment is important regarding the second half of the 1990s, but in the first half armed forces were nonexistent as an organized interest group. In order to define and defend their interests, the armed forces had to be organized in the first place, but the decision to create and organize armed forces was again a decision between different alternatives. It followed from deeply ingrained ideas of the state among Baltic decision makers that were not determined by necessity. Rather, as shown in the preceding chapters, they reflected a particular, historically contingent understanding of the state. That this understanding of the state corresponded with the subsequent self-interests of the armed forces is not surprising: it reflects the complementary relation between the modern nation-state and the armed forces (Krippendorff 1985, 275).

All the same, security policy is often taken for granted and delegated to those representatives of the state who deal with security on a professional basis and claim expert knowledge. Because security is usually represented as being detached from everybody's everyday experience, this delegation of authority is often argued to be in everyone's interest. Democratic interference and participation in foreign and security policy are thus not exactly encouraged. The executive representatives of security prefer to see themselves as tightly knit groups of experts with privileged access to knowledge and information, on the basis of which they justify their claim to keep legislative and public interference at bay. The U.S. Defense Department, for example, is certainly no exception in including in its publicly accessible reports "only what [is] absolutely required by law." It even "seems proud of its failure to inform" the general public (Walt 1991, 228). Equally habitually, populations accept this procedure and by so doing contribute to their own political incapacitation.

Excluding the public from as much information as possible is conventionally justified on the basis of three principles (Mouritzen 1998b, 84–85). First, a state's bargaining position in foreign and security policy may be weakened by openness and by the revelation of internal disunity (principle of bargaining). Second, the state's supreme interests rule out open debate (principle of supreme interest). Third, foreign policy is too remote and abstract for the average citizen, who thus may be neither interested in assessing foreign policy decision making nor qualified to make informed choices as to foreign and security policy (principle of remoteness). As a consequence, security policy is more often than not elite business. However, even within the elite, at any given point different conceptions of national security issues compete with one another, and "what scholars and policy makers consider to be national security issues is not fixed but varies over time" (Katzenstein 1996, 10).

David Baldwin, stressing the power dimension of security, argues that security as a concept is an important issue not least because it "has been used to justify suspending civil liberties, making war, and massively reallocating resources during the past fifty years" (1997, 9). Kalevi Holsti adds that

> probably few concepts employed in statecraft and in the study of international politics have as vague referents as do *security* or *national security*. The terms have been used and abused by many governments to justify external aggression and the stifling of internal opposition. Robespierre, Napoleon, Kaiser Wilhelm, Joseph Stalin, and Senator Joseph McCarthy and some of his colleagues, to mention just a few, have justified purges; restraints on the freedoms of speech, press, and assembly; character assassination; and even mass murder in the name of "national security." Most governments that have launched wars of aggression or significant military interventions abroad have similarly claimed that their policies were designed to defend or preserve national security. (1995, 84–85, emphasis in original)

According to James Der Derian's estimation, in the name of security

> peoples have alienated their fears, rights and powers to gods, emperors, and most recently, sovereign states, all to protect themselves from the vicissitudes of nature—as well as from other gods, emperors, and sovereign states. In its name, weapons of mass destruction have been developed which have transfigured national interest into a security dilemma based on a suicide pact. And, less often noted in international relations, in its name billions have been made and millions killed while scientific knowledge has been furthered and intellectual dissent muted. . . . From its origins, security has had contested meanings, indeed, even contradictory ones. (1995, 25, 28)

Despite or perhaps because of this vagueness, the concept—or the rhetoric—of security is being used inflationarily. It is a political corollary of the argument that the more issue areas are treated under the label of security, the more issue areas can be expected to be excluded, on the basis of the principles sketched earlier, from democratic codetermination. This possible outcome is one of the dangers of security's becoming "infinitely flexible" (Kuus 2003a, 45) once it is addressed in cultural terms or otherwise broadened. It has made some authors argue *against* security (Neocleous 2000). Still, other authors argue for a broadening of the security agenda because referring to an issue in terms of security assigns importance and priority to this issue (Booth 1997, 111). Importance, however, does not automatically mean that the issue is being tackled according to the principles of democracy and transparency. It has even been argued that articulating security means presenting an issue as an existential threat and that this procedure involves the claim to use extraordinary means to deal with the issue (Buzan, Wæver, and de Wilde 1998, 24). As always, such claims should be subject to verification or falsification by means of empirical analysis.

The concept of security was for a long time undertheorized. It still is, but during the 1990s the situation took a turn for the better. Scholars started systematically to criticize the traditional, narrow conception of security and to examine the consequences of the articulation of security. They also engaged in a theoretical investigation of what security, security issues, and security subjects could possibly be if the understanding of security were decoupled from the conventional focus on the nation-state and on military security. This investigation has often been described as one of broadening the approach to security (Krause and Williams 1996). Issue areas other than the military were allowed to be referred to as security issues, and referent objects other than the nation-state were included in the security discourse even though the nation-state and national security still enjoyed a leading position in many writings, realist and otherwise.

By sketching Stephen Walt's and Barry Buzan's positions here, I do not mean to suggest that the debate on security and security studies within the intellectual and political community since the late 1980s can in any meaningful way be reduced to these two positions if a comprehensive survey were intended. Rather, I consider the focus on Walt's understanding of security with its emphasis on military force and on Buzan's notion of a comprehensive security agenda to be important in understanding the construction of security in the Baltic states in at least two ways.

First, the construction of security in and for the Baltic states has unfolded partially through a tension between a narrow and a comprehensive understanding of security. The Estonian scholar Peeter Vares, for instance, follows Walt by stating that the political elites of the newly independent

Baltic states "had to create the *security* policies of their countries, but no local Balt had ever been involved in *strategic* studies" (1993, 5, emphasis added). His colleague Aare Raid, however, follows the alleged "predominant understanding of security today," namely "the coordination and integration of numerous functions carried out by the nation-state: the military, political, economic, environmental, social, cultural and judicial" (1996, 8–9).

Second, both Walt and Buzan adhere to a narrow understanding of the referent object of security, namely the nation-state, which is in accordance with most policymakers' view in the Baltic states. Thus, I neglect many important contributions to the debate on security. This procedure does not imply any evaluation of the intellectual merits of the neglected approaches both in themselves and as a contribution to the ongoing debate on what "security" may be. The section on Walt is longer than that on Buzan because it includes a cursory sketch of the evolution of security studies since World War II as Walt sees it.

In 1991, Stephen Walt proclaimed a "renaissance" of security studies. This proclamation logically presupposes the existence of something that could be revived. Barry Buzan disagrees and claims that the concept of security needs to be "habilitated." According to Buzan, "it cannot be rehabilitated because it has never been properly developed" (1991, 3). These divergent assessments follow from divergent understandings as to what security is and what security studies is dealing with. According to Walt, "security studies may be defined as *the study of the threat, use, and control of military force*" (1991, 212, emphasis in original).[1] And elsewhere: "Security studies seek *cumulative knowledge* about the role of military force" (222, emphasis in original). However, in order to avoid an overemphasis on military power both as a source of and as a threat to national security, Walt includes in security studies the analysis of "what is sometimes termed 'statecraft'—arms control, diplomacy, crisis management, for example" (213). In light of Walt's understanding of what security studies is about, it thus makes sense to speak about a renaissance. Issues of war and the use as well as the threat of the use of force have been studied by, among others, historians devoted to diplomatic and military history as well as philosophers. Security studies as an intellectual discipline applying rigorous methodological standards, however, has evolved fairly recently.

---

1. The subsequent parenthetical page references in the text refer to Walt 1991.

According to Walt, many studies written in the United States before World War II lacked systematic evidence and were historically uninformed. They often expressed the authors' personal experiences and reflected their personal as well as collectively held judgments and prejudices. This was especially true for the presupposed aggressiveness and hostile intentions of the Soviet Union: "Because the Soviet desire to expand was taken for granted, more attention was paid to deterring it than to verifying the assumption or explaining its origins" (215). Security studies was often located in think tanks and institutes with questionable distance from governmental institutions and were engaged in policy advice rather than in empirical and theoretical research. Politically ambitious researchers' proclivity toward becoming engaged in policymaking outside the academic ivory tower frequently coincided with policymakers' inclinations to select from scholarly work only what was needed in particular circumstances to legitimize their political decisions. As Walt, referring to Hans Morgenthau, writes, "active involvement in policy debates inevitably tempts participants to sacrifice scholarly integrity for the sake of personal gain or political effectiveness" (222). This sacrifice may in part be a consequence of the intimate connection between security studies and real-world issues. In security studies' early stages, the situation was even more difficult because of the lack of access to independent data, the lack of technological knowledge required to analyze the development of modern weapons systems, and the importance to the discipline's development (or a lack thereof) of civilian strategists within governmental agencies who were only occasionally interested in independent scholarly work.

During the Vietnam War, security studies was almost ostracized by academic departments. With the end of the war, the discipline reappeared on the academic agenda and matured both quantitatively and qualitatively. In addition to the end of the Vietnam War, Walt identifies several reasons for the discipline's renaissance. The collapse of détente and the deterioration of U.S.-Soviet relations in the late 1970s and early 1980s revived interest in U.S. national security policy. Independent research institutes provided scholars with a growing body of increasingly reliable data, thus reducing the notorious dependence on politicized data provided by governmental agencies. Financial support by independent foundations increased considerably, and the introduction of new journals promoted scholarly publishing in the field of security studies. The revival of security studies was finally facilitated "by its adoption of the norms and objectives of social science," leading to its recognition as an accepted discipline within the academic world (221).

Furthermore, the discipline's narrow emphasis on the role and use of force in international politics has resulted in its intellectual coherence. According to Walt, this consistency would be at risk if the agenda of security studies were to be broadened significantly. The broadening of the agenda is precisely what Barry Buzan suggests, however. It is not surprising, then, that Walt rejects Buzan's ideas by stating that "defining the field in this way would destroy its intellectual coherence" (213). This coherence can best be preserved by concentrating on the role of force in international politics. Introducing new nonmilitary objects of security studies would result in the discipline's excessive expansion and "make it more difficult to devise solutions to any of these important problems" (213). Walt acknowledges that "nonmilitary issues deserve sustained attention from scholars and policymakers, and that military power does not guarantee well-being" (213). However, from this assessment it does not follow that security studies should deal with these nonmilitary issues. Furthermore, "the fact that other hazards exist does not mean that the danger of war has been eliminated" (213).

In contrast to Walt, who sees the merits of security studies in its narrow focus on military force, Buzan regards this narrow focus precisely as a reason for the theoretical underdevelopment of the concept of security. According to Buzan, policymakers and analysts' equation of security with national power was not only a result of the dominance of realist thought patterns, but also a consequence of the political environment of both World War II and the ensuing Cold War. In this environment, security was dealt with in the narrow confines of strategic studies, which often were neither theory oriented nor independent of the national policy level. Buzan thus recommends decoupling security studies from strategic studies as a precondition for broadening the security agenda and separating security from power according to the realist model of international politics. Although in the case of security the discussion is, in Buzan's words, "about the pursuit of freedom from threat" (1991, 18),[2] threats neither emanate exclusively from nor can be banned exclusively by means of military force. In Buzan's interpretation, security is about "the ability of states and societies to maintain their independ-

---

2. The parenthetical page references in the text from this point refer to Buzan 1991.

ent identity and their functional integrity" (18–19). Basically, security is about survival, but it includes "a substantial range of concerns about the conditions of existence" (19). A longer quotation from Buzan clarifies the scope of this range of concerns:

> The security of human collectivities is affected by factors in five major sectors: military, political, economic, societal and environmental. Generally speaking, military security concerns the two-level interplay of the armed offensive and defensive capabilities of states, and states' perceptions of each other's intentions. Political security concerns the organizational stability of states, systems of government and the ideologies that give them legitimacy. Economic security concerns access to the resources, finance and markets necessary to sustain acceptable levels of welfare and state power. Societal security concerns the sustainability, within acceptable conditions for evolution, of traditional patterns of language, culture and religious and national identity and custom. Environmental security concerns the maintenance of the local and the planetary biosphere as the essential support system on which all other human enterprises depend (19–20).

Because of the intimate relationship between the nation-state, security, and the military, a possible consequence of a broad understanding of security may be the militarization of nonmilitary security issues, such as environmental security. This argument does not speak against the concept of environmental security in itself, but it calls attention to the dangers inherent in too naïve an expansion of security. Security-political relations reflect power relations within societies and the power of interest groups, the military among the most influential. These power relations may be decisive in answering, for example, the question of whether environmental security "will lead to a militarization of environmental politics, or rather help to demilitarize security thinking" (Brock 1991, 407). Moreover, security policies in the five sectors cannot be pursued independently of one another. Security thus may not simply be seen as the sum of the cumulative securities in these sectors because, for example, "what is being followed in the name of military security in the short term may eventually go against the civil security" (Chaturvedi 1996, 33), not to mention the environment. What is being followed in the name of security in one of the five sectors may exclude what the security of (one or several of) the other sectors requires. All too often, critics say, political decision making still reflects the nation-state or the military logic of security.

Categorizing security requires saying whose security we are talking about, for "security is a derivative concept; it is in itself meaningless. To have any meaning, *security* necessarily presupposes something to be secured; as a realm of study it cannot be self-referential" (Williams and Krause 1997, ix, emphasis in original). Both Walt and Buzan assign to *national* security the role of the most important referent object of security studies: Walt implicitly by not discussing alternative referent objects such as individual, international, human, or global security; Buzan by claiming that states are "the principal referent object of security because they are both the framework of order and the highest source of governing authority" (1991, 22) and by explicitly subordinating individual security to national and international security (1991, 54). Although Buzan expands the conventional understanding of security considerably and offers a realist critique of the neorealist approach, his research agenda is still a narrow, conventional one because he insists on and prioritizes the nation-state as the referent object of security.

Because, as Buzan argues, "the nature of security defies pursuit of an agreed general definition" (1991, 16), and because such a general definition is not even desirable, it is more appropriate in the present context to discuss the construction of what political and academic elites are referring to as security rather than the construction of security proper. This procedure does not relieve the scholar from thinking about what security is or what it could possibly be, but for the time being a less ambitious approach may be in order. Here, approaches based on language theory have proven to be useful starting points for analysis. They reflect an understanding of language that is based on the assumption that "to engage in a speech act is to give meaning to the activities which make up social reality. Language thus no longer describes some essential hidden reality; it is inseparable from the necessarily social construction of that reality" (George and Campbell 1990, 273).

Following a linguistic approach to the study of security, Ole Wæver and associates have introduced an approach that treats security as a self-referential practice of articulation. Here, what matters is both security agents' definition of a person, an object, or a development as a threat and a political audience's acceptance of this designation—a process that the authors call *securitization*. Without acceptance, there is no securitization, but only a securitizing move. With acceptance, however, securitization implies and justifies a distinctive political behavior culminating in the use of extraordinary means. What does not matter, then, is a threat's real existence because, save for rare exceptions, there is said to be no way for a community and for those speaking on behalf of it to know objectively whether a threat really exists. Thus, knowledge underlying the process of securitization is knowledge in the sense that the community knows that someone or something is a threat because this threat perception is socially constructed as reality (Wæver 1995; Buzan, Wæver, and de Wilde 1998, 21–29).[3]

As outlined earlier, different notions of the sectors to which the security logic should be applied compete with one another, so that the priority on the military sector is increasingly being challenged. Different conceptions of what has to be secured (the individual, the society, the government, the state, the nation, a particular set of norms and values, and so on) are also contesting with one another, so that the emphasis on the nation-state is increasingly being criticized. Different strategies of how to achieve the security of the either explicitly defined or taken-for-granted object(s) coexist as well. As with every construction, the construction of what policymakers refer to as security, the object(s) and sector(s) of security, and the strategies with which policymakers try to secure these objects are inseparably connected with power relations within society. These power relations are at least as important to explaining which conceptualization of security gets the upper hand, as is the theoretical maturity and intellectual coherence of the particular conceptualization. Indeed, one-dimensional and intellectually undemanding constructions of national security with the nation-state as the referent object and the military aspect of security as the prioritized issue area have long dominated the security agendas and official interpretations of national interests in many states. They were based on short-run self-interests, simple and Manichaean perceptions of the enemy, zero-sum thinking, and military, even nuclear, means with which to keep the alleged enemy

---

3. In their most recent work, Buzan and Wæver no longer seem to be interested in the intersubjective dimension of securitization. Now they define successful securitization traditionally in terms of "visible outcomes such as war, mass expulsions, arms races, large-scale refugee movements, and other emergency measures" (2003, 73). By so doing, they have cleansed from their approach one of the ingredients that were most challenging from a conventional international relations point of view and most promising from a critical point of view because it assigned to the audience codetermination in security policy.

at bay (all linked to domestic power relations and interest groups). "National interests, defined essentially in terms of power, intimately related to the concept of security, which in turn was understood in the sense of preventing the potential adversary from invading one's space" (Chaturvedi 1996, 23). Certainly, simplistic readings can be very powerful.

One popular way of constructing security is to say that "there are no alternatives" (Jundzis 1995, 570; Klaar 1997, 19), although it does not require much theoretical sophistication to see that there are always alternatives. Another way of assigning to competitive views an inferior status in terms of rationality is to claim the monopoly of logic for oneself (Arnswald 2000, 99). Such statements are obviously under the suspicion of ideology. In fact, ideology "frequently takes the form of 'commonsense'—ideas that are sufficiently 'taken for granted' as to be beyond the realm of rational debate" (Jackson 1989, 51). In addition, and by no means mutually exclusive, postulating that there are "no alternatives" often results from "the anxieties associated with conflict," referred to in social psychology as rigidification: "People start to have a kind of tunnel vision with regard to the issues in conflict. They see only a limited range of possibilities for resolving the conflict. They lose their creative potential for conceiving a range of options which might make the conflict a constructive experience in which both sides might profit. Rigidification and tunnel vision are often associated with excessive anxiety" (M. Deutsch 1987, 40). Refusing to permit the existence of alternatives may be an expression of the inability or unwillingness to think creatively in alternatives. Assigning to the competitors the task to explain, justify, and legitimize their "deviating" claims, while at once absolving oneself from legitimizing one's own claims because they are "logical", "natural", "obvious," or "without alternatives," is a convenient means to demonstrate the alleged superiority of one's own intellectual position.

Another useful means for naturalizing one's own claims is geography. The Russian Federation is a direct neighbor of Estonia and Latvia, and with its Kaliningrad exclave also of Lithuania. In writings inspired by geopolitics and realism, this geographical proximity *in itself* makes Russia a threat to the Baltic states because, in Hans Mouritzen's words, *"power and incentives wane with distance from states' home base"* (1998a, 4, emphasis in original). According to Walt, because "the ability to project power declines with distance, states that are nearby pose a greater threat than those that are far away" (1994, 23). Buzan adds that "because most political and military threats travel more easily over short distances than over long ones, insecurity is often associated with proximity" (2000a, 2). These assessments can be supported historically and empirically, but the equation of proximity with a threat should not be naturalized. Geography in itself has little explanatory power; the meaning assigned to geography is socially constructed in order, among other things, to "disguise arguments about politics, religion and culture" (Mazower 1999, xiv). The effects of geography presuppose what Alexander Wendt calls "structures of shared knowledge" (1995, 73), translated into political and security-political meaning assigned to geography by political agents for various purposes. Although often seemingly justified in the light of historical experience, it is inapposite to derive threat perceptions directly from "geographical realities" (Ruhala 1988, 120) because the meaning assigned to these realities may be subject to change. Otherwise the evolution of the Nordic subsystem from a zone of almost permanent warfare to a security community would not have been possible. In this case, geography has not changed, but contiguity is no longer represented as threatening.

In the Baltic states, however, geography still appears strongly to legitimize security policies. Note, for example, Zaneta Ozolina's statement that "Latvia's unique geographical location makes an army absolutely necessary" (1996, 50). Geography is used to provide legitimacy for a particular kind of policy, here the buildup of armed forces, that is then said to be without alternatives. Thus, although geography in itself may be neutral, references to and the articulation of geography and its use for political purposes are not. Especially in combination with the word *unique*, the articulation of geography often is an antidiscursive and thus an antidemocratic practice of domination that locates political decision making outside the realm of public discussion and adds to the lack of democracy that still characterizes decision making in foreign and security policies in the Baltic states (and elsewhere) (Albrecht 1986, 152–73; Czempiel 1996).

Security-political meaning assigned to geography was remarkably constant in the Baltic states throughout the 1990s. These representations made themselves to some extent independent of the factual developments, and no example shows this rhetorical stability more clearly than the representation of the Russian Federation's Kaliningrad region. Ground forces deployed here showed a marked decline in number from approximately 50,000 in 1992 to 10,400 in 1999 (and slightly up again to 12,700 in 2000). As to military equipment, the figures were still substantial, but a downward trend was obvious in most categories. From 1995 to 2000, the number of main battle tanks decreased from 870 to 816; ACVs from 980 to 850; mortars from 410 to 345; attack helicopters from 52 to 20. The Baltic Fleet's submarines declined from 9 to 2; principal surface combatants from 23 to 6; patrol and coastal combatants from 65 to 26; mine countermeasure from 55 to 13, and so on. The operationability and even floatability of parts of the remaining vessels were questionable. Naval aviation under the command of the Baltic Fleet decreased from 195 combat aircraft in 1995 to 68 in 2000, whereas the number of armed helicopters increased slightly from 35 to 41.[4] In particular, the personnel figures call into question the aptness of the frequent—almost ritualistic—complaints of excessive militarization of the Kaliningrad oblast, which were part and parcel of the representations of the military developments in the Baltic Sea region throughout the 1990s.

With respect to Kaliningrad, the georhetoric usually saw the combination of Russian troop presence in Kaliningrad and the region's specific geographical location in itself—that is, its spatial separateness from Russia proper—as a military threat to the security of the neighboring states. Regardless of the reduction in manpower and military equipment, the Kaliningrad oblast was during the 1990s constantly referred to in terms of large, high, or excessive militarization (McCausland 1996, 53; Motulaite 1996, 166; Herd 1999, 201). Language has made itself independent of factual developments and is often adhered to for political purposes. For example, the Lithuanian government in its Basics of National Security unmistakably represents the "specific geopolitical environment [as being] hardly predictable due to existing militarized territories and states of unstable democracy" in the chapter on "potential risks and dangers" to Lithuanian security. It claims that "[m]ilitary capabilities in close proximity to Lithuanian borders" are an external risk to Lithuanian security. This formulation clearly, albeit implicitly, refers to the Kaliningrad oblast, the demilitarization of which is said to be one of Lithuanian foreign policy's main tasks.[5]

4. For the figures, see the IISS yearbooks on the military balance (1992, 98; 1995, 113–20; 1999, 116; 2000, 120–26). The downward trend seems to have stopped, and the figures for 2004 are fairly similar to the figures for 2000 (IISS 2004, 107–8).

5. Further implicit references to Kaliningrad include "attempts to impose upon Lithuania dangerous and discriminatory international agreements" and "military transit through Lithuania," which are seen as "external risks, challenges and potential challenges and political dan-

As early as 1994, the Baltic Assembly, in its Resolution Concerning the Demilitarization of the Kaliningrad Region and Its Future Development, called the oblast's demilitarization an "essential element for the security process in central Europe and the entire Continent" (1995, 64–65). However, the Assembly also included in its resolution a call for the restoration of old— that is, German—place-names in Kaliningrad, thereby addressing "a most sensitive issue . . . in a most provocative manner" (Wellmann 1996, 173). The Assembly itself thus provoked a negative Russian response to its call for demilitarization. It can be argued that the call in itself was justified at that time because of the temporary increase in troops stationed in the oblast in the process of the Russian troop withdrawal from central Europe and the German Democratic Republic.[6]

Consider also the geographical metaphor of "gray zones." In the Baltic Sea region, the end of the superpower antagonism resulted in a kind of instability that is often referred to in terms of unpredictability and gray zones. As compared with the rigidity of the Cold War regime, however, the current instability is positive and should be addressed in terms of flexibility, openness, and opportunities. In particular, the "gray zone" metaphor misses the point and still reflects Cold War terminology, if not thinking. Throughout the 1990s, it was used in Estonia, Latvia, and Lithuania primarily to describe the geographical location, perceived as uncomfortable, between NATO and Russia (Kristovskis 2000, Section Security Options; Lachowski 2000, 270). It presupposes (at least) two poles between which a gray zone can be located. Yet, as I show later in this book, it is a distinctive mark of the post–Cold War period that in terms of politics and security the Baltic Sea region is increasingly represented in its entirety, without dividing lines, spheres of influence, or, thus, gray zones—however ambiguously, conflictually, and hesitantly this representation and its transformation into politics unfold. "Geography is destiny?" (van Ham 1998)—hardly. To paraphrase Alexander Wendt (1992), geography is what states (and other actors) make of it.

Defending official policies may include tactics that go far beyond argu-

---

gers by the geopolitical environment." See Republic of Lithuania, Basics of National Security, 19 December 1996 (amended as of June 1998), pt. 2, chap. 9, first sec., and pt. 1, chap. 5.

6. The IISS gives a figure of 103,000 troops deployed in Kaliningrad oblast in 1993 (1993, 104). Some Russian politicians sharply condemned the resolution as interference in Russian internal affairs.

mentation. In Peter Berger and Thomas Luckmann's words, these tactics may include "therapy" as well as "nihilation"—that is, techniques that aim at converting the deviants to the official conceptualizations by convincing them that they are simply "wrong" ("therapy") or by "liquidat[ing] conceptually everything outside the same universe" ("nihilation" or "negative legitimation"). Basically, "nihilation *denies* the reality of whatever phenomena or interpretations of phenomena do not fit into [the official] universe." It may take the form of either giving the deviant phenomenon an

> inferior ontological status, and thereby a not-to-be-taken-seriously cognitive status [or] *incorporat[ing]* the deviant conceptions within one's own universe, and thereby . . . liquidat[ing] them ultimately. The deviant conceptions must, therefore, be *translated* into concepts derived from one's own universe. In this manner, the negation of one's universe is subtly changed into an affirmation of it. The presupposition is always that the negator does not really know what he is saying. His statements become meaningful only as they are translated into more "correct" terms, that is, terms deriving from the universe he negates. (1967, 132–33, emphasis in original)

For example, the Lithuanian Parliament, on 14 January 1999, passed a law according to which defense spending had to be increased to 2 percent of GDP by the year 2001. However, in the spring of 2000 the New Union Party challenged this official political strategy and opted for investments in education and science. These claims were supported in opinion polls by 80 percent of the Lithuanian population and by almost one hundred thousand signatures demanding a change of the state budget law. Furthermore, the New Union at that time enjoyed its highest ratings in public-opinion polls and thus appeared to be a serious rival to the ruling parties at the general elections scheduled to take place in October. Parliament (by a sixty-four to twelve margin) and government rejected the New Union claims because cutting defense expenditures would allegedly deteriorate both Lithuania's NATO prospects and the state of the Lithuanian defense system. Some commentators on the claims, however, used tactics such as nihilation, character assassination, and custody based on kinship and by so doing gave the heretical claim a negative ontological status. For example, they reminded their audience that the father of one of the chief protagonists behind the campaign and a New Union leader had been a KGB colonel, thereby indicating that the son of such a father could not possibly be up to something good for Lithuania; if his father had not been a KGB colonel, the son would surely not have made such "incorrect" demands (Goble 2000b).[7]

Before I dig deeper into the construction of security in the Baltic states by means of two case studies on the Russian troop withdrawal and the writing of security in national security documents in the next chapters, a brief look at the constructors of security is in order. Although these constructors often disappear in the technical and impersonal language of security documents and equally often hide behind ostensible *Sachzwänge*—that is, circumstances beyond their personal influence and control—it is nevertheless identifiable human beings who make the decisions that are subsequently referred to as and translated into national security policy. When the Baltic states regained independence, it was clear to many observers that they were not ready for it. Janis Penikis notes that "no one is ever ready for independence" (1996, 23), but perhaps Terry Eagleton's statement that "the best preparation for political independence is political independence" (2000, 7) is more to the point.

In Estonia, the first free elections since the collapse of the interwar democracy took place on 20 September 1992. They brought to power a coalition in which the Isamaa (Fatherland) Party played the decisive role. Isamaa's main slogan at that time was "Plats puhtaks!" ("A Clean Break with the Past!"—literally "Clean the Place!"), a slogan that was to become influential, among other things, in selecting the personnel for the ministries: "the people who came to power not only had no Communist background, they (if one may say so) had no background at all; no experience in governing" (Tallo 1995, 127). Peeter Vares and Mare Haab come to similar conclusions: "The political elites of the three Baltic republics [*sic*] are being formed rather rapidly out of comparatively young (30–35 years of age), energetic and politically ambitious people, who, as a rule, do not have adequate preparation for active policy making. It is rather their party adherence than academic training and competence that determines their political career" (1993, 299).

---

7. Likewise, parliamentary speaker Landsbergis called the initiators, the New Union, a "mouthpiece of the strategic interests of a foreign country" and the initiative "Russia's policy" (*RFE/RL Baltic States Report* 2000c). See also Tracevskis 2000a, 6, and *RFE/RL Baltic States Report* 2000d.

It is no surprise, then, that the resulting policies resembled, as Olav Knudsen has put it, "foreign policy pluralism" (1993, 62). Mette Skak writes about the incipient Lithuanian diplomacy in terms of "infantile disorders" (1996, 196). Such disorders did not, however, prevent Lithuanian diplomacy from coming to an agreement with the Russian Federation—and its equally unorganized foreign policy negotiators (Baev 1996, 27–47)—on Russian troop withdrawal one year earlier than its Estonian and Latvian counterparts. Indeed, Skak assigns to Lithuanian diplomacy "sufficient tactical skills to prove their value as a resource for security policy" (1996, 196). However, the issue was not only one of infancy or pluralism, but also one of power because "transition also means a struggle for resources" (van Zon 1995, 461). At stake at this point were both the question of the future orientation of the Baltic states' foreign and security policies and the actors' personal careers. The ideas of acting or aspiring foreign policy representatives as regards security policy were at that time hardly elaborate and homogeneous. In Lithuania, for example, in late 1991 these ideas ranged from alignment with Western states via balanced relations with both the West and the East to neutrality and regional security arrangements.[8] The evolution of a particular foreign and security policy opened or closed the doors to the foreign and security policy bureaucracy for individual actors. Individual careers and the translation of specific ideas into effective foreign policy thus were inseparably connected with one another.

The institutional preconditions for the development of a consistent foreign and security policy were in fact not encouraging. Because of a 1944 modification of the Soviet Constitution vesting each Soviet republic with the right to establish foreign and defense ministries as well as "national" troop formations, the Baltic republics, too, had established "foreign ministries" and the formal right to pursue foreign policies. Yet these "ministries" were never taken seriously by the central Soviet authorities. They were introduced primarily in order to justify the Soviet leadership's claim to membership for each Soviet republic in the UN. As early as 1944, Soviet foreign minister Vyacheslav Molotov had requested that "each of the sixteen Soviet republics (then including a Karelo-Finnish republic) be given a seat in the United Nations" (Skak 1996, 97). Afterward, it was demanded that only the

8. For surveys based on interviews in late 1991, see Petersen 1992, 32–34; Kerner, Stopinski, and Weiland 1993, 26–33.

Ukrainian, Belarusian, and Lithuanian Soviet Republics be represented in the UN.[9] The republics' "foreign ministries" were staffed with only five or six employees. They functioned primarily as control instruments with clear connections to the secret service, and the employees had only limited contact with Western states, mostly in cultural and trade matters (Vares 1993, 3). The ministries were only "ministries on paper," "purely fictional, as suggested by the deafening silence on their existence in the literature on Soviet foreign policy" (Skak 1996, 96, 28).

The lack of expertise characteristic of the incipient foreign and defense ministries of the Baltic states is therefore hardly surprising. A penchant for grand gestures—demanding too much, too soon—often made up for the obvious lack of qualifications on the part of the foreign representatives of the Baltic states (Vares 1993, 3). More serious than the initial lack of expertise has proven to be the competition between different foreign policy actors—foreign policy representatives of the popular fronts, parliamentary commissions dealing with foreign affairs, the ministries of foreign affairs, prime ministers, presidents, foreign policy advisers, and so on—resulting in unclear, competitive, and often contradictory foreign policy orientations (Knudsen 1993, 62–63; Vares 1993, 5). Early on, an unfavorable division of labor among the ministries preordained a specific understanding of security. In Latvia, for example, the Ministry of Defense—"initially overloaded with Soviet style bureaucrats with good formal but very few real qualifications for leading the development of a ministry of defense of a small state" (Clemmesen 1998, 238)—was responsible for security issues and unsurprisingly narrowed down security to military issues, while the Ministry of Foreign Affairs was preoccupied with establishing contacts to foreign countries. It was only in 1998 that both ministries "produced a *joint* paper on Latvian NATO policy" (Ozolina 1999, 24, emphasis in original).

To make things even more opaque, personal ambitions have often col-

9. At the Yalta Conference in February 1945, Molotov had vainly asked for a seat in the UN for the Lithuanian Socialist Soviet Republic (in addition to the Ukrainian and Belarusian seats). One year later, at the Paris Conference, the Soviet leadership again demanded a seat for the Baltic republics; again, the demand was rejected (Clemens 1991, 298; Bollow 1993, 14; Misiunas and Taagepera 1993, 126). During the UN's founding period, representatives of the last independent prewar governments of the Baltic states had also tried to achieve a seat for the Baltic states in the world organization. They, too, had failed (Bollow 1993, 13, 18).

lided with formal responsibilities. A case in point is the situation in Lithuania. Before the approval of the Constitution of the Republic of Lithuania in a referendum on 25 October 1992, assigning to the president the main role in foreign and security policy, decision making in these areas was based on the Provisional Basic Law passed by the Supreme Council on 11 March 1990. Purely legally, the chairman of the Supreme Council was subordinated to the Presidium of the Supreme Council in foreign and security matters; his or her formal foreign and security powers were not considerable. De facto, however, the then chairman of the Presidium, Vytautas Landsbergis, "gradually concentrated in his hands enough power to make all major foreign and security decisions" on the basis of his formal responsibility to negotiate with foreign states, the strong support in Parliament that he enjoyed until the end of 1991, and his personal ambitions (Nekrasas 1996, 66). All of this would not have been too serious if the foreign and security policy representatives of the Baltic states had been allowed a probationary period. This, however, was not the case. As shown in chapter 4, immediately after the recognition of independence, the first security policy decisions of long-term importance had to be made. Likewise, immediately following the recognition of independence, the negotiations on the withdrawal of Soviet troops from the territories of Estonia, Latvia, and Lithuania began. These negotiations and the subsequent writing of security in national security documents are analyzed in the next chapters.

7

The Russian Troop Withdrawal

FROM 1991 TO 1993–94, no other issue kept the foreign and security policy elites in the Baltic states as busy as the negotiations on and the process of the withdrawal of 120,000 Russian (former Soviet) troops from the territories of Lithuania, Latvia, and Estonia. As shown in chapter 4, the withdrawal was understood as mandatory in light of constructing Baltic nation-states. It was also seen as mandatory in light of constructing security. From the point of view of the political decision makers in the Baltic states, the presence of the Soviet/Russian army was a matter of foreign troops having crossed the border and established themselves illegally in states occupied by them.

The members of the Soviet/Russian army probably did not share this view. For most of them, the issue was simply one of having crossed an intra-Soviet administrative line and been transferred from one Soviet republic to another. After the international recognition of the independence of the Baltic states, they nevertheless found themselves abroad, as confirmed by the Soviet State Council's recognition of Baltic independence on 6 September 1991, and deeply involved not only in Baltic-Russian relations, but also in the domestic political struggles in the Baltic states. For example, representing the Russian army's presence on Lithuanian territory as a threat is said to have served the purpose of declaring a permanent national state of emergency with which the conservative parties sought to mask the deficiencies of the government's reform policy and to immunize themselves against any kind of criticism (Christophe 1997, 302).

Regardless of internal political struggles, the presence of what most Baltic policymakers interpreted as an occupation army was perceived as a threat to the security of the Baltic states. For example, in November 1991 Latvian vice minister of foreign affairs Martins Virsis was quoted as saying that the "Soviet troop withdrawal is the most crucial security requirement"; the chairman of the Foreign Affairs Committee of the Latvian Supreme Council,

Mavriks Vulfsons, was quoted as saying that "the longer the Soviet army stays in Latvia, the greater the polarization of politics in Latvia"; and the Defense and Interior Affairs Commission secretary, Mikhail Stepitchev, was quoted as saying that the defense issue in Latvia was one of "interior defense [against] Soviet troops and Soviet people who are fighting to maintain their privileges."[1]

As shown in chapter 5, the reserve exhibited by major portions of the Soviet army in the years leading to the independence of the Baltic states hardly gave reason for the interpretation that the presence of the Russian forces in itself posed a direct threat to the Baltic states' security. At that time, however, some decision makers feared "a potential armed conflict with Russian troops still located in Estonia" (Haab 1994, 149), and others feared that parts of the Russian army, especially the officers, stationed in the Baltic states could make themselves independent of Moscow and get out of control (Allison 1993, 49). In October 1991, for example, the officers of the Baltic MD wrote to Marshall Evgeni Shaposhnikov, the commander in chief of the armed forces of the Commonwealth of Independent States (CIS), "informing him that they would ignore any order he might issue telling them to withdraw from the Baltic states to Russia" (Donnelly 1992, 37–38).

The completion of the withdrawal of the main body of the Russian troops from Lithuania on 31 August 1993 and from Latvia and Estonia one year later showed that an act as militarily and strategically challenging, as technically complicated, and as politically and psychologically demanding as the pullout of the Russian troops could be organized in a peaceful way even though it strained Baltic-Russian relations considerably. The issue was militarily challenging because, from the point of view of the Russian military, both "retaining a military presence on the strategically significant Baltic coastline [and] maintaining lines of communication with the Kaliningrad region" were at stake (Lukic and Lynch 1996, 364). It was politically and psychologically demanding because the Russian troop withdrawal meant rearriving at where Peter the Great had started his expansions some centuries ago and acknowledging Russia's downgrade to the status of a regional power. It was technically complicated because there were no facilities on the territory of the Russian Federation that could possibly digest the sol-

---

1. All cited in Petersen 1992, 25. For privileges regarding housing and education, see Karklins 1994, 121.

diers returning from the Baltic states (in addition to those forces withdrawn from Poland and the former German Democratic Republic).

The negotiations on the Russian troop withdrawal were temporarily an obstacle to the development of closer cooperation among the Baltic states. Russia's "game to play [Lithuania] off against the other Baltic partners" (Lachowski 1994, 578) was facilitated by the Lithuanian government when it aimed at and reached a separate agreement on the withdrawal of Russian troops from Lithuanian territory, "thereby undermining the united Baltic front" (Lieven 1994, 78).[2] From a strictly national point of view, this strategy proved to be successful, and Russia completed the troop pullout from Lithuania one year before the withdrawal came to an end in Estonia and Latvia. In Lithuania, the situation was substantially more relaxed than in Latvia and Estonia. To be sure, in the beginning maximalist claims by the Lithuanian government collided with equally maximalist demands on the part of the Russian negotiators. The former wanted the Russian troops to be withdrawn immediately and unconditionally; the latter wanted to drag the troop withdrawal process on and on and to optimize the withdrawal conditions. The former regarded the troop withdrawal as a precondition for negotiating other controversial issues; the latter regarded negotiations on other issues as a precondition for the troop pullout. The former wanted the Russian government to pay compensation for the damage done in more than fifty years of the Soviet army's presence; the latter wanted the governments of the Baltic states to pay compensation for more than fifty years of Soviet investments in military and traffic infrastructure.

The Lithuanian government's view was presented by Vytautas Landsbergis in his capacity as chairman of the Lithuanian Supreme Soviet as early as 5 October 1991 (Skak 1996, 200) and confirmed in a referendum on 14 June 1992: 68.9 percent of all eligible voters supported the Russian troop withdrawal.[3] In order to make the Russian negotiators sign a timetable on troop pullout, the Lithuanian government in September 1992 agreed that the issue

2. Lithuania's " 'solo' action" is an example of what Dzintra Bungs has called "disharmony in intra-Baltic relations. . . . Apparently neither Latvia nor Estonia was informed beforehand that such an accord was in the making" (1998, 90).

3. In the literature, the figures on the referendum are not entirely clear. The correct interpretation seems to be that approximately 76 percent of all registered voters participated in the referendum; approximately 90 percent of those voting polled for troop withdrawal and com-

of compensation could be negotiated later. By so doing, it removed a major stumbling block to the completion of the troop withdrawal (Stankevicius 1994–96, chap. 1, sec. 2 [13]).[4] The second reason for the Russian troops' early pullout from Lithuania seems to have been a reasonable limitation of the Lithuanian-Russian agenda. Neither the Kaliningrad transit issue nor the border issue nor the question of social guarantees for Russian military pensioners burdened the negotiations in an insurmountable manner. The issue of Russian transit to and from the Kaliningrad oblast through Lithuania entered the agenda only after the completion of the troop withdrawal from the Baltic states (Butkevicius 1993a, 174).[5] Likewise, an agreement on social guarantees for retired military personnel was signed only after the completion of the troop withdrawal, on 18 November 1993 (Norkus 1998, 142). Regarding the Lithuanian-Russian border treaty, the Lithuanian authorities showed a more relaxed attitude than their Estonian and Latvian counterparts, resulting in the 24 October 1997 border treaty, ratified by Lithuania in 1999 and by the Russian Duma in May 2003. The Lithuanian-Russian relations were further advanced by a readmission agreement signed in June 2003 and by the agreement on border-crossing rules that has been in force since 1 July 2003.

Finally, the number of Russian speakers in Lithuania was significantly lower than in Estonia and Latvia. This configuration was translated by the Lithuanian government into a reasonable citizenship law ("zero option"). The December 1991 Law on Citizenship granted Lithuanian citizenship to all those people who had been citizens before 1940 and to their descendants, but also to those permanent residents who, when the citizenship law was adopted, had a legal income and had been living in Lithuania at least ten years, irrespective of their ethnic origin. Additional requirements were the oath of loyalty to the state, a Lithuanian language test, and the abandon-

---

pensation, amounting to 68.9 percent of all those entitled to vote. Thanks to Grazina Miniotaite for detailed information on this issue.

4. On 13 June 2000, the Lithuanian Parliament approved legislation binding every Lithuanian government to demand compensation from Russia for five decades of occupation by the Soviet Union. For Latvia, see Clemens 2001, 222.

5. For some time, the transit proceeded on the basis of an intergovernmental treaty regulating the Russian troop withdrawal from the former German Democratic Republic and Poland using the Mukran-Klaipeda ferry service.

ment of any other citizenship (Norkus 1998, 147–48; Jurkynas 1997, 126). This citizenship law could not be used as a pretext for maintaining a Russian military presence in Lithuania. Although the Russian military considered free access to the Kaliningrad oblast through Lithuanian territory one of the most important issues during the withdrawal negotiations, the Lithuanian citizenship law did not give the Russian military any pretext to intervene militarily in order to guarantee both nondiscrimination against the Russian-speaking minority and—from the military's point of view certainly much more important—free access to the Kaliningrad oblast (Baev 1996, 162).

On 8 September 1992, several months after a May 1992 Lithuanian proposal that the Russian side had ignored, Lithuanian defense minister Audrius Butkevicius and his Russian counterpart Pavel Grachev, after roughly one year of negotiations that were accompanied by the beginning of the Russian troop withdrawal, signed a timetable on troop pullout. According to the timetable, the last Russian soldier would have to leave Lithuania on 31 August 1993. This agreement was a substantial deviation from the attitude hitherto assumed by the Russian military, which aimed to protract the troop withdrawal as long as possible. In September 1992, 35,000 Russian troops were still in Lithuania. In December that year, Lithuania's defense minister said that he expected the remaining 15,000 soldiers to leave by 31 August 1993. The troop numbers indicate a continuous Russian troop withdrawal from Lithuania regardless of repeated announcements by the Russian president and Ministry of Defense officials indicating the opposite.

The Russian government's basic decision to withdraw the troops completely from the Baltic states was occasionally met with "indirect resistance" (Lukic and Lynch 1996, 365) by the Russian military. Although the military seems to have been "in basic accord with the Russian Government as far as the broad outlines of policy towards the three Baltic states [were] concerned" (Lukic and Lynch 1996, 363), the political and military leadership sometimes exhibited divergent attitudes toward the precise procedure and minutiae of the troop withdrawal, just as did the military leadership in Moscow and the local authorities. For example, although the Russian Parliamentary Committee of International Affairs voted on 22 September 1992 to declare the agreement on troop withdrawal from Lithuania null and void, the military personnel on the spot continued withdrawal. Although the Russian government halted the pullout on 29 October on the basis of a presidential decree that confirmed a 21 October declaration of the Defense Ministry, the military personnel in Lithuania proceeded with the withdrawal (Sharp 1993, 605; Wettig 1993, 12; Skak 1996, 152).[6]

As in Lithuania, the Russian troops in Estonia and Latvia were on the whole withdrawn continuously regardless of various official statements indicating or announcing the opposite. But what may look in retrospect like a smooth and frictionless process of Russian troop pullout was in reality neither. For example, between February and December 1992 the Russian military sent an additional 2,630 military personnel to Latvia (Vares and Haab 1993, 293). According to Jane Sharp, Russian foreign minister Andrey Kozyrev in March 1993 reportedly "warned the Baltic states that gross violations of human rights against ethnic Russians would require the dispatch of vast numbers of Russian 'peace-keeping forces' in a 'new Yugoslavia'" (1993, 605). Summarizing Russia's policy toward the "near abroad" in 1994, Zdzislaw Lachowski notes that "Russia is embarking more and more aggressively on peace-enforcement activities in its direct neighborhood" (1995, 780). In light of Russia's intervention in Transdniestria in 1992, the combination of "peace enforcement" and the "Russians abroad" issue was particularly explosive (Baev 1996, 39). In early May 1994, Russian defense minister Grachev threatened to reinforce Russian troops in Estonia in the case of a continuation of what he deemed Estonian unwillingness to make concessions regarding the rights of retired Russian officers (Lachowski 1995, 779). In Pavel Baev's assessment, in the case of Estonia "it would not be an exaggeration to say that on the political level everything was made ready for a Russian 'peace intervention' in support of the 'popular demand' in Narva (which was not that difficult to organize) for reunification with Russia" (Baev 1996, 163).[7]

6. The additional 722 Russian troops brought into Lithuania from February to December 1992 changed the overall picture only slightly (Vares and Haab 1993, 293). Lithuanian defense minister Butkevicius nevertheless in the summer of 1992 announced the formation of an operative group with which to prevent, if necessary by force, the stationing of new Russian military units in Lithuania. Like the suspension of Russian oil and gas deliveries to Lithuania, the October decree has to be seen as interference in the Lithuanian election campaign, wherein the Russian government apparently hoped to contribute to the replacement of the conservative government by the more flexible and compromise-oriented LDDP (Christophe 1997, 324–25).

7. Baev's bracketed assumption seems to be questionable. In December 1990, only 7 percent of Narva's inhabitants preferred incorporation into the Russian Federation, whereas 36 percent opted for Narva as a regular city in Estonia and 39 percent for autonomy or some other

However, reluctance on the part of some Baltic negotiators to accept anything deviating from their original positions also rendered finding a mutually acceptable compromise difficult. Estonian negotiators rejected some Russian proposals as to the deadline of the troop pullout, only to complain later that there had been no Russian proposals in the first place. For example, a January 1993 proposal by the Russian delegation to complete the troop withdrawal by the end of 1994 under condition that sufficient housing be constructed was rejected by the Estonian delegation, who then complained about the lack of "a strict schedule for the withdrawal." In November 1993, the Russian delegation suggested a deadline of 31 August 1994, but the Estonian negotiators declined to discuss this proposal because of a lack of authority to do so.[8] Furthermore, burdening the negotiations with too many controversial issues aggravated the tensions inherent in the talks. A major obstacle to a compromise was the dispute over the Latvian-Russian and the Estonian-Russian borders that were being negotiated at the same time. From September 1991 to September 1994, the withdrawal of the former Soviet troops to Russia was indeed "the long-standing central conflict in the area," but, more important, the conflict continued for three years "significantly below the level of open violence" (Baranovsky 1995, 246), and political means were the prevailing methods of conflict settlement.

Because the Russian government was in no legal position to refuse withdrawing the troops, two arguments were presented to justify the slow process of the troop pullout. First, the lack of housing for returning soldiers was often mentioned. To accommodate 120,000 returning troops (in addition to those troops returning from Poland and the former German Democratic Republic) was indeed a technical challenge, the dimension of which some Baltic negotiators seem to have underestimated. Reasoning that social

---

kind of special status (Gorohhov 1997, 128). Estonian sovereignty in a confederation with the Russian Federation was preferred by 79 percent of Narva's inhabitants at the referendum on independence on 3 March 1991 (Kaplan 1993, 216).

8. See *Estonian Review* 1993b (for the quotation) and 1993d. As early as September 1992, the Russian delegation offered 1994 as the deadline for the troop withdrawal (*Estonian Review* 1992a). Although the Russian delegation in December 1992 proposed a draft agreement, On the Withdrawal of Russian Forces and on Temporary Conditions for Their Presence on the Territory of the Republic of Estonia, the three Baltic states in July 1993, at the occasion of the G7 Summit in Tokyo, emphasized "the Russian government's refusal to present draft agreements including timetables" (*Estonian Review* 1992b, 1993c).

problems may continue forever in Russia, Latvian legislators were reluctant to accept the lack of housing as an argument (Karklins 1994, 121).

Second, the Russian authorities used the Russian-speaking minorities especially in Latvia and Estonia as a reason to maintain troops in both states as long as possible. The Russian military had introduced the human rights situation of the Russian-speaking minorities outside the Russian Federation as a vehicle with which to legitimize its activities in Transdniestria in mid-1992. Making the troop withdrawal from the Baltic states contingent on the improvement of the human rights situation of the Russian-speaking minorities in Latvia and Estonia did not, however, entirely produce the desired results. It is not surprising that the Baltic states' representatives perceived "Moscow's current policy as a continuation of Russian (or Soviet) imperial policy" (Sergounin 1997, 331). But Western observers and some moderate nationalists in Russia were also disturbed by the identification of the violation of the rights of the Russians and of "those identifying themselves ethnically and culturally with Russia" in the 1992 Draft Military Doctrine as a possible casus belli (Baev 1996, 39). Although the military leadership had invented the "Russians abroad" issue, the political leadership continued to adhere to the issue even when the military had again lost its interest in it. In the summer of 1993, for example, Yeltsin and Foreign Minister Kozyrev condemned the Estonian Law on Aliens as "apartheid" and "ethnic cleansing" (Baev 1996, 16?)

Frequently, the reference to the human rights situation in the Baltic states did not correspond with the actual situation of the Russian-speaking minorities in Estonia and Latvia as assessed by international organizations. An official of the Russian Foreign Ministry frankly explained in July 1992 that the West is "highly sensitive to this issue, in contrast to us" (cited in Karklins 1994, 122). It is not surprising, then, that Russian authorities often referred to the human rights situation in Latvia and Estonia in order both to damage the Baltic states' reputation abroad and to justify the slow troop pullout. This reference had not much to do with a real interest in the Russian speakers' situation in Latvia and Estonia. It had even less to do with the Russian speakers' wish to have the Russian Federation's authorities speak and act on their behalf (Maley 1995, 5). This tactic hardly produced the desired results, although the Baltic states found themselves "under international scrutiny" (Zhuryari 1994, 79), and their reputation was temporarily stained. The Russian authorities' intentions were easily discernible, and they did not succeed in keeping their aim secret from international organizations that were monitoring Baltic-Russian relations. The growing ritualization of the accusations and the rhetorical exaggerations did not increase the credibility of the complaints, either. The domestic factor—Foreign Minister Kozyrev's aspirations to appear even more nationalist and extremist than Vladimir Zhirinovsky and to outpace the latter in rhetorical aggressiveness—contributed to a Russian self-dequalification on the issue, just as the Russian military's abuse of human rights in Chechnya did (Baev 1996, 40–41).

In addition, although international organizations often hesitated to name Russia directly, the internationalization of the "Russians abroad" issue through appeals to the UN and the Conference on Security and Cooperation in Europe (CSCE, which became the OSCE in 1994) resulted in strengthening the point of view that the Russian troops had to be withdrawn. At the CSCE Summit in Helsinki on 19 July 1992, the Russian delegation did not veto section 15 of the summit's final declaration, which called "on the participating States concerned to conclude, without delay, appropriate bilateral agreements, including timetables, for the early, orderly and complete withdrawal of . . . foreign troops from the territories of the Baltic States." The UN General Assembly, on 15 November 1993, called for "the early, orderly and complete withdrawal of foreign military forces" from the Baltic states, carefully avoiding calling Russia by name.[9]

At this point, approximately 20,000 Russian troops were still left in Estonia and Latvia. Although supported by international organizations, the Baltic governments were in a relatively weak bargaining position because the remaining Russian troops in their territories still outnumbered their own armed forces by far. The Baltic governments had to accept compromise agreements because they lacked the means to compel the Russian authorities to perform the troop withdrawal in the way they had originally demanded, and they had to accept parts of the Russian claims.

However important the involvement of international organizations, the United States, and other Western states may have been, other factors were more important for the Russian decision to withdraw the troops from the

9. CSCE Helsinki Document, 9–10 July 1992; UN document 4/48/18, 15 November 1993, as cited in Lachowski 1994, 580.

Baltic states. First, as noted earlier, the Russian military was interested in free access to the Kaliningrad oblast, but the Lithuanian policy on citizenship did not deliver any pretext for a military intervention in Lithuania. Furthermore, the Lithuanian leadership exhibited a unique willingness to co-operate with Russia at that time. The Estonian policy on citizenship, however, might have been used as a "justification" for a Russian military intervention, but the Russian military had neither strategic interests in Estonia (especially in the northeastern part, with its concentration of Russian speakers) nor troops stationed in the Narva area. A military intervention thus would not have been possible without cross-border deployment. Moreover, the military authorities in Moscow had no particular interest in the navy and in the submarine bases in Liepaja (Latvia) and Paldiski (Estonia). Limited access to Paldiski and the ballistic missile early-warning station in Skrunda (Latvia) could be negotiated peacefully (Baev 1996, 162–63).

Second, the shift of regional emphasis from the Baltic Sea to the Caucasus helps explain the Russian troop withdrawal from the Baltic states. Third, in late 1991 the Soviet armed forces were "without a master: highly armed . . . but without a political mission" (Schröder 1996, 132). It was in the Russian political leadership's interest to achieve as soon as possible political control over the potential for violence concentrated in the hands of the military leadership—the more so because there was growing unrest among the body of officers. In order to transform them into reliable, efficient, and loyal armed forces of the Russian Federation, the buildup of which was hesitatingly decided in April 1992, a "healthy core" was required. The Russian Ministry of Defense regarded the troops withdrawn from the Baltic states as at least "less damaged" than other parts of the former Soviet armed forces (Wettig 1993, 21–23).

Fourth, technical considerations may also have played a certain role. The introduction of new currencies in the Baltic states made the presence of Russian troops increasingly expensive. Unpredictable, fiercely anti-Russian Baltic volunteers challenged the presence of the Russian troops in a way that threatened a violent escalation of the conflict. Attempts to reestablish military control might have clashed with the buildup of a multitude of locally (if at all) controlled paramilitary forces (Park 1991, 262; Lieven 1996, 176). The prospect of becoming involved in a violent conflict must have inconvenienced a Russian leadership that was aiming at the image of both a trustworthy partner of the West and, based on its support for the Baltic strivings for independence in 1991, a friend of the Baltic peoples. Fifth, interventions by international organizations may also have influenced the Russian political and military authorities' decisions. Yet "western influence produced success only when the Russian leadership came to the conclusion that the troop withdrawal was in its own interest" (Wettig 1993, 23).

In retrospect, more important than the technical details of the troop withdrawal were its peaceful progress and termination only two years (Lithuania) and three years (Estonia and Latvia) after the Baltic states achieved independence. More important than the fact that several agreements negotiated in connection with the troop pullout remained unsigned by the Russian side was the fact that Russia all the same withdrew the troops according to the agreements (Stankevicius 1994–96, chap. 2, sec. 3, [44] and [46]). The troop withdrawal was a tremendous achievement that qualifies the critical assessments of the efficiency of the incipient foreign and security policy elites in the Baltic states and the Russian Federation referred to in the preceding chapter. If the elites are blamed for the mistakes they made in an early stage, they should also be given some credit for the successful treatment of this very sensitive issue, which numerous authors often identified as a potential source for a violent escalation of Baltic-Russian relations from 1991 to 1994 and, paradoxically, even afterward (discussed later in this chapter). The governments of the Baltic states refrained from their earlier maximalist demands, directed to some extent at the domestic audience, and entered into compromise agreements that were certainly far from what they considered optimal at that time. In all three states, the initial demand to withdraw the Russian troops immediately was converted into accepting a two- and three-year period for the troop pullout. In Latvia and Estonia, the troop withdrawal was preceded by agreements on social-security guarantees for Russian military retirees; in Lithuania, such an agreement followed the completion of the troop withdrawal. In Latvia, Russia was permitted to use the ballistic missile early-warning station in Skrunda for a fixed period; in Estonia, it was allowed to use the nuclear submarine training base in Paldiski for a shorter period. The three Baltic states adjourned the negotiations on Russian compensations for the damage done in the years of Soviet troop presence.

At that time, the troop withdrawal agreements were certainly contro-

versial: according to the former Latvian minister of defense, Latvia, by signing the agreement, was "sacrificing its own vital interests" (Jundzis 1995,
563).[10] However, with regard to constructing a nation-state, the withdrawal
of Russian troops was a necessary precondition for effective sovereignty.
With regard to constructing security, the Russian troops' pullout marked the
end of what was regarded as a state of occupation and removed the greatest
obstacle to the development of national security policies, including the writing of national security concepts. With regard to security community building, the peaceful process of the Russian troop withdrawal, in addition to
nonviolently restoring independence, could have been another step toward
the emergence of a mutual-response community. The success of the negotiations could have been used as a point of departure for the evolution of
benevolent and good-neighborly relations between the Baltic states and the
Russian Federation. As one observer noted at that time, "with the troops finally gone . . . there should be prospects for a new and more constructive
phase in relations between Russia and the Baltic countries" (Bildt 1994, 78).
As shown in chapter 11, however, the potentiality for the development of relations based on mutual responsiveness and trust inherent in the way both
sides constructively dealt with the troop withdrawal issue has as yet hardly
been used. Furthermore, after the conclusion of the Russian troop withdrawal, the issue contributed in at least three ways to a deterioration of
Baltic-Russian relations and therefore to insecurity.

First, the completion of the withdrawal of the main body of Russian
troops on 31 August 1994 did not mean that all Russian troops had left the
territories of the Baltic states. Especially the ballistic missile early-warning
station in the Latvian town of Skrunda and the presence of Russian soldiers
there were permanently politicized and securitized by many observers even

10. The Estonian-Russian agreement was criticized for a variety of reasons. First, it was
signed in contradiction to the Estonian Law on Conducting Foreign Affairs because the draft
agreement "had not been officially approved by the Estonian Government prior to the signing
procedure" (Haab 1995, 56). Second, the agreement was said to include provisions in violation
of Estonian laws (56). Third, the agreement was an object of intense dispute as to the question of
whether too many concessions had been made (Tallo 1995, 129). Fourth, critics said that signing
an agreement on withdrawal of occupation troops would legalize their presence retroactively
(Tammerk 1994, 3).

after August 1994. It was surely one of the strategic installations on Latvian territory that the Russian authorities considered important.[11] According to a February 1994 Latvian-Russian agreement, Russia was permitted to control the Skrunda installation until 31 August 1998, with 599 Russian military specialists and 199 civilian technicians to operate the station, plus 89 military security guards and their families. An important role for monitoring and observing compliance with the agreement was assigned to the OSCE (Viksne 1995, 79; Lejins 1996, 47; Zaagman 1999, 21–23). The stationing of nearly 1,000 specialized Russian military personnel encountered severe skepticism on the part of many Latvian observers regarding the Russian authorities' willingness to stick to the 1994 agreement and to close Skrunda according to schedule—the more so because the abandonment of the radar station would result in a stretch of 2,500 kilometers between the British islands and Greenland being uncovered by ground-based Russian surveillance systems for the time being (Puheloinen 1999, 60; Rogov 2000, 27).

Many commentators believed that international observance was insufficient to provide for the withdrawal of the Russian servicemen from Skrunda in 1998 and for the demolition of the installation in 2000. However, Edgars Skuja, the deputy director of the Latvian Foreign Ministry's Political Department and Latvian representative on the international commission supervising the dismantlement of the facility, emphasized at the time that the "international attention means that the contract will be kept and everything will happen on time" (cited in Kahar 1997b, 1). Some authors have also speculated about the Russian authorities' ability to replace the installation with a new antiballistic missile facility in the Belarusian Gancevichi by 1998. If Russia were not able to replace Skrunda, "problems [might] arise in the future with Russia possibly asking for an extension of the time limit for the old Skrunda radar station" (Lejins 1996, 47). Yet, according to the then Latvian foreign minister, Valdis Birkavs, "Russia was very precise in completing the

11. Regarding the importance that the Russian authorities assigned to the facility, it is revealing that President Yeltsin erroneously included Skrunda in a 5 April 1994 decree on *permanent* military bases in the former Soviet republics (Lachowski 1994, 581). The Russian authorities were not the only ones to consider Skrunda important: the adviser to the Latvian government, Eriks Tilgass, reported in November 1991 that both "the United States and NATO were said to include the facility in their long-term planning documents for post–Cold War Europe" (Petersen 1992, 32).

terms of the agreement" (cited in Medearis 1998, 1). On 31 August 1998, the Skrunda station was closed down. The Latvian government allowed sixteen Russian military personnel to stay there until the end of 1998 because of a lack of housing in Russia.[12] In October 1999, the dismantlement of the facility was completed and with it the almost ten years' process of Russian troop withdrawal from the Baltic states.

Second, with the exception of the Russian specialists in Paldiski referred to earlier, all Russian military forces stationed in Estonia were officially declared on 31 August 1994 to have been withdrawn. Paradoxically, some of the troops withdrawn from Estonia in compliance with the troop withdrawal agreement were apparently seen as the "main concrete military threat to Estonia" (Haab 1998, 111). In particular, several of the officers of the 8,000-strong Seventy-sixth Guards Airborne Division stationed in Pskov are said to "have formerly served in Estonia and are well aware of the local Estonian conditions" (Haab 1998, 111). This division indeed had a certain tradition of being referred to as a threat to Estonian security. In February 1993, for example, the Estonian minister of defense said that this division "could invade Estonia on sixteen minutes notice."[13] The Russian military, however, seems to have refrained from a military operation against Estonia in 1993 because, among other things, it would have involved cross-border deployment with recourse to forces stationed in the Leningrad MD (including the Seventy-sixth Airborne Division in Pskov). There were no Russian troops based in the Narva region, and in the whole of Estonia only small contingents were left at that time. It has to be noted, however, that the stationing of considerable troops near the borders is not in accordance with the spirit of security community building, which requires the demilitarization of borders.

Third, after the withdrawal of the main body of Russian troops, some authors began to focus on the Russian military retirees in the Baltic states as a threat to security. According to Vejas Gabriel Liulevicius, "prior to the pullouts, the army demobilized thousands of younger officers to take advantage of the new situation. While the term 'retirees' conjures up pictures of aged

---

12. Both the overall ecological state of Skrunda and the radar station's noxious potential effects on those who had been exposed to the electromagnetic radiation for years are unknown. At the moment of the dismantlement, Skrunda residents were financially or medically assisted neither by the Russian nor by the Latvian authorities (Cengel 1999, 8).

13. *Estonian Review* 1993a.

veterans, harmless and infirm, the reality is much more threatening. Latvians and Estonians fear these young demobilized officers were left to be an active 'fifth column' " (1995, 394). Juris Dreifelds pointed out at the time that "there is a fear among many Latvians that these retirees, some as young as forty, could become a destabilizing force used by Russia in the future for direct military functions, sabotage and for agitation among the Russian-speaking population. Most of these retirees are bitter about the breakup of the USSR and the emergence of an independent Latvia" (1996, 172). According to Rein Helme,

> the possibility of political manipulation of some contingents of that people [the Russian-speaking population] to cause internal instability still remains a real risk factor. It is often alluded, in this connection, to thousands of retired Russian militaries in Estonia. The list of those contains data on the 10,517 retired military up to August 26, 1994. Some have affirmed, that many of them possess weapons. It is possible, but not enough to operate immediately as an armed unit. (1997, 108)

Renatas Norkus added that "shortly before the departure of the Russian troops, there were approximately 22,000 and 10,500 retired Russian officers living in Latvia and Estonia respectively. Consequently, a widely shared fear prevailed in the two countries that this group could serve as a propaganda resource and a potentially active 'fifth column' for Russia" (1998, 142). And Talavs Jundzis identified as an "important threat source . . . separate groups within Latvia, who continue to work against its independence and for the return of the communist system or of another totalitarian regime. Predominantly among their members are former Soviet military persons and their dependents who, at the time of the Soviet Union, entered Latvia in large numbers as military retirees" (n.d., 5).

This peculiar mélange was composed of deep-seated prejudices, threat perceptions, images of enemies, half-truths and truths, as well as anxiety as a result of, among other things, past victimization. The intention to naturalize individual threat perceptions is obvious. It is almost impossible either to divide the cluster in its pieces or to dissect its constituent parts regarding their objective contents. It is also hardly necessary because, first, language constructs rather than reflects what is considered reality, and, second, it is the cluster in its entirety rather than its isolated ingredients that makes a

sober and nonemotional treatment of the issue difficult. Of course, the Soviet military stationed in the Baltic republics had a very bad reputation—
and rightly so—not only because it was perceived as an occupation army. It
also acted "as a state within a state" (Dreifelds 1996, 172) and was involved
in illegal activities. The environmental legacy of the Soviet troop presence is
alarming (Ozolina 1996, 39–40; Clemens 2001, 222). Members of the Soviet
army were also treated preferentially with respect to many things—for example, housing and education (Karklins 1994, 121–22). Some Soviet military
retirees are said to have become "part of the backbone of the Soviet loyalist
movement in Latvia" (Lieven 1994, 205). Furthermore, the Russian army stationed in the territories of the Baltic states consolidated this image in many
respects. It is said to have violated international norms and to have been involved in illegal business activities in the Baltic states. It also continued military maneuvers in 1992 and 1993 on Baltic territory (Vares and Haab 1993,
292; Vares 1994, 141).

It is also obvious that the independence of the Baltic states challenged
the Russian soldiers' hitherto convenient and privileged Baltic world. As a
consequence, many authors assign to the withdrawn Soviet/Russian troops
a deeply rooted hostility to the Baltic states and peoples. According to
Dreifelds, for example, "there is a widespread bitterness in military and
other circles [in Russia] toward the Baltic republics for their apparent role in
rupturing their previously comfortable world" (1996, 180). Anatol Lieven
adds that "the Balts were rightly seen as the cutting edge of the disintegration of the Soviet Union. Among some officers, this had led to a particular
hatred of the Balts" (1994, 203). As a corollary of these assessments, Olav
Knudsen speculated at the time that "the main challenge of Baltic security
policies is not going to be how to get the Russians out, but how to keep them
from coming back in once they are out" (1993, 66).[14] Furthermore, a 1994
opinion poll apparently confirmed these presumptions. The study analyzed
the main threat perceptions of 615 upper-level military officers in selected

14. Here, Lieven is more to the point: "it is not clear to me why some Western 'experts' automatically assume that [withdrawn Russian troops] are passionately anxious to come straight
back again, when an attack on the Balts would clearly involve major economic—if not military—retaliation from the West, catastrophic loss of export earnings, and therefore a massive
internal financial and economic crisis. If you are prepared to risk this to come back in, then why
go out in the first place?" (1996, 176).

Russian military districts. Of those asked, 49 percent identified Latvia as among their chief enemies, 47 percent Lithuania, and 46 percent Estonia (SINUS Moskau 1994, 37). At the same time, it should be noted that the results of the opinion poll were made public in the very month in which the Soviet (and later Russian) military presence in Latvia and Estonia came to an end (with the exceptions mentioned earlier). However deeply rooted the Russian military's hostility to the Baltic states, though, it proved insufficient to prevent the Russian troops from leaving. They took everything with them that was not nailed down, and what they could not take with them, they demolished. But they left, and they did so without firing a shot.

The military pensioners who remained in the Baltic states exhibited and continue to exhibit only scarce inclinations to act as someone's "fifth column" or to perform activities that called into question Latvian, Lithuanian, or Estonian independence.[15] They showed also no proclivity toward encouraging others to act on their behalf. In addition, many of them are suffering from advanced age and poor health. In Estonia, only approximately one-third of the military pensioners were aged fifty and younger; 42 percent were sixty-one years old or older (Norkus 1998, 142). The overall magnitude of the Russian military pensioners should also not be exaggerated. The approximately 22,000 and 11,000 former Soviet officers who were allowed to remain in Latvia and Estonia, respectively, formed less than one percent of these countries' populations. Accordingly, in spring 1994 only 2.2 percent of Latvian citizens and 1.8 percent of Latvian noncitizens considered "people unloyal to Latvia" the most threatening factor to Latvian security and independence (Ozolina 1998, 133). Likewise, Estonian respondents "usually [did] not view the some 10,000 retired Russian military residing in Estonia as threatening" (Aalto 2000, 76).

Thus, there is a gap between the scholarly and officially cultivated image of the enemy, on the one hand, and the general public's perception, on the other hand. In the terminology of the securitization approach to the study of security, it can thus be said that the securitizing actors failed to convince the general public that the presence of the military pensioners was indeed a threat to national security and that the issue needed to be moved beyond the normal and established political rules of the game. Securitiza-

---

15. The Latvian government, however, banned a military veterans' organization from openly agitating for the restoration of the Soviet Union (Ozolina 1996, 44).

tion takes place only if and when the audience accepts a securitizing move on the part of securitizing actors as such. By refusing to subscribe to the point of view that the military pensioners were a threat to national security, the Latvian and Estonian public also refused the government the right to handle the issue through extraordinary means. By doing so, the public served as a desecuritizing corrective against those actors who assigned to the issue urgency and priority.

If we agree with Ole Wæver that a "period of desecuritization [is] the ideal condition for a security community" (1998, 93), then the desecuritizing attitude of the Latvian and Estonian populations can be seen as an important element in security community building. The military pensioners issue thus can be classified as a failed attempt at securitization or, strictly speaking, as no securitization at all. As a consequence, the Latvian and Estonian governments had to deal with the issue within the framework of normal politics instead of, for example, according to a "throw them out" logic (which would also hardly have been accepted by international organizations monitoring Baltic-Russian relations). Like the case for the whole complex of the Russian troop withdrawal, a search for compromise took the place of radical, maximalist demands. In Latvia, the issue was thus solved in the 30 April 1994 Latvian-Russian Accord on the Social Security of Russian Federation Military Retirees and Their Families Who Reside on the Territory of the Republic of Latvia, and it was solved in Estonia in the 26 July 1994 Estonian-Russian Agreement on the Russian Military Pensioners in Estonia.[16]

16. For details, see Norkus 1998, 143. Haab notes that the Estonian-Russian agreement violated a number of Estonian constitutional provisions. The ratification of the agreement by the Estonian Parliament in late 1995 thus was in contradiction to the Estonian Constitution; nonratification, however, would have undermined Estonia's credibility in international politics.

8

National Security Documents

THE NEGOTIATIONS on and the process of the withdrawal of Russian troops from the territories of the Baltic states dominated the security policies of Estonia, Latvia, and Lithuania until the main body of troops was pulled out in August 1993 in Lithuania and one year later in Estonia and Latvia. Only after the objects of suspicions in connection with the presence of the Russian army did not materialize could an earnest conceptualization of national security start. These conceptualizations resulted in Lithuania in the Basics of National Security of the Republic of Lithuania, in Latvia in the Security Concept of the Republic of Latvia, and in Estonia in the National Security Concept (see appendix, documents 1–4).

Besides carrying forward the historical narrative, the textual-contextual analysis of the evolution of the national security concepts and the discussion of the documents in this chapter has three aims. First, the chapter and the appendix contribute to the authenticity of the analysis by including excerpts from the documents with a focus on risk assessments and threat perceptions. Second, I discuss parts of the documents in light of both their internal coherence and selected scholarly work on security, perceptions, and conflict. Third, the chapter delivers the textual background for the following chapters. National security documents can be expected to "end up with unduly [sic] generalizations and simplistic, unspoken assumptions relying on ideological bias and subjacent, maybe even subconscious, agendas" (Jæger 1997, 4). However simplistic and generalized they may appear, they deserve close attention because they are "privileged textual representations of the state's security policy" (Jæger 1997, 5).

Lithuania

In Lithuania, designing a national security concept was no simple task. By order of the government, the defense and intelligence ministries started as early as November 1991 to work on a formal document outlining what should in future be referred to as the national security for Lithuania (Petersen 1992, 32). This task was temporarily made difficult by the bizarre simultaneity of an LDDP government, a president sympathizing with the LDDP, and a Ministry of National Defense that continued to be led by a Sajudis defense minister until September 1993. Assigning to Parliament the task of drafting a security document furthermore resulted in delays as it opened up the procedure to party politics. This proceeding is not wholly a bad thing because it situates security policy firmly within the parliamentary process rather than assigning it to the bureaucrats in the security establishment, whose democratic legitimacy is often questionable. It is, however, time-consuming. An ad hoc working group appointed by the president presented to Parliament the first draft of a national security concept in the summer of 1993. The parliamentary opposition, for its part, thereafter presented an alternative concept in mid-1994. At the end of 1994, following the discussion of the two drafts in Parliament, a new ad hoc working group, representing all parliamentary factions, was inaugurated. In the autumn of 1995, Lithuanian policymakers came to an agreement on the text of the national security concept, but Parliament did not accept it. On 19 December 1996, finally, the Law on the Basics of National Defense and its annex the Basics of National Security (see documents 1 and 2 in the appendix) were adopted and came into effect on 8 January 1997 (Bajarunas 1995, 20; Linkevicius 1995, 101; Jæger 1997, 18–19).

In part 1, chapter 2, first section, the Basics of National Security define as the "main objects of national security human and citizen's rights, fundamental freedoms and personal security; the cherished values of the nation, its rights and conditions for a free development; state independence; constitutional order; the integrity of the state's territory; and environment and cultural heritage." Certainly, the Basics of National Security indulge in a very expanded understanding of security, which almost makes the law practically inapplicable. This approach to security, arguably, reflects the self-perception of "small states with big worries" prevalent at that time

(Heinemann-Grüder 2002). Yet the document in fact displays a loquacity that almost amounts to securitizing everything. Moreover, in part 2, chapter 9, first section, the Basics define "external risks, challenges, and potential challenges and potential dangers conditioned by the geopolitical environment" in such a comprehensive manner that any effective foreign and security policy almost necessarily has to violate them.

The enumeration of "external risks, challenges, and potential challenges and potential dangers conditioned by the geopolitical environment" widens considerably—and potentially open-endedly—the range of issues with possible relevance to security. Calling on every citizen to defend this broad range of security objects leads to a situation in which "everyone is expected to defend everything with every possible means" and *"everything* [appears] to be a question of national security" (Jæger 1997, 22, emphasis in original). Furthermore, given the political function of the articulation of security, this very broad definition expands the government's sphere of responsibility and helps restrict democratic control and participation along a broad range of issues. The expansion of responsibilities of the state's authorities may also overstretch the government in the sense that it becomes involved in too many fields at the same time. This overstretch may, among other things, collide with civil society, complicate finding a balance between loads and capabilities, and render difficult the implementation of an effective policy that is indeed in accordance with the law. For example, the ban on "direct or indirect interference into Lithuania's domestic affairs" was not in accordance with Lithuania's membership in the EU. "Military capability in close proximity to Lithuanian borders" actually defined a possible presence of NATO troops in Poland as an external risk, a challenge, a potential challenge, or a potential danger to Lithuanian security. The ban on "spreading propaganda and disinformation" is open to the interpretation that any statement deviating from what the government considers information may be considered a risk to security, so the ban may serve as a means of domination with which to prevent official standpoints from being challenged by competitive views. As long as Lithuania was prevented from joining NATO or the Western European Union (WEU) or both, the interpretation of "preventing Lithuania from obtaining international security guarantees" as a risk to Lithuanian security turned against both alliances. "Attempts to impose upon Lithuania dangerous and discriminatory international agreements" is again open to interpretation and defies general definition. Moreover, the ban on "investment of capital with political goals" and its specifications pave the way to political interference into economic affairs.

Regarding the internal dimension of risks and domestic crises, the document also swarms with unclear terms and formulations. For example, it raises the following questions: At which point is a rise in unemployment, a decline in production volume, and a decrease in the gross national product (GNP) to be considered "critical"? And critical to whom or what? What is precisely meant by "backwardness" of the economy? What does "the Nation's immunity and sense of identity" mean and what are "national values"? When does the differentiation in wealth approach a "critical" level? Furthermore, the definition of "disregard for long-term national goals in the State policy" as a potential internal risk declares these "long-term political goals" sacrosanct and beyond political competition and thus restricts the freedom of opinion. In the final analysis, the document makes use of generalizations and formulations that are almost meaningless in their lack of specification and open to political interpretation.

Latvia

Zaneta Ozolina differentiates between four phases in the evolution of the Latvian security policy (1999, 18–26). She identifies as the first phase the period from 1988 to 1990, during which time the Latvian Popular Front and the Latvian Writers' Union defined security primarily in terms of demilitarization of the Latvian republic's territory. Initially, demilitarization was identified with the withdrawal of Soviet troops and was not an expression of criticism of the military in general; just the opposite: the desirability of territorial military formations had been stressed as early as June 1988. On 18 December 1988, the Latvian Popular Front adopted an official document titled "On Attitudes toward the Armed Forces of the USSR" that reflected the disastrous conditions in the Soviet army and the ongoing discussion among dissidents on the possibility of stationing Latvian conscripts in the Latvian republic.[1]

In general, however, security issues did not rank paramount on a Lat-

1. This demand was blocked by the central authorities with reference to the success of the Soviet conscription system—deploying conscripts outside their home republics—in World War II (Clemens 1991, 243).

vian Popular Front agenda dominated by economic, cultural, and political issues. One year later, the Second Congress of the Latvian Popular Front devoted an entire chapter of its program to demilitarization. Now, the issue was not narrowly confined to Soviet demilitarization but argued in the more general context of détente. Furthermore, the congress problematized the presence of Soviet troops in Latvia in a more comprehensive manner. The most important passages of the Popular Front's resolution are quoted in chapter 5 and need no repetition at this point. Although the resolution did not generate the wished-for result of a dialogue between Moscow and Riga, it laid the foundation for the Latvian position on the issue of the withdrawal of the Russian troops after independence.

The second phase of Latvia's security policy, from 4 May 1990 to August 1991, was marked by the growing tensions between the central Soviet authorities and the Latvian Popular Front as a result of the Declaration of Independence by the Latvian Supreme Council on 4 May 1990. Declaring de jure independence and at the same time a transition period to full independence was, however, a sensible move that helped ease the tensions. Likewise, endorsing neutrality as a security policy option not only reflected the popular appreciation of neutrality at that time, but could also be seen as a nonprovocative stance toward the Soviet Union. Furthermore, it was in accordance with recommendations by Western observers (Bitzinger 1991). The nonaggressiveness of the civil resistance program, based on a comprehensive proposal elaborated and disseminated by the Latvian Popular Front in December 1990, could also not be doubted (Jundzis 1995, 553–54; Ozolina 1999, 20). However, the Latvian republic's authorities began the buildup of republicwide armed forces in October 1990 with customs and militia posts. Under the impression of a change from a perceived potential threat emanating from the Soviet Union to an actual threat in January 1991, the improvisational creation of voluntary territorial defense forces and the inauguration of the Department of Public Safety by the Council of Ministers followed. Reacting to assaults against unarmed customs and border guards, the document "On Immediate Self-Defense Actions to Be Taken in Connection with the Violent Actions of Soviet Armed Forces on the Territory of the Republic of Latvia" was approved on 24 May 1991. It aimed to mobilize all means and personnel available and "ordered the preparation of a draft law setting out the procedure whereby a state of emergency could be declared, as well as a draft law on the establishment of the National Guard; the establishment of a nonmilitary defense center attached to the presidium of the Supreme Council; and the establishment of a border guard division within the Security Department of the Latvian Council of Ministers, complete with regional divisions, border posts, and a training facility" (Ozolina 1996, 32).

The third phase of Latvia's security policy evolution lasted from 20 August 1991 to 31 August 1994. The first date marked the declaration of full independence by the Latvian Supreme Soviet and the international recognition of Latvia's independence in the aftermath of the failed coup d'état in Moscow; the second date marked the completion of the withdrawal of the main body of Russian troops from Latvia's territory. Latvian foreign and security policy representatives during this period aimed to strengthen Latvia's independence by having it join international organizations such as the UN and the CSCE, by creating the legal foundation for Latvia's security and defense system, by negotiating the troop withdrawal with the Russian Federation, and by establishing both a Ministry of Defense (on 13 November 1991) and national armed forces.[2] As this list of priorities indicates, this period saw an increased narrowing of the understanding of security that resulted from, among other things, the preoccupation with the Russian troop withdrawal and the perceived fragility of the recently established independence. The 1994 Latvian Defense System Concept, for example, regarded as the main sources of threats to Latvia the presence of an alien army in the country; the uncontrolled operations of foreign espionage and counterespionage units; the presence and extremist activities of various Communist, imperialist, and other anti-independence organizations; a high level of crime involving the activities of armed, international, and organized criminal groups; economic instability and the country's dependence on foreign energy resources; and the use of the demographic situation by anti-independence activists in their own interests (Ozolina 1999, 22).

After the completion of the troop pullout, Latvia entered the fourth phase of its foreign and security policy evolution. Within less than a year, Parliament adopted the Foreign Policy Concept (on 7 April 1995) and the Cabinet of Ministers adopted the Latvian National Security Concept (on 13

2. The Latvian armed forces were split into the Home Guard (Zemessardze), established by a Supreme Council's Law on the Home Guard on 23 August 1991 and responsible to the Supreme Council until the end of 1994, and the regular armed forces, formed by order of the minister of defense on 13 December 1991.

June 1995). These concepts define the maintenance and preservation of independence, territorial integrity, language, and national identity as the main goals of Latvia's foreign and security policy. They emphasize, among other things, Latvia's desire to become fully integrated into the EU and NATO. The National Security Concept, moreover, identifies as the main threat to Latvian security efforts by other countries to destabilize Latvia's internal situation rather than an outright foreign military aggression. At the same time, it holds that Latvia poses no threat to any other country or to minorities within Latvia. However, although "theoretically well-written and practical in nature," the concept

> has for several reasons become virtually irrelevant. First of all, the [National Security Concept] document was prepared without the input of broad swathes of the political elite. Second, there was virtually no public discussion of the document after its preparation. Third, the very process by which the concept was adopted was illustrative of the fact that Latvia still has not developed adequate systems for taking decisions. The document was vetted by the National Security Council on 22 May 1995, and less than one month later, on 13 June, it was accepted by the Cabinet of Ministers. This surprising speed of action eliminated any opportunity for discussion, analysis, and amendment of the document. (Ozolina 1996, 36)

The plans to supplement the document with an annual national security plan failed. Recognizing the failure of the National Security Concept, President Guntis Ulmanis in spring 1997 ordered a new law on national security, which was approved by the Cabinet of Ministers on 6 May 1997.

The document (see document 3 in the appendix) is problematical in several respects. First, it creates the same set of problems as its Lithuanian counterpart both by inviting state institutions to become involved in too many areas at the same time and by securitizing a plethora of issues. Second, stating that "in establishing its security, Latvia does not threaten any other state" is a declaration of intentions that, like the whole document, ignores the character of perception and misperception in international politics discussed earlier in connection with the security dilemma and the relationship between different security communities. One's own conviction that the means with which one aspires to increase one's security do not threaten anybody does not necessarily result in another's belief in one's lack of aggressiveness. Intentions and perceptions may be identical, but a unilateral declaration of goodwill hardly suffices to convince others, especially if the provisions included in the remainder of the document may be read as pointing in a somewhat different direction. For example, as long as the leadership of the Russian Federation interpreted NATO enlargement in general and the membership of the Baltic states in particular as a threat to Russian national security, Russian leaders saw the Latvian government's aspirations for membership as threatening quite regardless of the benevolent tone of the overall principles of the Latvian security concept.

Third, aspiring to develop a legislative basis "for the prevention of any type of conflict" reveals an untenable understanding of conflict. Not only are conflicts an inevitable consequence of limited resources, but they are also "an essential part of human life and . . . fundamental to the capacity of people to develop a world which is more suitable to human needs than the world of today" (Birckenbach 1997, 17). Conflict may contribute to the establishment of unity and cohesion within a group and is by no means necessarily dysfunctional. Furthermore, it may serve as a unifying bond between antagonists and thus help reduce antagonism (Coser 1956). Aspiring to prevent conflict therefore ignores the creative and constructive potential inherent in conflict and paves the way for dealing with it in a suppressive manner. In connection with the expanded understanding of security outlined in the document, such an understanding of dealing with conflict runs the risk of a restrictive and suppressive engagement of the state security authorities in too many spheres of the society. This is true in particular because the identification and neutralization of "antigovernment oriented factions or individuals" in part IV.7 strictly speaking render oppositional activities impossible and protect any particular ruling party's particular points of view.

At the very least, this provision is an invitation to abuse and restrict oppositional activities because it assigns to state security institutions the task to protect not only the state and its underlying norms and values, but also the particular government in power by identifying and neutralizing antigovernment factions or individuals. Finally, taking into consideration both the existing and the future military potential of Latvia, *deterrence* seems hardly to be an appropriate label for Latvia's defense basis. In the Latvian Security Concept, the expectation of international assistance in case of a foreign aggression is confined to "political and economic pressure" and does not include military assistance. As long as Latvia did not enjoy international military security guarantees, the isolated Latvian armed forces could in fact not exhibit the capabilities to function as a credible military deterrent. An emphasis on "political resolution of military problems" was a logical consequence and a more promising way to deal with perceived threats.

Estonia

If "the military threat constitutes the prime risk factor on Estonia's national security agenda" (Raid 1996, 13) and if "military instruments and force are still considered the crucial elements in the defense of Estonian independence" (Haab 1995, 39), then the absence of a draft national security concept until October 2000 was as logical as the existence in Estonia, since May 1996, of the Guidelines of the National Defense Policy, defining national defense as "an inseparable part of ensuring national security." According to the guidelines, national defense aims "to guarantee the preservation of independence and sovereignty of the state, the indivisible integrity of its land, territorial waters, and airspace, its constitutional order, and the vitality of its people." Reflecting the preoccupation with military threats at that time, Estonia, according to the guidelines, "understands that the main sources of danger threatening state security are aggressive imperial aspirations and political and/or military instability."[3]

Reflecting five more years of peaceful development, diminishing precariousness of independence, and increasing self-confidence in international affairs, the security document of 6 March 2001, the National Security Concept of the Republic of Estonia (see document 4 in the appendix), apparently aims to dedramatize and desecuritize Estonia's direct political and military environment by downplaying both military risks and risks posed by outside political pressure. The authors of the concept were even criticized in the Estonian media for too strong an adaptation to the soft security discourse cultivated in the West, especially before 11 September 2001 (Kuus 2003b, 18). However, the document also emphasizes risks stemming from what is called the incomplete evolution of the post–Cold War Euro-Atlantic security framework—meaning, less cryptically, Estonia's exclusion from NATO.

---

3. Guidelines of the National Defense Policy of Estonia is available at http://www.vm.ee /eng/nato/def.policy.html.

The document's somewhat relaxed tone—downplaying specific military and political risks, and stressing economic, cultural, and so-called new security risks—should be read in connection with the discussion in chapter 11 on representations of Russia in security talks in Estonia. The change in the representation of Russia from a military threat to a cultural threat was in accordance with Estonian policymakers and scholars' enthusiasm for Samuel Huntington's (1997) idea of a clash of civilizations. For example, a significant factor in the cultural identity of the Baltic countries is said to have been *"living on the border* of Western civilization":

> We maintain that the most decisive role in the collapse of the Soviet Union was played . . . by the civilizational conflict between the Russian-Soviet Empire, the "New Byzantium" of the 20th century, and the Baltic and other East-European nations, representing the Western traditions of individual autonomy and civil society. . . . For Estonians and other people with a Western mind-set, living under the Soviets meant a "clash of civilizations" inside the mind of every single individual, the loss of personal integrity, and even the loss of the right to an authentic life-world. (Lauristin 1997, 29, 37, emphasis in original)

Huntington's idea of a clash of rather than a dialogue between civilizations, "avidly welcomed by cultural nationalists the world over" (Halliday 2000, 49),[4] has become "a key conceptual basis of speeches, policy analysis and academic research on foreign affairs" (Kuus 2002, 307) in Estonia. Furthermore, by making "the nebulous and ubiquitous category of culture" the basis of security policy, "security has become infinitely flexible" (Kuus 2003a, 45) and has left the narrow sphere of conventional security policy in a way that most of those scholars arguing for a broadening of the security agenda would not appreciate. Rather than expanding democratic codetermination in security policy, the document may exclude the public from an ever-increasing range of cultural issues because these issues are referred to in terms of security.

National security documents, however, are not only security docu-

---

4. The Estonian minister of foreign affairs has written a preface to the Estonian-language edition of Huntington's *Clash of Civilizations* and, like the prime minister, attended the book's presentation to the public. Thanks to Merje Kuus Feldman for this information.

ments, but also political statements. It may be speculated at this point that the representation of Russia as a military threat to Estonian security was abandoned in official documents primarily because it was not in accordance with NATO's representation of Russia. Because the goal of Estonian accession to NATO was approved by the Estonian Parliament in May 1996 and because NATO representatives reportedly "had urged Estonia to adopt a security policy concept,"[5] it is a reasonable assumption that the document was written in the light of NATO's self-understanding and self-representation.

The disappearance of military threats and of the representations of Russia as a threat from the Estonian security policy documents can be explained by two mutually supportive ideas. First, following an explanation that focuses on instrumentality, it can be argued that the shift of emphasis reflects a failure of the earlier approaches to membership in NATO. Estonian decision makers seem to have understood at that point that an invitation to NATO would hardly follow if they continued to represent Russia as a threat to Baltic and European security. Rather, the chance of becoming a NATO member was inversely proportionate to the extent to which they referred to Russia as a threat to the Baltic states in particular and to Europe in general. Thus, it is true that the governments of the Baltic states tried to use their perception of a threat emanating from Russia as an argument for NATO membership (Oldberg 2003, 274), but it is equally true that they received an invitation to NATO only when they abandoned their representation of Russia as a threat.

Continuing to present the Baltic states as threatened by and as a bulwark against presupposed Russian aggressiveness would have been counterproductive. The prospects of becoming involved in a military conflict between the Baltic states and Russia seems to have been one of the reasons why NATO refused too direct and too ambitious an engagement in the Baltic region in the first half of the 1990s. And during that decade, NATO represented Russia increasingly as a partner. Recommending oneself as a bastion against a partner neither makes sense nor is a promising preaccession strategy. The Baltic states' rhetorical adaptation of representing Russia as unpredictable rather than outright hostile followed, as did advertising themselves in terms of protecting Europe from uncertainty and unpredictability (Saudargas 1999, 78). Moreover, Baltic decision makers advertised their strivings after membership in terms of a general integrationist or reintegra-

5. *RFE/RL Baltic States Report* 2001b.

tionist policy, strictly following the rhetorical patterns given by NATO. Beginning in the mid-1990s, they adapted themselves to official NATO parlance by emphasizing NATO's character as a community of values rather than a community of defense and as a political rather than a military organization (membership to which is aspired to for political rather than military reasons) and by emphasizing their own role as a contributor to rather than a beneficiary of security. This change was both an expression of the idea of the return to the West and a rhetorical adaptation to NATO's anticipated set of expectations.

The second reason for the disappearance of the depiction of military threats and representations of Russia as a threat from Estonian security policy documents at the end of the 1990s can be revealed through discourse analysis. Discursive analysis, in one of its variations, is interested in rules governing articulation. It explores shared understandings that govern the use of language and differentiate that which can legitimately be said from that which cannot (Wæver 2002). It thus investigates "the relationship between the rules and conventions of specific 'language games' or 'forms of life' and their socio-historical and cultural meaning" (George and Campbell 1990, 273). But language games also reflect power relations. Wanting to become a member of NATO and the EU clearly required an adaptation to the language games played in and by these two organizations. Furthermore, because the process of transformation in the Baltic states was one of adaptation to the Western type of political, economic, and social organization, the discursive structures within which this transformation unfolded were by and large constituted in and by the West. Challenging the Western language games by adhering to different games was counterproductive. Sticking to the structure of discourse prevalent in the Baltic states throughout the 1990s would thus have resulted in isolation from the Western world.

9

The United States and Baltic Security

Friendship in a Historical Perspective

CAN ACTORS EXTERNAL TO A REGION play a significant role in regional security community building, with "significant" meaning a role other than that of serving either as a model or as a negative reference point? Mutual representations of friendship can establish bonds between actors internal and actors external to a region. As noted in chapter 3, referring to one another as friends is standard procedure in international politics. It is also standard procedure in U.S.-Baltic relations, and there is no reason to treat such references a priori as mere rhetoric. Analyzing the substance of references to friendship should therefore be a part of the analysis of security policy, as should exploring the extent to which friends may influence one another with respect to basic thought patterns underlying foreign and security policy. I analyze the U.S.-Baltic friendship in this chapter and in chapter 10, focusing on historical events here and exploring current affairs in the following chapter.

As also sketched in chapter 3, assuming the role position of a friend may follow from a Kantian structure of the system, but on the foreign policy level it may also be assumed contrary to the perceived structure: others may be represented as friends precisely because the system is not understood as one following a Kantian logic. Under these circumstances, the probability that friends keep their promises may to some extent be influenced by the degree of symmetry among them, with *symmetry* meaning a "mutual relation of parts in respect of magnitude and position."[1] I do not suggest here that symmetry is an indispensable precondition for the evolution and endurance of

---

1. *The Shorter Oxford English Dictionary on Historical Principles* (Oxford: Clarendon Press, 1973), 1:2220.

friendship patterns. Without symmetry among parties, however, obligations resulting from friendship are also asymmetrical. Asymmetrical obligations create different degrees of incentives to enter a friendship relation and to maintain it, and they render a regression to a second-level reasoning more likely in individual cases even when the friendship is in principle based on a third-degree internalization with mutuality as its foundation. The assumed role pattern is symmetrical, but if the friends and their capabilities are not symmetrical, the consequences resulting from accepting the role of a friend will also be asymmetrical. If mutual representations as friends are cultivated contrary to the perceived structure of the system—in other words, if they are a foreign policy strategy rather than an expression of the structure of the system—asymmetry in friendship relations is likely to make these relations quite unstable at the policy level.

Baltic decision makers' references to the United States as a friend can thus be explained as a politicorhetorical proceeding, the aim of which is to motivate the U.S. government to apply policies toward the Baltic states as if the United States were a friend.[2] In this case, cultivating bilateral friendship patterns appears to follow paradoxically from the perception of the international system as a Lockean kind of anarchy in which the use of violence cannot in principle be ruled out. Such a perception can be found in, for example, part 1 of the 1999 *White Paper of the Ministry of National Defense of the Republic of Lithuania,* stating that the absence of an immediate military threat at that time provided "a breathing space" during which time credible armed forces could be developed. In particular, the rule of mutual aid inherent in friendship had relevance to the Baltic governments as long as they were excluded from NATO. Patterns of friendship may develop from systems of collective defense, but they may also evolve independent of formal alliance responsibilities. If taken seriously and internalized to the third degree, they render formal alliance commitments indeed unnecessary. Owing to the asymmetry

---

2. The kind of practice that produces friends "involves treating others as if one not only respected their individual security concerns but also 'cared' for them, a willingness to help them even when this serves no narrowly self-interested purpose. In treating Alter in this way Ego is casting Alter in the role of friend, and, given the symmetry of the role, taking the same role for himself" (Wendt 1999, 341–42). For lack of a possibility to *treat* Alter in a specific way, Ego replaces treatment of Alter with a specific linguistic designation: *calling* Alter a friend and rhetorically supporting Alter's policies.

in U.S.-Baltic relations and to the differences in the degree of obligations resulting from assuming the role of a friend, however, regression to second-level internalization (i.e., an instrumental understanding of friendship in terms of self-interest) cannot be excluded in any given situation. These differences give the inherently symmetrical role of friendship a profoundly asymmetrical bias—quite regardless of Baltic decision makers' sincerity and dedication to their role as a friend, which I shall not at all call into question here. Yet whatever kind of aid the United States could expect from the Baltic states, it would be negligible in comparison with what the Baltic governments hope for from their American friend.

References by U.S. decision makers to the Baltic states as friends are more difficult to explain and cannot be understood without taking into consideration domestic politics: the Baltic American community is said to remain "a well-organized and effective political force in U.S. politics," preventing any president from "appearing to have 'sold out' the Baltic states and to have contributed to a 'new Yalta' in the context of [NATO] enlargement" (Asmus and Nurick 1996b, 128). Stressing that the Baltic states "are becoming good economic partners and reliable friends" and emphasizing "the friendly relations that have been continuously maintained . . . since 1922"[3] between the United States and the Baltic states thus not only could have been indicative of a moral ingredient of U.S. foreign and security policy (Asmus 1998, 24), but also in part could have reflected the influence of domestic pressure groups. This influence should, however, not be exaggerated. Even at the end of the 1990s, it was still argued that U.S. support might not mean Baltic NATO membership (Krickus 1999, 19). In April 2001, after several years of intense lobbying, twenty U.S. Senators still opposed NATO enlargement, and forty-eight were undecided or without opinion, whereas only thirty-two were in favor (Lithuanian-American Community 2001).[4]

There appears to have been at least five reasons for the somewhat lim-

3. White House Office of the Press Secretary 1996; Charter of Partnership among the United States of America and the Republics of Estonia, Latvia, and Lithuania, signed in Washington, D.C., on 16 January 1998, Preamble (*U.S. Information and Texts* 1998).

4. The September 2001 attacks on the World Trade Center and the Pentagon, the subsequent change in priorities on the U.S. security policy agenda, and U.S.-Russian rapprochement in light of the "war on terror" helped transform the issue of NATO enlargement into one of secondary importance and eased the Baltic way into NATO.

ited influence of the Baltic émigré communities on U.S. decision making. First, with the restoration of Baltic independence, the main political task of the émigré organizations—reminding the world of both the illegality of the Soviet occupation and the de jure independence of Estonia, Latvia, and Lithuania—came to an end, resulting in what can be called an identity crisis for these organizations, bereaved of their main mission (Plakans 1998, 644). Second, whereas some of the organizations favored a collective (Baltic) approach, others prioritized an individual approach (Estonian, Latvian, or Lithuanian). For example, the Association for the Advancement of Baltic Studies (AABS) aimed at "an explicitly 'Baltic' consciousness" (Plakans 1998, 644). Representing the concept of area studies, the AABS favored the representation of a coherent Baltic region rather than one of individual states with individual histories and cultures—a perception that was popular in the United States anyway and was translated into policy recommendations by, for example, Charles Kupchan, arguing in the context of NATO enlargement that "the American people and their representatives are not likely to respond favorably when asked to extend ironclad defense guarantees to countries they could not locate on a map" (1996, 271).

Increasing the visibility of the Baltic states thus called for the prioritization of a collective rather than an individual approach. However, the enduring tensions between these approaches limited the émigré organizations' penetrating power, whereas speaking with a single voice would have benefited their case. This limitation was politically reflected especially in the debate on taking either an individual or a collective path to NATO. Estonian officials' call for intense lobbying on the part of the Estonian American organizations, "including cooperation with other ethnic organizations," accordingly encountered dissent because many of the groups were in favor of prioritizing Lithuania, which was for some time considered the Baltic state best prepared for NATO membership (Huang 2001, 32). The Joint Baltic-American National Committee (2001), representing the Estonian American National Council, the American Latvian Association, and the Lithuanian American Council, thus somewhat competed with the national émigré organizations, but also united with them in the advocacy of Baltic NATO membership. By stressing the "vital interest of the United States" in Estonian, Latvian, and Lithuanian NATO membership, these organizations advocated an instrumental interpretation of the U.S. support in terms of self-interest, and the substance of friendship declarations was undermined.

Third, as the U.S. media's interest in Baltic affairs waned immediately after 1991, the émigré organizations lost an important means to influence public opinion and to exert pressure on political decision makers (Plakans 1998, 646). Fourth, by disregarding "the realities in place" in the Baltic states, thereby revealing the limits of "long-distance assistance," some exile organizations limited their influence on political decision making in the United States as well as in the Baltic states. For example, a comprehensive draft document for Baltic state security and defense infrastructure developed by the Baltic Foundation is said to have "completely ignored the Latvian constitution and all its laws and concepts. Since it additionally envisioned a force structure beyond the dreams of even the former Soviet officers, it could not be seriously considered and all the time and effort had been wasted" (Zalkalns 2002, 47).

Finally, the émigré organizations' political objective—unconditional and immediate NATO membership for the Baltic states—did not match with the more sophisticated and nuanced, yet at once more ambiguous and undecided approach to Baltic security pursued by the Clinton administration (see chapter 10). Because of many emigrants' personal experiences with the Soviet policy toward the Baltic states from which they had escaped, parts of the émigré organizations were said to be backward looking, anti-Communist, and anti-Soviet (Venclova 1995, 362). They seem to have translated this attitude into profound skepticism toward Russia and its potential to develop along democratic lines. This skepticism, however, was not in accordance with the Clinton administration's policy. Thus, although U.S.-Baltic relations displayed a somewhat sentimental character,[5] the U.S. government did not seem to lose its critical distance from the Baltic states, as some authors suspected would happen when several Baltic American diplomats and officers' assumed leading positions in the Baltic states (Lieven 1996, 177).

Patterns of friendship and amity may result from "longstanding historical links" (Buzan 1991, 190). Yet history is often represented in such a manner that it seems to support the notion of friendship. This procedure is not surprising because "statesmen and society actively shape the lessons of the past in ways that they find convenient" (J. Snyder 1991, 30). There is no rea-

---

5. The term *sentimental character* was suggested by one of the referees of the draft manuscript.

son to suspect that friendship patterns should deviate from this general rule. It is thus important to keep in mind that friendship (as well as nonfriendship and prefriendship) does not necessarily have to have anything to do with history as explored by historians. This lack of connection, however, does not automatically make it fragile or bound to dissolve when under pressure.

No event in the history of U.S.-Baltic relations has been quoted and referred to more often than the U.S. nonrecognition of the Soviet Union's occupation and annexation of Estonia, Latvia, and Lithuania in 1940. For example, the Charter of Partnership among the United States and the Republics of Estonia, Latvia, and Lithuania, signed on 16 January 1998, confirms in its preamble the narrative of state continuity and emphasizes the U.S. policy of nonrecognition by "recalling that the United States of America never recognized the forcible incorporation of Estonia, Latvia, and Lithuania into the USSR in 1940, but rather regards their statehood as uninterrupted since the establishment of their independence, a policy that the United States has restated continuously for five decades."[6] On 14 June 2000, at the occasion of the sixtieth anniversary of the U.S. nonrecognition of the Soviet takeover of Estonia, Latvia, and Lithuania, the U.S. Senate in a somewhat self-congratulatory mood adopted Concurrent Resolution 122, in which it "recognizes the 60th anniversary of the United States nonrecognition policy of the Soviet takeover of the Baltic states and the contribution that policy made in supporting the aspirations of the people of Estonia, Latvia, and Lithuania to reassert their freedom and independence" (U.S. Senate 2000).

Baltic political decision makers have reciprocated. For example, at the occasion of the signing of the U.S.-Baltic Partnership Charter, the Estonian president, Lennart Meri, emphasized that the United States "with its bipartisan support for [the] nonrecognition policy . . . was a true friend of the Baltics in a time of need, acting as a beacon of hope throughout the long, dark, and cold years of the Soviet occupation." The then Estonian foreign minister Toomas Ilves added that "the USA is one of Estonia's greatest and steadiest friends in the world." Lithuanian president Valdas Adamkus joined in the song by stating in a 14 July 2000 letter to U.S. president Bill Clinton that the U.S. policy of nonrecognition "gave strength and hope to hundreds of thousands of Lithuanian people under oppression and in

6. *U.S. Information and Texts,* no. 003 (21 Jan. 1998): 14. See document 6 in the appendix.

exile."[7] And the then Latvian foreign minister Valdis Birkavs characterized U.S.-Baltic relations thus: "Geography has not made us neighbors, but history has made us friends" (1997b). Cheap talk perhaps, considering that "[t]he United States [had] not protested strongly against the incorporation of the Baltic states into the Soviet Union, even though later on it claimed that it had never formally agreed to it" (Korbonski 1995, 131). The U.S. policy of nonrecognition and the maintenance of legations of the pre-1940 Baltic states in Washington throughout the Cold War, however, contributed to the continuous perception of illegality of the incorporation of the Baltic states into the Soviet Union and therefore to the perception of the Baltic claim to independence as legitimate, even though it is notoriously difficult to measure this contribution.

Indeed, in 1940 the U.S. government reacted immediately. For example, on 15 July, roughly one month after the occupation had begun, the United States issued Executive Order 8484, froze all Lithuanian assets in the United States, and made their transfer to Soviet Lithuania dependent on a license from the secretary of the Treasury. Sumner Welles, acting secretary of state, is quoted as stating on 23 July 1940 that the United States condemned "devious processes" where "the political independence and territorial integrity of the three small Baltic Republics . . . were to be deliberately annihilated by one of their more powerful neighbors" (cited in Gvosdev 1995, 17). Two months later, Secretary of State Cordell Hull, referred to the Atlantic Charter while stating "that the United States wished 'to see self-government restored to those peoples who have been forcibly deprived of it' " (Gvosdev 1995, 18).[8] The secret National Security Council document NSC 20/1 of 1949 stated that "we cannot really profess indifference to the further fate of the Baltic peoples. . . . It should therefore logically be considered a part of U.S. objectives to see these countries restored to something at least approaching

7. For Meri, see White House Office of the Press Secretary 1998b and Ilves 1998; for Adamkus, see *Newsfile Lithuania* 2000.

8. Lennart Meri said that Roosevelt should be remembered as the person who signed the Yalta Agreement, but also as the person who "laid the foundations for the U.S. policy of not recognizing the annexation of the Baltic Countries" and as "the author of the Atlantic Charter." Jüri Uluots, the last prime minister of the first Estonian republic, is said to have been aware of the Atlantic Charter (2004, 7–8).

a decent state of freedom and independence" (cited in Mantenieks 1990, 169).

Based on the June 1932 Stimson Doctrine of Nonrecognition of Forcible Seizure of Territory, the U.S. position on the Baltic states was reiterated—and inevitably increasingly ritualized—in numerous statements from 1940 to 1991. President Gerald Ford, for instance, shortly before departing for Helsinki to sign the CSCE Final Act in 1975, emphasized that the Helsinki Accords did not change at all the U.S. position on the illegality of the occupation of the Baltic states and did not legalize their incorporation into the Soviet Union (Mantenieks 1990, 169). Baltic émigré organizations such as the Baltic American Committee had indeed been afraid that the Final Act could legalize Soviet rule over Estonia, Latvia, and Lithuania (Thomas 2001, 127). On 18 November 1983, President Ronald Reagan assigned to the then established Baltic States Service Division at Radio Liberty the task of "reinforc[ing] the distinct identities of the Baltic States and separat[ing] them from the rest of the Soviet Union" (Clemens 1991, 299). President George Bush, six years later, confirmed the U.S. position at the occasion of the designation by the U.S. Congress of 14 June 1989 as Baltic Freedom Day (Mantenieks 1990, 169). Furthermore, the Baltic states were not consigned to invisibility in official maps issued by the U.S. government. Rather, the U.S. nonrecognition of the occupation of the Baltic states has been stressed in most maps published since World War II (Plakans 1998, 643).

At the Tehran Conference in November and December 1943, during which time the Baltic states were occupied by German troops, President Franklin Delano Roosevelt indeed argued in favor of self-determination for the Baltic states, elections, or a plebiscite. However, what "seemed to be a matter of principle in 1941–1942 was explained to Stalin as vote-getting expediency when Roosevelt met him at Tehran in 1943. Perhaps this was not Roosevelt's intention, but it comes across this way in the Russian minutes of their meeting, sandwiched between more formal tripartite meetings" (Clemens 1991, 297). Thus, Roosevelt seems to have represented the Baltic issue primarily in light of the 1944 presidential elections: in order to gain the support of Baltic and other émigré communities in the United States, he was more interested in maintaining the impression of a real U.S. interest in Baltic independence than in actually restoring independence. Roosevelt furthermore "said that he fully realized the three Baltic Republics had in history and again more recently been part of Russia and added jokingly that when the Soviet armies re-occupied these areas, he did not intend to go to war with the Soviet Union on this point." He is also quoted as having been interested in "the world opinion receiving some expression of the [Baltic] people, perhaps not immediately after their [the Baltics'] re-occupation by Soviet forces, but some day." He personally was confident "that the people would vote to join the Soviet Union."[9]

Informal arrangements among Roosevelt, Winston Churchill, and Joseph Stalin at Tehran preordained the postwar order by assuring the Soviet Union of, among other things, the restoration of the 1941 Soviet western border (including the Baltic states and the eastern part of Poland) (Loth 1987, 41). Marc Trachtenberg specifies that it is "clear, more generally, that Roosevelt was willing to accept Soviet control over certain areas—the Baltic republics and the eastern part of prewar Poland—no matter how the populations in question felt about rule by the USSR" (1999, 5). Throughout World War II, Roosevelt seems to have been disposed to see the Soviet Union as acting primarily in self-defense. At the same time, he perceived Germany "as the incomparably greater threat to national security" (Leffler 1994, 26)—an assessment he did not change even after the Soviet march into eastern Poland, its attack on Finland, or its seizure of the Baltic states.

In January 1945, a memorandum circulated by the deputy director of the State Department's Office of European Affairs is again worth quoting because it unmistakably makes clear the gap between the pragmatic position effectively taken by the U.S. government and the moralistic language in which it presented the Baltic issue then and later: "We know that the three Baltic States have been reincorporated into the Soviet Union and that nothing we can do can alter this. I would favor using any bargaining power that exists in connection with the foregoing matters to induce the Russians to go along with a satisfactory United Nations organization and the proposed Provisional Security Council for Europe to deal with . . . other trouble spots. I favor . . . our recognition of these areas as Soviet territory" (cited in Gvosdev 1995, 18). This position was not, in principle, changed by the replacement of Roosevelt by Harry S. Truman or by the hardening of the U.S. position on the Soviet Union in late 1945 and early 1946. According to Trachtenberg,

---

9. *Foreign Relations of the United States, The Conferences at Cairo and Tehran (1943)* (Washington, D.C.: U.S. Government Printing Office, 1961, 595), as cited in Bollow 1993, 3.

There was still no wish to commit American power in any serious way to the rolling back of Soviet influence over eastern Europe. America would never risk war, for example, to prevent the Communists from getting control over Czechoslovakia; even the threat of an American intervention would never be used to neutralize the threat of a Soviet intervention there. But the line would be drawn around the periphery of the area that had been consigned to the Soviets, and there was a growing willingness to defend that line if necessary with military force. (1999, 40)

From the Baltic states' point of view, the problem was that they were on the wrong side of that line. Did U.S. foreign policy during the transition years from 1988 to 1991 actively help them cross that line? During the 1988 presidential campaign, Vice President George Bush addressed the Lithuanian American community in a letter, stating that condoning the Soviet occupation of Lithuania would be "to turn our backs on our own values and heritage" (cited in Gvosdev 1995, 19). Accordingly, in September 1989 the State Department issued a declaration on the U.S. policy of nonrecognition (see appendix, document 5). Thus, official U.S. representatives still emphasized strongly the U.S. nonrecognition policy. The center of the two-pillar strategy characteristic of the U.S. policy toward the Baltic states after World War II—officially and rhetorically stressing the nonrecognition of Soviet control, while at the same time "not allow[ing] the non-recognition . . . to disrupt the basic pattern of U.S.-Soviet relations" (Gvosdev 1995, 18)—increasingly switched over to the latter pillar. Walter Clemens noted at that time: "As prospects for Baltic independence soared, Washington seemed to pull back its support or at least remain silent" (1991, 300).

This is not the place to explore the foreign policy priorities of the Bush administration from 1988 to 1991 in detail. Suffice it to say that the Baltic issue surely did not rank high on the agenda because the imminent dissolution of the Soviet Union, in addition to being a tremendous challenge in itself, jeopardized a main objective of U.S. foreign policy at that time—namely, the Strategic Arms Reduction Talks (START)—and threatened to result in a loss of Soviet control over nuclear weapons (Garthoff 1994, 461–72, 553–58). Furthermore, the negotiations on the unification of the two German states kept the Bush administration busy. The U.S. government saw the success of the Soviet reform process associated with Gorbachev and his team as a precondition for European security, transatlantic relations, and U.S. security. It is easy, with the benefit of hindsight, to blame the Bush administration for believing this, but at that time no one knew what would happen if Gorbachev failed and was replaced by someone else. Gorbachev's failure might have resulted in a much more traditional Soviet approach to the Baltic issue than the approach he actually exhibited as outlined in chapter 5. The Bush administration consequently placed its support of Gorbachev and the Soviet reform policy above all other issues at that time. Bush defended his careful and reserved attitude toward the independence of the Baltic states at a press conference on 6 April 1990 by pointing out "how important [the U.S.-Soviet] relationship is in arms control and on the peace of an emerging democratic Europe" (cited in Gvosdev 1995, 27).

In addition, beginning on 13 February 1990, the Bush administration was preoccupied with the negotiations on the accession of the German Democratic Republic to the purview of the German Federal Republic's Basic Law (Two Plus Four). As a consequence, the U.S. National Security Council found itself void of Baltic experts. The Lithuanian declaration of independence of 11 March 1990 thus inconvenienced the foreign policy establishment in Washington and suffered in this respect from "bad timing." It is true that only one hour after the declaration of independence, White House spokesman Marlin Fitzwater appealed to the Soviet government "to address its concerns and interests through immediate, constructive talks with the Government of Lithuania." Two days later, however, Bush addressed not the Soviet government, but rather the Lithuanian government in urging the latter "[to work] it out with the Soviets to achieve what they want" (both cited in Gvosdev 1995, 22) and thus returned to his nonadventurous policy. The influential American Committee on U.S.-Soviet Relations supported Bush's cautious position on the Baltic independence by labeling Lithuanian independence a " 'parochial' interest" (Vardys and Sedaitis 1997, 170).

Obviously, the Bush administration had no interest in undermining Gorbachev by supporting the Baltic states prematurely, too strongly, or too openly. Likewise, it had made clear to the Soviet leadership, and Mikhail Gorbachev is said to have realized, that "Bush would not take advantage of [Gorbachev's] problems in the Baltics and Eastern Europe" (Ekedahl and Goodman 2001, 124).[10] As long as the success of Gorbachev's reforms was

---

10. According to Ekedahl and Goodman, then U.S. secretary of state James Baker even stated that "Washington would not object 'if the Warsaw Pact felt it necessary to intervene' in

seen to be more important to U.S. interests than Baltic independence and as the condition of possibility for that independence, an outspoken and active support for the Baltic case could indeed have undermined both objects. Furthermore, from the U.S. point of view, the Baltic states should regain their independence in a manner that would not create obstacles to friendly future relations between them and the Soviet Union. Therefore, the initial U.S. response emphasized the necessity of resolving the Baltic-Soviet issue peacefully and by means of negotiation arriving at mutual agreements, and it downplayed the role the United States could possibly play: "The prevailing view within the State Department was to let matters in the Baltic States 'unfold at their natural development,' with a strong emphasis on having the matter resolved between the Lithuanians and the Soviets themselves, without any outside mediation or interference by the United States" (Gvosdev 1995, 23–24).

This restrained position was suggested because in general "domestic reform and *glasnost* in the Soviet Union are American interests, but not ones about which the United States can do much" (Nye 1988, 406). Just the opposite: U.S. interference in Soviet internal affairs could very well have supported conservative and status quo-oriented opponents to the reform process by giving the impression that Gorbachev's course was primarily serving U.S. rather than Soviet interests. As a result, the Bush administration maintained a low-profile course on the Baltic issue and in principle a Soviet Union/Russia-first policy. For example, it criticized as "damaging" the 22 March 1990 U.S. Senate Concurrent Resolution 108 affirming U.S. support for Lithuanian self-determination and the 4 April 1990 House of Representatives Concurrent Resolution 289 supporting Lithuanian independence and demanding the administration to reestablish diplomatic relations "at the earliest possible time" (Gvosdev 1995, 26).

The U.S. contribution to Baltic independence was thus a fairly indirect one of avoiding too direct an involvement in what it considered Soviet internal affairs. Far from having been priority number one on the Bush administration's foreign policy agenda and far from having been amalgamated with U.S. interests, the Baltic states' independence happened to be a by-product

---

Romania" (2001, 124). His suggestion is said to have been called "stupid" by Soviet foreign minister Shevardnadze.

of U.S. foreign policy rather than an object in itself. As Clemens observes, "like Roosevelt during World War II, Bush evidently took no deep interest in national self-determination for the Baltic peoples. His major concerns were approval ratings from American voters and improved Soviet-U.S. relations" (1991, 303). This policy is easy to criticize as a case in which "consigning Balts to Soviet overlordship was a small price to pay for Moscow's coopera- tion in other matters" (Clemens 1991, 303; see also Brill Olcott 1990, 30–31). Yet between 1988 and 1991, carefully designing the U.S. position on the Baltic states seems to have been as mandatory as developing a position to- ward Gorbachev's reform process in the first place. At times, the principle of national self-determination of the Baltic peoples collided with other U.S. in- terests that were deemed more important; it was subordinated to them and thereby devalued. From 1988 to 1991, the representatives of the Baltic inde- pendence movements indeed often had reason to feel offended by the U.S. position. However, they probably misunderstood that a stronger U.S. in- volvement would not necessarily have promoted their case.

The U.S. administration welcomed Baltic independence once it was achieved, and although it did not hasten to recognize that independence,[11] it interpreted independence as the realization of a long-standing U.S. foreign policy objective. More to the point, it was the policies not undertaken by the U.S. government at that time that helped reestablish independent Baltic states. It was the policy of not supporting Baltic independence at the ex- pense of the reform process in the Soviet Union; of not placing Baltic inde- pendence over Soviet reform policies; of not getting involved in Soviet internal affairs too directly; of not recognizing the independence too early. The representatives of the Baltic popular movements were probably disap- pointed with the U.S. lines of argumentation and the resulting (non)policies. A State Department spokesperson argued, for example, that U.S. recogni- tion of Baltic independence was not at all necessary because the United States had recognized Baltic independence as early as 1922 and had never withdrawn from this recognition. In accordance with the traditional view on sovereignty discussed in chapter 4, the spokesperson continued by saying that diplomatic relations could not be reestablished as long as the Baltic

11. The United States was the thirty-seventh country to recognize the independence of the Baltic states on 2 September 1991, the same day Gorbachev announced the intention to do so. Again, the U.S. objective was not to stab the Soviet leader in the back.

states had no effective control over their territory and no ability to enter into and fulfill international obligations (Vardys and Sedaitis 1997, 168–69). All Sajudis leaders and staff members interviewed by Richard Krickus were indeed "convinced the Bush Administration would recognize an independent Lithuania if the Sajudis state [sic] elected enough of its candidates to the Lithuanian Supreme Soviet to restore the state which had been crushed in 1940" (1993, 180).[12]

Likewise, U.S. vice president Dan Quayle's reaction to repeated interventions by Soviet paratroopers in Estonia, Latvia, and Lithuania in order to capture Baltic deserters from the Soviet army did not support the notion of friendship. He reportedly declared that "an army had a so-called 'right' to capture its 'deserters' " (Gvosdev 1995, 28). In 1989, however, President Bush had designated Lithuania as an occupied country to which the 1949 Geneva Convention, assigning a protected status to citizens of a country occupied by foreign military forces, was to have been applied, strictly speaking. Bush nevertheless did not comment on the conscription of Baltic draftees into the Soviet army.

Furthermore, from a Lithuanian point of view the inclusion of the city of Klaipeda in a list of "Soviet" cities in the U.S.-Soviet Civil Aviation and Maritime Agreement (Gvosdev 1995, 32) must have been totally unacceptable. Moreover, President Bush, in the spring of 1990, refused economic sanctions against the Soviet Union in response to the Soviet economic embargo on Lithuania, thereby ignoring a nonbinding U.S. Congress resolution, and he mildly called the Soviet embargo "another unfortunate step" (Vardys and Sedaitis 1997, 170; see also Garthoff 1994, 424; Gvosdev 1995, 29–30). Parliament chairman Vytautas Landsbergis of Lithuania was at the very least humiliated by being received by Bush in December 1990 only after several requests and as a private person "because the State Department chose not to recognize the Landsbergis regime as a valid government" (Gvosdev 1995, 35). Lithuanian prime minister Kazimiera Prunskiene in April 1990 on her visit to the United States had to enter the White House through the tourist entrance (Vardys and Sedaitis 1997, 171).

Likewise, the U.S. response to the January 1991 assaults in Vilnius and

12. Thus, although the U.S. position supported the narrative of state continuity, Sajudis members had hoped for an *explicit* recognition of independence on the part of the United States immediately after the declaration of independence on 11 March 1990.

Riga was soft and conciliatory. When Bush called Gorbachev on New Year's Day 1991, he did not mention the increasing tensions in the Baltic republics (Ekedahl and Goodman 2001, 252). This position reflected both the preoccupation with the beginning UN offensive against Iraqi forces in Kuwait and President Bush's interest in "build[ing] a lasting basis for U.S.-Soviet cooperation," confirmed by the president in his 29 January State of the Union address to Congress and the American people even after the January assaults on Vilnius and Riga (Garthoff 1994, 447). Condoleezza Rice, an adviser to the president, is also said to have "resisted Baltic Americans' request for U.S. pressure against Soviet intervention" (Clemens 2001, 218). When Soviet forces were preparing their intervention in Lithuania, the U.S. Department of Agriculture announced that "it had already allocated $900 million of the $1 billion in credits that had been made available in December. Sources knowledgeable on this issue felt that the Administration was rushing the assistance through" (Gvosdev 1995, 34). Likewise, the White House announced that it would not stop export credits. Of course, Bush described the events in Lithuania as "deeply disturbing," but he "did not impose even symbolic sanctions," although both houses of Congress passed resolutions recommending "that the president consider economic sanctions" (Garthoff 1994, 446; see also Vardys and Sedaitis 1997, 179). Rather than announcing effective U.S. countermeasures against the Soviet aggression, the U.S. ambassador to Moscow, Jack Matlock, in January 1991 warned Gorbachev only that "armed intervention to crush Lithuania would sever movement toward détente and trade in East-West relations" (Clemens 2001, 51).

The Bush administration is said to have claimed "that the policies undertaken by the United States government were instrumental in leading to the re-establishment of the independent Baltic states" (Gvosdev 1995, 17). The analysis presented here points in a somewhat different direction, however: the Bush administration's position on Baltic independence was distinctly reserved and confined to rhetorical support at the most general level. As a consequence, the "Baltic states were disappointed by the manœuvring of the U.S. Administration between Moscow and the Baltic capitals in the *perestroika* years. They realized that the Baltic states were far beyond the sphere of interest of the United States" (Vares and Haab 1993, 286). The Baltic issue was certainly subordinated to maintaining "business as usual" relations with the Soviet leadership in order not to throw obstacles into either Gorbachev's reform or arms control negotiations. If the Baltic peoples themselves managed to restore their independence, so be it; if they managed so peacefully and through negotiation, all the better; the U.S. government would not hesitate to interpret the achievement as its own success. For the time being, however, the Baltic issue was seen primarily as an obstacle to the development of a new world order, which was preferably to be based on as little change as possible.[13]

During the restoration of Baltic independence, U.S. nonpolicies and low profile on the issue were in play rather than active U.S. support of the Baltic case. Ultimately, U.S. noninterference and rhetorical restraint may nevertheless be seen as a condition for the possibility of Baltic independence. More important in the present context than speculating whether a more active U.S. involvement would have accelerated Baltic independence or, on the contrary, encouraged stronger resistance on the part of the Soviet government (or its domestic critics) is the conclusion that interpreting the U.S. position on the Baltic independence issue in terms of "friendship" as outlined in chapter 3 would mean stretching the concept beyond recognition: the U.S. government did not actively come to the assistance of the Baltic independence movements in January 1991. Nor did the Bush administration amalgamate U.S. interests with the interest of the Baltic peoples as formulated by their legitimate governments at that time. The Baltic states' self-determination was certainly considered legitimate, but furthering the Baltic case was not equated with U.S. interests. Adhering to U.S. interests triumphed over adhering to the general principle of self-determination.

When it comes to the issue of the Russian troop withdrawal from the Baltic states, the U.S. role and its influence on Russian political and military decision makers are also often exaggerated. This is not surprising for at least three reasons. First, it reflects the "old American foreign policy error to exaggerate the effect the United States can have on others" (Russett 1993, 131). Second, in general, "when the other behaves in accord with the actor's desires, he will overestimate the degree to which his policies are responsible for the outcome. . . . Actors almost always feel responsible for exerting influ-

13. Korbonski argues that in addition to an extreme decentralization in the U.S. foreign policy decision-making process from the late 1980s onward, the younger regional specialists who entered the bureaucracy had one thing in common: "they were quite conservative in their outlook, disliked change and had only one goal—maintaining international stability at all cost" (1995, 134).

ence when the other acts as they wish" (Jervis 1976, 343, 345). Third, from the point of view of constructing friends, it is obvious why Baltic security constructors refer to the United States in the most flattering words. As outlined in the beginning of this chapter, references to the United States as a friend are explainable as a political-rhetorical proceeding, the aim of which is to motivate the U.S. government to apply policies toward the Baltic states as if the United States were indeed a friend.

Similar to "security" as a speech act (Buzan, Wæver, and de Wilde 1998, 27), the articulation of "friendship" does not necessarily require an actor to refer to someone else as a friend. The logic of friendship may be implicitly present in a text. For example, the "friend" may be both hidden and unmistakably present in statements such as the following: "America played the decisive role in persuading Russia to pull out of the Baltic states" (Lejins 1998, 29). As shown in chapter 7, however, foreign influence on the Russian military and on Russian political leaders was limited. The cited statement thus tells us more about the author's intention to construct the United States as a friend by means of exaggerating the effect of U.S. foreign policy than about the procedure of the Russian troop withdrawal. Had the Russian leadership come to the conclusion that the continued presence of Russian troops in the Baltic states were in its political, military, or strategic interest, then the "quiet, unofficial U.S. reminder (in the press) that U.S. aid was linked to a complete Russian pullout, or agreement on it, from all the Baltic states" (Lachowski 1994, 579) would hardly have sufficed to change its mind. The reminder by the U.S. Congress that aid to Russia was contingent upon Russian troop withdrawal from the Baltic states was neither quiet nor unofficial. However, by October 1992, when the U.S. Congress passed the $24 billion Freedom Support Act,[14] a timetable on troop withdrawal from Lithuania had already been signed by the Russian and Lithuanian defense ministers, and the Russian delegation to the Russian-Estonian talks had offered a deadline for troop pullout.

---

14. According to Vares and Haab, the act contained three conditions: "*(a)* the President must present a report to Congress on the progress of troop withdrawal from the Baltic states before aid is issued; *(b)* Russia will receive only 50 per cent of the aid if the President is unable to report either considerable progress or a working schedule for withdrawal; and *(c)* if Russia has not withdrawn its entire army from the Baltic states or concluded negotiations to do so within a year, no further aid will be granted" (1993, 292).

U.S. diplomats' involvement in the negotiations on troop withdrawal in general and on three specific issues in particular (the Kaliningrad transit issue, the Skrunda ballistic missile early-warning station, and the Paldiski nuclear submarine training base) induced the Baltic governments to accept compromises that were in part considered detrimental to their security and interests at that time. The former Latvian defense minister, for example, argues that:

> Latvia signed the agreements [on the Russian troop withdrawal] under strong pressure from the U.S. and other Western states, in the process sacrificing its own vital interests: 1) agreeing to allow to remain in Latvia a substantial number of former Soviet military personnel and their dependents . . . many of whom are hostile to Latvia's independence; 2) agreeing to legalize the presence of a military installation for an unnecessarily long time and under risky conditions; 3) not getting an agreement on compensation for ecological damage and other losses; 4) not receiving payment for equipment taken from the Latvian army at the time of the 1940 occupation. (Jundzis 1995, 562–63)

Latvia is said to have agreed to the Skrunda accord, as outlined in chapter 7, by "succumbing to U.S. pressure" (Lukic and Lynch 1996, 365; see also Garthoff 1994, 787), and Russia is said to have been left, at President Clinton's insistence in January 1992, "with only a reduced term for Skrunda" (Lejins 1998, 29). Which side of the coin an author stresses depends to some extent on the position he or she takes, yet what is often described in the literature either as U.S. pressure on Latvia or as U.S. pressure on Russia seems to have been both, thus indicating that the U.S. administration was indeed looking for a compromise, undermining attempts at one-sided appropriations of the United States as a friend.

Likewise, regarding the Kaliningrad transit issue, "U.S. diplomats had urged Vilnius to come to some accommodation with Moscow on this issue, arguing that any agreement with Russia was better than none" (Liulevicius 1995, 396). This argument is said to have been a "dangerous argument . . . since the territorial sovereignty of an independent country is at stake" and because it could reduce Lithuania to being a "corridor for regular Russian military transit" (Liulevicius 1995, 396). Yet in 1994 only one percent of all goods transferred through Lithuania from the Russian mainland to Kaliningrad was military, with "only a fraction of this" being personnel and weapons (Oldberg 1998, 7). Moreover, clearly, the package deal of transit rights in exchange for most-favored-nation status was important to Lithuania at that time because Russia was still the most important trading partner. This discussion anticipates the tension between, on the one hand, a traditional understanding of security emphasizing sovereignty, territorial integrity, and military security prevalent among Baltic decision makers and scholars, and, on the other, an innovative understanding of security focusing on cooperation, permeability of borders, and nonmilitary security evolving among U.S. diplomats at that time. We reencounter this tension in the next chapter.

10

The Clinton Administration's Policy Toward the Baltic States

THE CLINTON ADMINISTRATION inherited from the Bush administration the nonexistence of a policy toward the Baltic Sea region in general and toward the Baltic states in particular. In this chapter, I analyze chronologically the Clinton administration's most important foreign policy initiatives toward the Baltic states and incorporate them into the overall U.S. policy approach to Europe's north. This procedure is suggested because the Baltic states have increasingly been incorporated into a wider U.S. policy toward northeastern Europe. I furthermore examine how the U.S. policy in northern Europe relates to security community building in the region. Reflecting the basic approach to security and security communities pursued in the whole book, I deal primarily with the policy's underlying thought patterns, especially with respect to security, rather than with its material aspect. Finally, I explore the profundity of U.S.-Baltic friendship declarations. The chapter's aim is not to rationalize thought patterns and policy retrospectively, but rather to show the evolution of the U.S. policy toward northeastern Europe in its complexity, including ambivalences, contradictions, and unresolved tensions.

Apart from the rhetorical cultivation of a special relationship between the United States and Estonia, Latvia, and Lithuania as outlined in the preceding chapter, there are at least three reasons for paying special attention to U.S. conceptualizations of security for the Baltic states and the Baltic Sea region. First, the Clinton administration's mapping of northeastern Europe was very similar to the understanding of the Baltic Sea region introduced in the first chapter as the most appropriate one in the present context. It essentially was a policy toward Estonia, Latvia, Lithuania, and the northwestern parts of Russia, including the Kaliningrad region. Poland, northern Germany, and the Nordic states (Denmark, Finland, Iceland, Norway, and Sweden) were included in this policy but given relatively little attention. As a corollary of this mapping of northern Europe, the U.S. policy inevitably influenced security community building in the region and thus cannot be ignored. Second, Baltic political decision makers and analysts attached and continue to attach importance to U.S. engagement in Europe in general and in the Baltic Sea region in particular to an extent that cannot be observed with regard to other external actors (Lejins 1998; Norkus 1999). It can therefore be assumed that the United States had and continues to have a greater capability than others to exert influence on the political decision making in the Baltic states.

Third, the United States is said to have "strong security links with the region" (Duke 1996, 183). Indeed, especially the Clinton administration directed considerable attention to northeastern Europe. Security and cooperation in the Baltic and Barents Sea regions were officially represented as a "priority for the United States" (U.S. Embassy in Stockholm n.d.). In the words of U.S. deputy secretary of state Strobe Talbott, one of the chief designers of the U.S. policy in Europe's north, "for Americans, the fate of the Baltic states is nothing less than a litmus test for the fate of this entire continent" (2000a). This statement sounds as moralistic as several of the official statements referred to in the preceding chapter, but Ronald Asmus presents an argument that is more in line with the analysis presented in chapter 9. In a policy paper coauthored with Robert Nurick, Asmus (who in 1996 was with the Santa Monica–based think tank the RAND Corporation) stressed that as a result of the pressure exhibited by Baltic American interest groups, enlarging NATO to east-central Europe "without a credible strategy on the Baltic issue" would in all likelihood be politically impossible for any U.S. president (Asmus and Nurick 1996b, 128).[1] The "Baltic issue" therefore was in more than one respect "one of the most delicate questions facing the Alliance" (121) and especially the U.S. government. It may thus be argued that the U.S. policy toward the Baltic states was in part a corollary of one of the basic weaknesses of NATO's enlargement policy and rhetoric during its early stages—namely, the failure to pay adequate attention to the consequences resulting from either including the Baltic states in or excluding them from NATO. It is also obvious that the U.S. policy toward the Baltic

---

1. Asmus and Nurick 1996b is subsequently cited by page references in the text.

states reflected U.S. domestic considerations. In the categories introduced in chapter 3, a second-level internalization of friendship in terms of instrumentality can be suggested here rather than a third-level internalization in terms of legitimacy.

Even in retrospect, one of the most striking aspects of the article by Asmus and Nurick is the absence of an explicit role assigned to the United States in building Baltic security. Just the opposite: the Nordic states are called upon "to find additional resources" and "to further their own political, economic and military involvement in, and cooperation with, the Baltic states," among other things, because "the risk of an undesirable Russian reaction to Western security assistance would . . . clearly be much greater if, for example, non-Nordic countries and leading NATO powers such as the [United States] or Germany assumed this role" (133).[2] Likewise, the authors argue for investments in Russia's Kaliningrad oblast but dissuade the United States (and Germany) from taking the lead. They in fact reduce the U.S. role to encouraging its European allies to accelerate the process of EU enlargement, to include Estonia in the first round of enlargement, and to be committed to creating "the conditions that will eventually make Baltic membership in NATO possible" (135). This position is certainly much less than the Baltic governments expected from their American "friend." Indeed, according to Asmus and Nurick, the United States had shown only "limited" support for Baltic membership in NATO (123).[3] As building blocks for a Baltic strategy, the authors recommended encouraging political and economic reform; Baltic defense cooperation; Nordic-Baltic cooperation; EU en-

2. Nordic policymakers instantaneously rejected the role assigned to the Nordic states. Obviously, they feared that NATO could shift the responsibility for the consequences of its own enlargement policy to them. Moreover, being incorporated into NATO's enlargement strategy as an "essential building block" of an "Alliance strategy to safeguard the Baltic states" (132) ran counter to the image of self-determined foreign and security policies carefully cultivated in the Nordic states, in particular Sweden. Being instrumentalized by NATO in order to do a job that NATO was unwilling or unable to do was indeed as unattractive a prospect as both the possibility of Russian reactions against Nordic involvement in the Baltic states and the risk of finding themselves "dangerously isolated in a future disagreement with Moscow" (133).

3. Talbott, while commenting on the U.S.-Baltic Partnership Commission meeting in Tallinn on 7 June 2000, confirmed the lack of support for Baltic NATO membership by saying, "I hope, by the way, that over time there will be solid and indeed increasing support in the United States for the three Baltic States becoming not just partners, but also allies" (Talbott 2000b).

largement to (at least one of) the Baltic states; an open-door strategy on NATO membership; and what is called "dealing with Moscow" (129–39). The latter would include, first, making sure that the Western states take "legitimate Russian grievances" seriously; second, reacting flexibly to arrangements proposed by the Russian authorities that take account of the burdens and the objective difficulties resulting from the withdrawal of Russian troops from the Baltic states, Poland, and the former German Democratic Republic; and, third, incorporating Russia into the "emerging web of multilateral security cooperation wherever possible" (137–38).

Ronald Asmus was later promoted from analyst at RAND to U.S. deputy assistant secretary of state for European affairs in charge of northeastern Europe. It is therefore not surprising that the Clinton administration's policy toward Europe's north included many ingredients of his recommendations. The RAND policy paper may thus be seen as a point of departure and a rough and preliminary edifice for the emerging U.S. strategy for Baltic security, the more so since the paper anticipated NATO's 1997 enlargement decision by concluding that the Baltic states "are unlikely to be in the first tranche of new Alliance members" (121).

Indeed, the Baltic states were not invited to begin accession talks with NATO at the meeting of the North Atlantic Council on 8 July 1997 in Madrid, a decision anticipated by U.S. secretary of defense William J. Perry in October 1996 with reference to their lack of capability to take on NATO's Article 5 responsibility (Kozaryn 1996). Instead, they were offered a charter of partnership with the United States, which became the Charter of Partnership among the United States of America and the Republics of Estonia, Latvia, and Lithuania (hereafter U.S.-Baltic Partnership Charter). Baltic commentators welcomed this offer because "the opportunity of security consultations on a bilateral basis with the [United States] is a significant security asset for three small states that have been more than once pawns in big power politics" (Lejins 1998, 31). Consultations, however, are not equivalent with security assistance in case of need. The Baltic governments, therefore, did not want the charter to be seen as a substitute for Baltic NATO membership, but rather as an interim solution during what they considered a transition stage until they were finally invited to membership. They appreciated the charter, signed in Washington, D.C., on 16 January 1998, because it tied the Baltic states closer to the United States. Furthermore, the charter referred explicitly to security and understood it in a relatively narrow—that is, military and defense-oriented—manner that was in accordance with the view on it prevailing in the Baltic states. The charter was a further development of the U.S. Baltic Action Plan presented by the U.S. assistant secretary of state to the ambassadors of the Baltic states in Washington on 29 August 1996. Like its successor, the Baltic Action Plan included U.S. support for Baltic integration in Western security institutions and stressed the importance of good relations with all neighbors, including Russia (Ferm 1997, 551).

The primary objectives of the U.S.-Baltic Partnership Charter seem to have been encouraging the Baltic governments to continue their political and economic reforms; assuring them of U.S. assistance in integrating the Baltic states in Western institutions; and signaling the Russian Federation that NATO's Madrid declaration should not be seen as the final say with respect to NATO enlargement (Winner 1998, 52). As stated in the preamble, the charter is "a political commitment declared at the highest level." It includes in its Principles of Partnership, in classical Westphalian parlance, a "real, profound and enduring interest [by the United States of America] in the independence, sovereignty, territorial integrity, and security of Estonia, Latvia, and Lithuania." Furthermore, the United States, among other things, "welcomes the aspirations and supports the efforts of Estonia, Latvia, and Lithuania to join NATO"; reiterates that NATO enlargement is an ongoing process; and affirms that NATO membership for the Baltic states may follow from their ability and willingness "to assume the responsibilities and obligations of membership."[4] At the same time, the charter did not commit the United States to "supporting Baltic entry to NATO by what in Congress is called a 'date certain' " (Zakheim 1998, 115). It may thus have been an exaggeration to say, as Latvian president Guntis Ulmanis reportedly did, that the charter provided a "strategic philosophy for the next century" (Chaise 1998, 8). From the Baltic governments' point of view, the main benefit of the charter appears to have been that the U.S. government for the first time officially committed itself to a specific policy toward Estonia, Latvia, and Lithuania— a policy that could not have said to exist in any formal sense before the signing of the charter (Nurick 1997, 57). The charter's presentation of security

4. Charter of Partnership, preamble: Principles of Partnership and Commitment to Integration. *U.S. Information and Texts*, no. 003 (21 Jan. 1998): 12–15.

cooperation follows a rather conventional reading of security and puts emphasis on military achievements and cooperation among governmental actors (see document 6 in the appendix).

According to an official publication from the Estonian Ministry of Foreign Affairs, the Baltic states "look on the charter as a confirmation of the United States' special attitude and interest toward the Baltic States." The White House, however, in its summary of the U.S.-Baltic Partnership Charter, stressed that the Baltic states would not be treated in a specific way, but like any other state: they "will have to meet the same criteria and standards [for membership in NATO] expected of other states."[5] The original idea behind including in the charter's Principles of Integration a section stating that the partners "believe that, irrespective of factors related to history or geography, [international] institutions should be open to all European democracies willing and able to shoulder the responsibilities and obligations of membership, as determined by those institutions" seems to have been a way of invalidating Baltic suspicions that NATO could discriminate against the Baltic states because of their former incorporation into the Soviet Union or their spatial contiguity to Russia or both. But from the allegedly uniform treatment of all candidate countries does not follow a special or even a preferential treatment of the Baltic states. The Estonian minister of foreign affairs nevertheless interpreted the charter as "a document of the future that supports Estonia, Latvia, and Lithuania's participation in the transatlantic integration processes of the 21st century" (Ilves 1998). The White House, however, in the summary of the U.S.-Baltic Partnership Charter referred to earlier, made clear that the charter "in no way pre-commits the United States to Baltic membership in NATO."

Patterns of friendship may evolve independently of formal alliance responsibilities. If taken seriously and internalized to the third degree, they make formal commitments unnecessary. Acting like friends according to Wendt's friendship pattern is, however, unlikely as long as security guarantees are explicitly ruled out. According to the White House's summary, the charter "does not offer 'back-door' security guarantees. The Baltic governments understand, and have said so publicly, that such guarantees can only come through NATO membership." A Lithuanian member of Parliament,

5. *Estonian Review* 1998; White House Office of the Press Secretary 1998c.

Stasys Malkevicius, thus may have been on shaky ground when reportedly stating in connection with investments of the U.S. company Williams International in the Lithuanian oil sector that the United States would defend Lithuania like "it defended Kuwait against Iraq" (Tracevskis 1998, 3).

The U.S. government did support the buildup of national armed forces in the Baltic states, but this financial support was far from impressive. For example, as part of the United States Foreign Military Financing Program, the Baltic states' military organizations received up to $18.2 million in 2001 ($6.5 million, $5.35 million, and $6.35 million for Lithuania, Latvia, and Estonia, respectively, in contrast to $4.7 million given to each state in the year 2000) (Mladineo 2000, 6). The section on security cooperation in the fact sheet issued by the White House following the signing of the charter revealed both a fairly traditional understanding of security and the financial limitations of the U.S. support (see appendix, document 7). Likewise, following a conventional approach to security, the U.S. government, in a joint communiqué of the U.S.-Baltic Partnership Commission on 7 June 2000, explicitly "welcome[d] the progress made by Estonia, Latvia, and Lithuania in approaching the goal of raising their defense budgets to 2% of their GDP" (U.S.-Baltic Partnership Commission 2000). By so doing, the U.S. administration underscored the traditional linkage of security with a certain amount of investment in the military.

Up to this point, the discussion does not support the claim asserted in the previous chapter that the U.S. administration broke with conventional patterns of thinking about security and included innovative elements in its policy toward northeastern Europe. Indeed, the Partnership Charter stuck to a very traditional way of addressing issues of security. It did so arguably in order to reassure the Baltic governments, who were facing both the temporary exclusion from NATO and the limited U.S. support for membership in the alliance. It can also be suggested that by adapting its view on security to the one prevalent in the Baltic states, the U.S. government retrospectively acknowledged and appreciated the Baltic peoples' reestablishment of independence. But perhaps *adapting* is not really the right word. It may indeed be said that the rather conventional reading of security adhered to by the U.S. government did actually reflect the views on security shared by U.S. decision makers at that time. The U.S. policy toward northeastern Europe was not ready-made, but evolved over time. The understanding of security changed in 1997, however, with the introduction to the public of the U.S. Northern Europe Initiative (NEI) by the then assistant secretary of state for European affairs, Marc Grossman. This is not to suggest that the NEI immediately replaced the approaches pursued hitherto. Rather, different (and not necessarily compatible) approaches were now being pursued at the same time, with the NEI tentatively challenging the more traditional approaches, but facing skepticism and irritation on the part of many analysts who misunderstood it and criticized that the initiative was too weak on the material side to be taken seriously.

Likewise, many policymakers in the Baltic states were puzzled about the basic thought patterns underlying the NEI.[6] They did not appreciate that the Baltic Sea region of all places was chosen to exemplify new approaches to U.S. foreign and security policy, and they felt that more traditional approaches would better serve Baltic security. Treating northern Europe and the Baltic Sea region as a laboratory for new policy approaches and theory testing is indeed problematic. As has been argued with respect to the Barents Sea region, diminishing a region and its inhabitants to objects of theory testing "will certainly face increasing criticism from the communities under 'testing'—and from the side of humanities in general" (Lehtinen 2003, 51).

The U.S. government, however, seems to have understood at that time that neither security nor prosperity in the Baltic Sea region would result automatically from NATO or EU enlargement or from bilateral military assistance programs. Instead, the realization of U.S. interests in the region required a more sophisticated approach that, among other things, understood the Baltic states as part and parcel of regional dynamics in northeastern Europe rather than in isolation. The U.S. administration is said to have begun to see that "increased and improved connections and cooperation on the political, economic, cultural and military level are critical for the security and stability of this region" (Winner 1998, 55) and beyond. Likewise, President Clinton stated retrospectively that "we launched the Northern Europe Initiative because we recognized . . . the importance of strengthening regional cooperation among the Baltic states, Russia, and all the countries bordering the Baltic Sea. Only in this way can we create the stability and prosperity that will lead to full integration of northern Europe, including

6. For NEI's basic thought patterns, see Möller 2000; van Ham 2000; Browning 2001.

northwest Russia, into the broader European and transatlantic mainstream" (Clinton 2000).[7]

From this view, it followed that the strategy for the Baltic states had to be integrated into a strategy for northeastern Europe as a whole. The latter strategy, however, had to be developed in the first place. It involved harmonizing three political regions and regional policies that during the Cold War had been treated rather separately and distinctly. First, the far north had been seen as a potential theater of nuclear war and a political space dominated by nuclear superpower rivalry, requiring policies specifically tailored to the nuclear setting (Young 1985–86; Jalonen 1988; Tunander 1989). Second, after Iceland, Norway, and Denmark had signed the Washington Treaty in April 1949, the Nordic states (with the exception of the U.S. air bases in Iceland and Greenland and Norway's importance to maintaining sea lines of communication between the European and American allies) "did not dominate American strategic thinking or concerns" (Zakheim 1998, 116). Third, the Baltic Sea region (here in its narrow understanding as the area directly adjacent to the Baltic Sea) had been regarded as a "strategic backwater," receiving "relatively little attention in U.S. policy" during the Cold War (Independent Task Force 1999, foreword).[8] As a consequence of the spatial alteration in the U.S. administration's mapping of northern Eu-

7. Legislation on the NEI does not encompass "all countries bordering the Baltic Sea." It includes Iceland and Belarus, but excludes Germany. According to the then NEI coordinator Conrad Tribble, the omission of northern Germany was a mere technical error: the administration pointed out the omission, and the relevant congressional staffers acknowledged the error, but it was decided not to open up the bill again because at that time it had already been launched into deliberation. In order to avoid opening up the law for more substantial changes, the technical corrections were renounced (personal communication, 17 August 2000). Belarus is included in the legislation because a few NEI projects in Lithuania also involve Belarus. Otherwise, it is not covered by the NEI. The legislation's reference to the NEI as a "framework agreement" is also incorrect. See Public Law 106-255, Cross-Border Cooperation and Environmental Safety in Northern Europe Act of 2000 (henceforth quoted as Public Law 106-255), sec. 6: Definitions, (1) and (2), available at http://www.state.gov/www.regions/eur/nei/nei_bill.html. This U.S. State Department Web site is a permanent electronic archive of information released prior to 20 January 2001.

8. However, the region was important as part of the "larger confrontation with the Soviet Union. If conflict erupted in Europe between NATO and the Warsaw Pact, NATO control of the Baltic Sea was seen as important to limit Soviet penetration into Germany and Denmark from

rope and the merging of three formerly distinct areas into one, the importance assigned to the Baltic states decreased from 1998 onward in parallel with a shift of emphasis from business and trade promotion to environmental and nuclear issues in the framework of the NEI.

After five years of indecision and apparent lack of interest, the U.S. policy toward the Baltic Sea region reacted, in the second half of the 1990s, to the emerging network building around the Baltic Sea rim by both launching its own initiatives and participating in several regional and subregional forums. To be sure, the processes of regionalization around the Baltic Sea since the early 1990s were evolving fairly independently of U.S. initiatives. It may also be called into question whether the U.S. government really appreciated the regional trends in Europe's north. For example, it did not pursue further the idea of "broader cooperative security arrangements specifically designed for the Baltic area" as suggested by policy advisors (Asmus and Nurick 1996b, 138). But because regionalization trends in the Baltic Sea region could undeniably be observed, the U.S. administration had to react in one way or another in order not to lose contact with and influence in the region. It may also be asked whether *initiative* was an appropriate term for the U.S. policy in the region. Because the U.S. government reacted to rather than launched policy developments in northern Europe, *noninitiative* would be a more appropriate label.

Since the mid-1990s, however, the United States has become a participant in a variety of functionally differentiated, albeit in part overlapping, (sub)regional networks in northeastern Europe with different objective areas and degrees of institutionalization as well as different membership compositions and principles. These institutions, specifically tailored to the requirements in northeastern Europe, include:

• the intergovernmental U.S.-Baltic Partnership Charter, with two bilateral working groups (on defense and military relations and on economic issues) and annual meetings of the U.S.-Baltic Partnership Commission;

• the Baltic Security Assistance Forum (BALTSEA), with the United States serving as a supporting nation;

---

the sea and to prevent Soviet submarines from exiting the Baltic to the North Atlantic area" (Sweedler 1994, 191).

• the Baltic Air Surveillance Network (BALTNET), developed from the U.S. offer of a regional airspace initiative to central European countries;

• the NEI (combining intergovernmental and nongovernmental activities and encompassing a wide range of issue areas);

• the intergovernmental, institutionalized CBSS, in which the United States has been an observer (as proposed by the Lithuanian government) since 1999;

• the Barents Euro-Arctic Region (BEAR) cooperation, partially overlapping but not congruent with the CBSS membership and with a different organizing principle (two parallel bodies, one representing national governments, the other one representing counties, oblasti, and republics in the Barents region, as well as one Sámi representative, with the United States being observer in the body representing national authorities); and

• the nonbinding declaration Arctic Military Environmental Cooperation (AMEC) signed by the U.S., Norwegian, and Russian defense ministers in September 1996 and confined to communication and cooperation among defense ministries.

The result is a "polity puzzle" encompassing "different kinds of cooperative processes and institutional frameworks" (Heininen 1999, 383) in northeastern Europe, a main objective of which seemed for some time to be cooperation with northwestern Russia or integration of northwestern Russia in low-politics networks or both (Heininen 1999, 398). From a U.S. point of view, it seems to have been equally important that through participation in as many networks, organizations, and institutions as possible, the U.S. government and departments made sure to be continuously kept informed of economic, political, and military developments in the region without becoming engaged too obviously, too directly, and too expensively—or too provocatively.

One of the most intriguing ingredients of the NEI in the present context was the absence of anything related to security in its traditional understanding. Following the official narrative, given in speeches by and interviews with prominent representatives of the U.S. policy toward northeastern Europe from 1997 up to the introduction in 2003 of the George W. Bush administration's Enhanced Partnership in Northern Europe (e-PINE)—in other words, U.S. policy representatives' advertisement of their policy and their version of how they wanted it to be seen by others—the NEI revealed a profound departure from a traditional understanding of interstate relations, regional cooperation, and security policy. Of course, it can be argued that an analysis relying on discourse, including policy manifestations and openly accessible sources, "leave[s] a lot of room for interpretation and this may lead to questionable research conclusions" (Bially Mattern 2000, 303) because it covers only the front-stage behavior of political decision making. This front-stage behavior, however, became remarkably coherent and constant with the arrival on the political scene of Strobe Talbott and Ronald Asmus, the main designers of the U.S. policy toward Europe and Russia. This consistency seems to reflect that "coherent images of phenomena are more likely to be formed if individuals have developed considerable expertise and experience relative to those phenomena" (Rosati 1995, 62). Thus, policymakers with long experience in policy analysis often develop consistent belief systems that are relatively insusceptible to rapid ad hoc changes and that they then transform into effective politics. Whatever else Mr. Talbott and Mr. Asmus are, they are certainly policy "experts" rather than "novices" (Rosati 1995, 62). Their message, thus, is not to be found in the language of it alone, but in its consistency.[9]

Furthermore, even if the language chosen to advertise a specific policy was initially picked out mainly to disguise "real" intentions and to veil what policymakers are saying on the back stage, it arguably shapes policy implementation in a way that the speakers might not have intended in the first place: "words and their meanings quickly escape their authors" (Barnes and Gregory 1997, 5). Language may exhibit a self-manipulative character, and policies might to some extent escape from their creators' intentions. The remarkably steady representation of the U.S. policy-in-the-making toward northeastern Europe since 1997 certainly influenced what the policymakers were learning to believe and to "know." Language may help cast belief systems—that is, a set of "propositions that policy makers hold to be true, even if they cannot be verified" (Holsti 1995, 273). Belief systems, in turn, are ar-

---

9. These claims are similar to but more cautious than those asserted by Ole Wæver, according to which "structures within discourse condition possible policies." Wæver holds that "although not every single decision fits the pattern to be expected from the structures used in the analysis, there is sufficient pressure from the structures that policies do turn within a certain, specified margin onto the tracks to be expected" (2002, 27).

ticulated through language and help shape political behavior.[10] This behavior may be more stable if supported by both beliefs and language rather than by language alone (Möller 2003b).

What, then, did the U.S. policymakers say about northern Europe? The U.S. government did not claim the leading role in northern Europe, but assigned this role to regional, especially Nordic, actors and the EU's Northern Dimension (which was launched at about the same time as the NEI). Nor did the U.S. government "mak[e] any unique claim to have a vision for the north" (Edelman 1999, 49; see also Tribble 2000). The initiative was indeed the rare case of a U.S. foreign policy adaptation to rather than a superimposition on a regional environment, and neither the lack of initiative nor the lack of originality devalued the NEI.

Initially, the NEI was launched to act as a guide for U.S. government agencies in northeastern Europe and to encourage the U.S. private sector to become more engaged in the region (Mulloy 1998). Subsequently, however, the U.S. administration developed a sense of its own role in Europe's north in terms of engaging Russia "in a positive and enduring manner" (Winner 1998, 57).[11] Indeed, according to the U.S. Department of State Bureau of European Affairs, northern Europe was considered "a promising area for broader U.S. efforts to integrate Russia into the West in a positive fashion. ... The U.S. goal is to demonstrate that integration and cooperation in the NEI region benefit Russia as well as its Baltic neighbors" (2000).

The six NEI priority areas were trade and business promotion, law enforcement, civil society, energy, the environment, and public health. As to trade and business promotion, the aim was to support U.S. companies interested in investing in and trading with the Baltic and Nordic states and northwestern Russia. As to law enforcement, the aim was to strengthen the rule of

10. It should be noted, however, that the impact of beliefs on behavior is often taken for granted in scholarly work but only seldom analyzed. Beliefs may influence behavior, but they are not necessarily the most important or even the only influence (Rosati 1995, 66–67).

11. The Independent Task Force sponsored by the U.S. Council on Foreign Relations defined as the main goal of the U.S. policy in northeastern Europe to help Russia make the "psychological adjustment" to see the Baltic region "not as a pathway to aggression but rather as an opportunity—that is, a gateway to greater cooperation and European integration" (1999, section: Is Northeastern Europe a Region?).

law and to help develop a functioning legal system in the Baltic states and northwestern Russia by, for example, fighting corruption. Regarding civil society, the Baltic-American Partnership Fund, cofinanced with the Soros Foundation, aspired to strengthen nongovernmental organizations (NGOs) and social integration in the Baltic states. With respect to energy, the U.S. government, under the NEI, supported the development and implementation of a Baltic regional energy investment strategy and a common regional electricity market as a precondition for economic growth in the region. Environmental challenges in the region, both nuclear and nonnuclear, were said to require regional approaches. Specifically, the treatment of nuclear waste in Russia's northwest was said to be of crucial importance for all states in the region. Finally, regarding public health, the focus was on fighting HIV and tuberculosis after the collapse of the health systems in the Baltic states and northwestern Russia. The NEI's objective areas were to

• integrate the Baltic states into a regional network for cooperative programs with their neighbors and support their efforts to prepare for membership in key European and Euro-Atlantic institutions;

• integrate northwest Russia into the same cooperative regional network [in order] to promote democratic, market-oriented development in Russia as well as to enhance Russia's relations with its northern European neighbors; and

• strengthen U.S. relations with and regional ties among the Nordic states, Poland, Germany, and the European Union. (U.S. Department of State Bureau of European Affairs 2000)

The NEI reflected a new understanding of international relations and security that was deliberately labeled "a new approach to diplomacy." Calling it "a new approach to security" would in all likelihood have been perceived as a provocation by the Baltic governments, which at that time were still interested in a very traditional approach to security. The White House Office of the Press Secretary (1998a) represented the regionalization of security in the Baltic Sea region as incompatible with the U.S.-Baltic Partnership Charter,[12] but the NEI included many ingredients that promoted a regional understanding of security. Among other things, it actively supported cross-border cooperation and regional linkages, and it understood borders as vehicles with which to tie communities together rather than separate them

---

12. The charter itself neither positively nor negatively refers to regionalization of security.

from one another. The initiative explicitly encouraged NGOs and private partners to participate in network building in the region. It thus promoted not only interstate cooperation or cooperation among governments, but also intersociety cooperation.

Thus, the NEI was an approach to foreign and security policy *"below* and *beyond* the nation-state level," as Sanjay Chaturvedi has suggested for a different context (1996, 36, emphasis in original). This feature, in turn, distinguishes security communities from zones of peace. Furthermore, the original concept of a security community did not exclusively refer to interstate relations, but to a sense of community among groups of people in general. These groups may be nation-states, but they may also be, for example, communities within nation-states and cross-border communities, adding to their respective national identities (if existent) further identities and we-feelings with respect to locality, regionality, common cultural heritage, and so on. By explicitly including the northwestern parts of the Russian Federation, the NEI aspired to avoid new dividing lines through the Baltic Sea region. This inclusive approach is an indispensable precondition for security community building in northeastern Europe. Thus, the NEI displayed many ingredients of security community building, which was much more important in the present context than the (rather negligible) amount of money allocated to the initiative.

The NEI did not have its own separate appropriation in the U.S. federal budget. Rather, it received money from several different sources, including the State Department and the Environmental Protection Agency. Only few of the NEI projects were financed solely by the U.S. government; most were cofinanced by governments in the region or by private NGOs or by both. According to Public Law 106-255, in the fiscal year 2001 not less than $2 million were to be used for NEI projects for eastern Europe and the Baltic states and not less than $2 million for assistance for the independent states of the former Soviet Union and related programs. These figures were certainly not impressive. Indeed, one of the reasons for launching the NEI was the necessity to use "increasingly scarce U.S. resources efficiently" (U.S. Department of State Bureau of European Affairs 2000, Section NEI: A New Approach to Diplomacy). Public Law 106-255 stressed that northern Europe "offers great opportunities for United States investment," and "partnership with other countries in the region means modest United States investment can have significant impact" (sec. 2, Findings and Purpose, [a] Findings 1 and 8). In other words, the NEI aspired to minimum U.S. financial input and maximal output.

However, more important than the financial input were the basic thought patterns underlying the initiative. By promoting cooperation in traditional low-politics areas such as environmental protection, the NEI intended to pave the way for cross-border cooperation in high-politics areas. Here, the issue is the familiar one of "spillover effect from low to high politics" (Vesa 1993b, 95). Furthermore, the "presence of environmental problems and the possibility of solving them by joint action" may "contribute to a sense of common interests and regional community" (Vesa 1993b, 95). Even within traditional high-politics areas, low-politics issues may be linked with high-politics issues, thereby obscuring the once permeable border. For example, the AMEC, although no ingredient of the NEI, implicitly links environmental issues with nonproliferation (Sawhill 2000, 16–18). Which of the two objectives is advertised for public consumption largely depends on the issue's popularity in domestic politics, on domestic power relations, and especially on the interest and influence (or a lack of it) of domestic pressure groups.[13]

Following the logic of conceptual change, security may also be made an affair concerning ordinary citizens by relieving it of its aura of secrecy, urgency, and high politics as well as by disassociating it from the state and the military. As Peter van Ham has observed, security and a sense of security may be linked with citizens' everyday experiences: "Especially since the development of this new security regime is not just an inter-state process (but involves regional authorities and non-state actors), its success will depend on whether ordinary people will consider it as a contributor to their sense of prosaic security and prosperity." The U.S. policy will be successful "if it [can] alter the sociopolitical environment in Europe's north to such an extent that it becomes almost irrelevant whether countries are full members of NATO or not, since their security would be assured by a dense network of

---

13. For example, there has always been skepticism in the U.S. Congress about the concept of environmental security. The creation of a position for deputy undersecretary of defense for environmental security during President Clinton's first administration has not changed this substantially. Advertising environmental protection in the language of nonproliferation thus may help make environmental protection palatable to the Congress. I am grateful to Steven G. Sawhill for sharing these insights with me.

daily cooperation based on complex interdependence and shared norms and ideas" (2000, 68, 78).

As a corollary of both the increasing insignificance of the military in intraregional and intersocietal relations and a growing sense of security among citizens resulting from daily cooperation in fields that are of practical relevance to their everyday lives, the NEI might perhaps have made military security cooperation unnecessary in the long run. The initiative is indeed said to have sought "to build a 'culture of cooperation' among the Baltic states, northwest Russia, the Nordic countries, Poland and Germany" (U.S. Department of State Bureau of European Affairs 2000, introduction). In addition to daily cooperation, language was important. Securitization basically means making an issue a security issue through the articulation of security. It thus means referring to an issue in terms of urgency and existentiality and thereby assigning to it a position above normal politics. If the political audience accepts this linguistic designation, the articulation of security justifies the use of extraordinary means to deal with the issue (Buzan, Wæver, and de Wilde 1998, 23–26).[14] If securitization works in one direction, there should be no reason to reject the possibility that it can also work in the other direction. Indeed, "in some cases security might be furthered exactly by the downgrading of security concerns" (Wæver 1998, 77). *Desecuritization* thus means referring to an issue in a nonsecurity and nonurgency terminology; it means downgrading an issue by not calling it a security issue.

Without any doubt, the designers of the NEI were perfectly aware of security, conventionally defined, in the Baltic Sea region and the perception of it on the part of the state actors in the region, in particular in the Baltic states, with their emphasis on discrepancies in military capabilities and a lack of security guarantees (Asmus 1998; Talbott 1998, 2000a). Yet by not referring to security on the official NEI Web site, by not referring to security even in the section on the environment (where security parlance almost suggests itself), the NEI did not treat northeastern Europe as a region requiring extraordinary means to tackle the problems. The existence of conflicts was acknowledged, but the initiative intended to resolve them with normal political

14. However, the articulation of security may also mobilize those policies that are possible within normal politics, but that would not have been mobilized had security not been articulated (Möller and Ehrhardt 2005, 53).

procedures prevalent in and among the democratic states and thus, among other things, without recourse to the use of or the threat of the use of force. The framework for cooperation in northeastern Europe, "in and of itself, is not going to eliminate our differences, but it was never intended to eliminate them. It was intended to give us a framework for managing them and building cooperation over time" (Asmus 1999). Thus, there is cooperation because there is difference and conflict. As suggested in chapter 3, living with differences rather than trying to eliminate difference is a way to conceive of security communities in a nondominative manner.

The NEI had initially stressed economic cooperation and business promotion. However, some developments during the second term of the Clinton administration pointed in a somewhat different direction. President Clinton, for example, commented on Public Law 106-255 by saying that there is "the need for continued international efforts to address the environmental dangers posed by nuclear waste in northwest Russia" (Clinton 2000). In one of its most detailed sections, the public law emphasizes that

> Northern Europe is home to significant environmental problems, particularly the threat posed by nuclear waste from Russian submarines, icebreakers, and nuclear reactors. . . . In particular, 21,000 spent fuel assemblies from Russian submarines are lying exposed near Andreeyeva Bay, nearly 60 dangerously decrepit nuclear submarines, many in danger of sinking, are languishing in the Murmansk area of Northwest Russia, whole reactors and radioactive liquid waste are stored on unsafe floating barges, and there are significant risks of marine and atmospheric contamination from accidents arising from loss of electricity or fire on deteriorating, poorly monitored nuclear submarines. (sec. 2[9] and [10])

This passage is important because the NEI's issue areas are defined more broadly than the AMEC program's issue areas. AMEC, on the one hand, is concerned with military communication and cooperation among defense ministries and thus includes the issue of nuclear waste from Russian submarines. The NEI, on the other hand, covered also civilian issues and could therefore both address the issue of nuclear waste from icebreakers and nuclear power plants and involve private actors. In Public Law 106-255, the issue of nuclear waste and environmental problems in northern Europe was addressed in the language of urgency and existentiality and therefore in the language of security. According to the law, addressing the environmental problems in northern Europe "remains vital to the long-term national interest of the United States." The terms *security* and *safety* were used interchangeably and apparently without much theoretical considerations (sec. 2 [13]). The consequences of addressing the environment in the language of "the national interest" and thus promoting the issue from low politics to high politics were not reflected in the law. Yet having Congress collectively refer to the solution of environmental problems as being vital to the long-term interest of the United States could have been an important step in imposing environmental security on the U.S. foreign and security policy agenda.[15]

Addressing environmental problems and nuclear waste in northern Europe in terms of urgency, immediacy, and the U.S. national interest may have helped to get support for the NEI in U.S. domestic politics. Similar to AMEC, the NEI's focus on environmental protection and nuclear waste treatment could also, if considered useful, have been advertised in terms of nonproliferation, thus giving the issue even more importance to U.S. national interest. Indeed, protecting the northern environment seems to have been too philanthropic and unselfish an objective because, as the Arctic Nuclear Waste Assessment Program concluded, "there is neither an indication [that] Russian dumping activities have increased radionuclide concentrations in American waters around Alaska, nor evidence that they pose any human health or environmental threat to the [United States]" (Sawhill 2000, 16–17). Addressing environmental problems and especially nuclear waste treatment coupled the NEI closer with AMEC and the Cooperative Threat Reduction Program (CTR) (Ellis 1997; Luongo 2001; Sawhill and Jørgensen 2001) and with U.S. strategic interests in terms of nonproliferation. This linkage may have been good news for the northern environment because the NEI, as stated earlier, was not limited to military environmental issues, but also covered civilian environmental issues such as nuclear-powered icebreakers and nuclear power plants in northern Europe. For the Baltic states, this development inevitably meant losing their own significance in the framework of the initiative's environmental protection issue area. For example, Public Law 106-255 did not refer to chemical weapons dumped in the Baltic Sea, which are arguably a greater and a more immediate danger to the

15. Steven G. Sawhill, personal communication, August 2000.

environment of the Baltic states and all Baltic Sea littoral states. This issue may have been totally unsalable to the U.S. domestic audience because neither U.S. strategic objectives such as nonproliferation nor the U.S. environment was threatened directly.

The NEI, then, was a considerable development of the U.S.-Baltic Partnership Charter. In particular with respect to security, it was a substantial departure from the charter. The Clinton administration's foreign policy regarding northeastern Europe indicated indeed a shift in emphasis from traditional interstate cooperation to the use of nonstate and nongovernmental actors. It intended to transform the conventional meaning of borders and to make them permeable; to minimize the U.S. government's involvement; and to demilitarize security both by following a comprehensive conception of security (largely without calling it security) and by generally downplaying the importance of military capabilities to security. Still, the implementation of this foreign policy was too ambivalent to evaluate the NEI as a complete and irreversible break with conventional foreign policy practices (discussed further later). Furthermore, the foreign-policy-in-the-making was too closely associated with the Clinton administration (and especially with Ronald Asmus and Strobe Talbott) to expect it to survive unchanged the replacement of the Clinton administration by an administration with a very different view on security. Indeed, the basic lines of thought underlying the NEI do not seem to be in accordance with the basic thought patterns on security of the George W. Bush administration.

At first sight, however, the U.S. policy toward northeastern Europe seems to have endured without much damage the first term of the George W. Bush administration and both its emphasis on military security and its interest in building rather than transcending boundaries. Indeed, by launching the e-PINE in October 2003, the Bush administration showed continued U.S. commitment to northern Europe. However, although rhetorically stressing continuity, it implemented profound changes. First, by including "cooperative security" on its agenda and by defining security in terms of "counterterrorism cooperation, control of the spread of weapons of mass destruction, border security, regional European challenges, and new threats" (U.S. Department of State n.d.a), the e-PINE now explicitly deals with conventional security issues. As Christopher Browning and Pertti Joenniemi have also observed, referring to terrorism means "importing into the region essentially non-regional security concerns" (2004, 249). It means that the

Baltic Sea region has lost to some extent its detachment from world politics, a detachment that arguably facilitated the development of a civilian understanding of security and the decrease in importance of security issues in the 1990s. The current trend is supported both by the Baltic governments' very audible support for the Bush administration's position on the "war on terrorism" and by the dependence of a continued U.S.-Russian rapprochement on successful cooperation in this very war. As a result, the foundations of the security relations in the Baltic Sea region seem to be more fragile today than before 11 September 2001 (Möller 2003c, 100).

Second, the language of threats in the e-PINE's sections on "cooperative security" and "healthy societies" indicates that the initiative's goal now is "to combat negatives, rather than to play on positives" (Browning and Joenniemi 2004, 249). In the original NEI, this focus on negatives could only be observed with respect to threats posed by nuclear waste. Third, by representing as NEI's "greatest success . . . Baltic State membership in NATO and the EU" (Conley 2003), the initiative is retrospectively reduced to and rationalized as a deliberate and target-oriented but also fairly traditional approach to the integration of the Baltic states in these organizations. As the analysis presented here has shown, it was not: the NEI was something substantially different from and more than just a disguised NATO and EU enlargement policy.

Fourth, the emphasis on the cooperation with Russia that was so prominent in the original NEI seems to have been replaced by an "eight plus one" approach (three Baltic states plus five Nordic states plus the United States). The deputy assistant secretary for European and Eurasian affairs in the U.S. Department of State, Heather Conley, indeed saw similarities and common interests only between "the United States and the *eight* states of the Baltic Sea region" (2003, emphasis added). Whatever the composition of these eight states is meant to be, it is different from the understanding of the Baltic Sea region prevalent in the Clinton administration's approach and does not seem to see the cooperation with and the integration of Russia as one of the most important policy goals. This approach, in turn, calls into question the basic lines of thought underlying the NEI.

It would perhaps have been naïve to expect the U.S. policy toward northeastern Europe to be unaffected in the long term by the changes in U.S. foreign and security policy after George W. Bush came to power. It should, however, also be noted that the Clinton administration had facilitated these changes by not decidedly and irreversibly breaking with older patterns of conducting and thinking of foreign and security policy. For example, one problem that the Clinton administration decided to address by largely ignoring it was that the notion of multiactor and demilitarized cooperation underlying the NEI did not relate easily to either the U.S. military assistance programs for the Baltic states or the rhetorical commitments to their integration into NATO. This ambivalence may have had two functions. It may have helped to increase the initiative's acceptability on the part of Baltic decision makers by addressing different actors in a different way and by emphasizing different aspects of the NEI according to who was addressed in a particular situation. This strategy may have induced everyone to welcome and support the initiative because each actor perceived its interests as represented by at least parts of the U.S. policy.[16] This ambivalence was also indicative of the Clinton administration's overall foreign policy style. This style aimed at preventing intra-administrative opponents from emerging by making sure that "different institutions in the Clinton administration could promote 'their' solution, without feeling that their advice had been frozen out." As a result, "U.S. attitudes and policies since the end of the Cold War have been marked by a series of ambiguities" (M. Walker 2000, 466, 471).

However, military assistance programs had helped construct traditional, territorial nation-states in Estonia, Latvia, and Lithuania, and these nation-states could be expected to pursue precisely the kind of conventional foreign and security policies the NEI aspired to transcend. The effective U.S. military assistance policy toward the Baltic states, although limited in scope, thwarted the understanding of security and cooperation underlying the NEI. This initiative, in its roundabout way through low politics and multiactor cooperation, may have had a strong impact even on conventionally defined security precisely by desecuritizing it and by dispossessing it of its aura of high politics. Integrating the Baltic states into a military alliance that the Russian Federation is not likely to join in the foreseeable future, however, not only supported the reading of the essence of state and military security prevailing among the Baltic governments, but also was precisely the opposite of creating a common northeastern European political space without dividing lines. It called into question the seriousness of the claim that the U.S. policy indeed aimed at integrating Russia in northeastern European co-

16. Pertti Joenniemi stressed this point in a personal communication, September 2000.

operation. Therefore, it was likely to reduce the incentives for the Russian authorities to engage in cooperation in the first place.

The Asmus/Nurick paper, the U.S.-Baltic Partnership Charter, and the NEI failed to offer convincing solutions to the basic contradiction inherent in the U.S. policy initiatives in northeastern Europe during the 1990s—namely, preparing the Baltic states for NATO membership while at the same time maintaining good relations with Russia on the basis of integrating it in as many networks as possible, except NATO. The contradiction was for some time a major stumbling block to the U.S. initiative's success. Essentially, if the NEI had taken seriously its own basic thought patterns with respect to security, it would either have convinced the governments of the Baltic states that their security can best be achieved both outside of NATO and according to the rationale underlying the NEI or have opened NATO in principle to all Baltic Sea littoral states—including the Russian Federation. For some time, however, the Russian government "justifiably object[ed] to the formal enlargement of a Western military bloc from which it would be excluded" (Kupchan 1996, 270–71). It even more strongly objected to the incorporation of what it regarded as former Soviet republics into NATO. As long as the government of the Russian Federation retained its disapproval of NATO membership for the Baltic states and as long as the Baltic governments refused even to discuss alternatives to NATO membership, reconciling the two views and maintaining good relations with Russia remained a challenge to U.S. policy that the U.S. initiatives found difficult to master. However, "promises [had] been given, expectations raised, prestige committed" (Wallace 2000, 475), so not delivering on the promises would have resulted in a severe loss in credibility and prestige. Rhetorically adhering to two objectives that for some time seemed to be mutually exclusive may thus have been the best interim solution available, for underneath the rhetorical surface cooperation may have evolved that ultimately would make the question of NATO membership indeed a question of secondary importance for both the governments of the Baltic states and the authorities of the Russian Federation.

Another challenge to the NEI was the emphasis on environmental protection in the NEI legislation during the second term of the Clinton administration. To be sure, the protection of the environment had always been one of the issue areas addressed by the NEI. The August 2000 public law, however, dramatized environmental protection in a way that called into question the basic thought patterns underlying the whole initiative. If we take into consideration the resistance to the concept of environmental security in the U.S. Congress, linking environmental protection explicitly with the long-term national interest of the United States in terms of safety, stability, and even security, rather than simply referring to environmental protection as a value in itself, may have facilitated collective support of the initiative in the U.S. Congress. At the same time, however, linking the NEI with U.S. national interests inevitably meant emphasizing state actors rather than nonstate actors, high politics rather than low politics, governmental rather than nongovernmental actors; in sum, it meant following a traditional understanding of interstate cooperation and security and regressing beyond the NEI's intellectual point of departure. This is all the more true because the legislative initiatives addressed nuclear waste in northwestern Russia and thereby entered into the core of Russia's military and energy affairs, with which regional and local actors, not to mention nongovernmental actors, were unlikely to be trusted (Palosaari and Möller 2004, 272–73).

The NEI, as interpreted here, had originally encouraged NGOs and private partners to participate in network building in the region. It promoted not only interstate and intergovernmental cooperation, but also intersociety cooperation. The initiative was an approach to foreign policy "below and beyond" the nation-state level and thus a potential component of security community building in northeastern Europe. The legislative developments sketched here, however, pointed in a somewhat different direction. This change of course may primarily have been a result of the U.S. legislative branch's skepticism about environmental security and did not necessarily reflect a deliberate change of intentions on the part of the U.S. government, the more so because it concerned only one out of six issue areas addressed by the NEI. It was, however, likely to influence U.S. policy implementation in a way that would be detrimental to security community building in northeastern Europe by making more difficult the reconciliation of U.S. interests with local interests (Möller 2002b). The current Bush administration apparently has decided to deal with this issue by deleting the protection of the environment from e-PINE's major issue areas and dealing with it only in the framework of the initiative's focus on "healthy societies" (U.S. Department of State n.d.b).

The last question to be discussed in this chapter concerns the profundity of the U.S.-Baltic friendship. I exclude the developments concerning the e-PINE from the discussion because they are too recent to allow more than mere conjectures. The basic theoretical conceptualization of international and intersocietal cooperation in the Baltic Sea region underlying the NEI differed significantly from the understanding of the security establishment in the Baltic states as analyzed in earlier chapters. In the Baltic conception, there was at that time relatively little appreciation of civilian readings of security and, understandably perhaps, even less appreciation of the Baltic states being used as "guinea pigs" for nonmilitary security "trifles" such as environmental security, which allegedly underestimated the seriousness of their security predicament. In the words of former Reagan administration deputy undersecretary of defense Dov Zakheim, the "warning of the need to take Soviet power in the North seriously," expressed by many U.S. military representatives in the late 1980s and early 1990s, "has carried over into the post–Cold War era, though the most vocal advocates of this view are not Americans, but rather the new Baltic states. On the other hand, with the passage of time, the United States itself, like the older Nordic states both within and outside NATO, has come to take a somewhat more relaxed view of the current power balance" (1998, 128).

Friendship, according to Alexander Wendt, is about national security. It explicitly includes the observance of two rules, the rule of nonviolence and the rule of mutual aid. On its third level of internalization, however, it includes a third rule—namely, amalgamation of interests: national interests are seen as synonymous with the friend's interests, international interests become national interests. Only on the third level can the observance of the rules be taken for granted. Only on the third level, too, may friendship reliably replace formal membership in a military alliance because the friend(s) are "known" to come to one another's assistance anyway—with the qualification that knowledge is always limited: the issue is one of probability rather than one of certainty. It is therefore the third-level internalization of friendship between the Baltic states and the United States that was of interest to and ultimately aspired to by the state actors in Estonia, Latvia, and Lithuania in order to ensure that the United States would actually behave according to the friendship rules. On the second internalization level, the friends follow the logic of instrumentality: they will observe the rules only if they consider it to be in their interest in a given situation. From a Baltic point of view, the logic of instrumentality was too weak a U.S. commitment to rely upon. As argued in the preceding chapter, friendship does not require symmetry between state actors, but symmetry increases the likelihood that friends will keep their promises. Because of the asymmetry in U.S.-Baltic relations and the resulting variations in the degree of obligations, the second-level internalization will probably remain of weight and give the inherently symmetrical role of friendship a profoundly asymmetrical bias in this case to the detriment of the Baltic states.

In chapter 9, I characterized the U.S. policy toward northeastern Europe as a "Soviet Union/Russia-first policy." This assessment, of course, is imprecise because it has always been a "United States–first policy." In other words, the U.S. policy toward the region has always followed U.S. interests in the region (and beyond). This position is still in line with friendship patterns, however, because friendship does not require the state actors to abandon their interests or to sacrifice them in favor of their friends' interests. Rather, friendship means incorporating others' interests into one's own set of interests. However, the reformulation of the U.S. policy toward northeastern Europe—from a military assistance program to a comprehensive approach involving international and intersocietal cooperation, network building in the region with as many and as diverse actors included as possible, a reduction in the role of both the military and state actors in providing security for and a sense of security on the part of the people inhabiting the region—means neither the abandonment of U.S. interests in the region nor their amalgamation with the Baltic states' interests as seen by their respective governments. Rather, it is a different way to pursue U.S. interests. These interests include strategic interests in terms of establishing nonproliferation; financial interests in terms of reducing the burden on the U.S. budget; politicoeconomic interests in terms of having access to information on economic developments in the region, primarily with respect to natural and energy resources; domestic political interests in terms of trying to gain the Baltic (and other) émigré circles' electoral support in the United States; military interests in terms of effectively concentrating expensive military efforts on what are considered geopolitical hot spots rather than wasting them in political settings where nonmilitary means have successfully been used in an effort to pacify, deescalate, and even solve conflicts and where the demilitarization of security made good progress during the 1990s.

As shown in this chapter and in the preceding chapter, the "United States–first policy" materialized in northeastern Europe in a variety of features. The U.S. nonrecognition policy on the Soviet occupation of the Baltic states did not effectively undermine the basic patterns of U.S.-Soviet relations during the Cold War. The existent frictions between them resulted from issues other than the Baltic question. In fact, active U.S. support for Baltic independence in the second half of the 1980s was modest. Gorbachev's reform course was prioritized and supported by not actively undermining it. The Baltic states' independence thus was a by-product of U.S. policy and was managed by the Baltic popular movements themselves—with active and involuntary support from the participants in the failed August 1991 coup d'état and from the central Soviet government itself, which, as chapter 5 has shown, refrained from using all means in its power to maintain the territorial integrity of the Soviet Union. Arms control and nonproliferation issues in the late 1980s and early 1990s were considered to be in U.S. national interests, whereas Baltic independence was seen as a threat to these very interests by making an agreement between Washington and Moscow more difficult. The U.S. policy toward the Baltic states, almost nonexistent until the mid-1990s, evolved very slowly, cautiously, and ambiguously. The U.S. domestic context and the influence of Baltic American interest groups were arguably more important to the development of the U.S. policy on the Baltic issue than was genuine interest in the Baltic states on the part of the U.S. government.

Without the deficiencies inherent in NATO's enlargement strategy, the Baltic states would in all likelihood have been of even less importance to the U.S. government. For example, U.S. financial support for Baltic military organizations was limited, just as was its support for the Baltic governments' wish to enter NATO as soon as possible. The U.S. government explicitly denied the interpretation of the U.S.-Baltic Partnership Charter as a U.S. security guarantee. Within the NEI, the northwestern parts of the Russian Federation were increasingly emphasized. This change of course is to be seen in light of U.S. strategic interests in nonproliferation rather than in environmental security or the environmental protection of northern Europe. Furthermore, U.S. foreign policy appeared increasingly reluctant—as indicated in the NEI—to follow the Baltic governments in their narrow understanding of security, their prioritization of the military aspect of security, and their adherence to the old-fashioned idea of membership in a military alliance as a security panacea.

In the final analysis, thus, U.S. policy has pursued U.S. interests rather than an amalgamation of U.S. and Baltic interests. If both sets of interests coincide, so be it. But in case of divergent interpretations of interests, U.S. interests gain the upper hand. A "willingness to help [the Baltic states] even when this serves no narrowly self-interested purpose" (Wendt 1999, 341) cannot be assumed. Reciprocal friendship among the United States and the Baltic states does not yet exist. Paradoxically, however, the conceptualization of security underlying the NEI in its original version, although tailored to meet U.S. interests in the region and not reflecting a fusion of U.S. and Baltic interests, may have benefited security community building in the Baltic Sea region and therefore the security of the Baltic states.

11

Representing Russia

The Politics of Negative Identification

THE PRESENT AND THE FOLLOWING CHAPTERS explore Baltic-Russian relations with particular emphasis on the tension that resulted from the simultaneity of the construction of Baltic nation-states and the construction of security. As shown in chapter 4, these constructions were not always mutually supportive. The resulting condition of tension can also be traced to Baltic-Russian relations especially in the first half of the 1990s, during which time the representation of Russia as a threat to the Baltic states was quite persistent. For most political actors in the Baltic states, the separation of Estonia, Latvia, and Lithuania from anything Soviet and post-Soviet seems to have been a necessary condition for the construction of the nation-state. Accordingly, Russia was represented as the Other that by virtue of its otherness helped consolidate the Baltic nation-states. However, constructing security in the shadow of the construction of the nation-state and the resulting politics of negative identification had consequences for both domestic politics in the Baltic states and international politics in the Baltic Sea region.

During the 1990s, most political actors in Estonia, Latvia, and Lithuania inhibited the development of friendship between the Baltic states and the Russian Federation by adhering to the image of Russia as the Other, and Others cannot be friends. On the one hand, international relations in the Baltic Sea region were empirically characterized by the absence of armed conflict. Accordingly, the representation of Russia prevalent in the Baltic states moved from enemy to rival. On the other hand, expectations of nonviolent conflict resolution, mutual representations of each other as postrivals, and corresponding politics between the Baltic states and the Russian Federation could not be observed. The representation of Russia in the Baltic states was stuck in rivalry, not because of the empirical evidence regarding the use of force throughout the 1990s, but rather because the nation-state-building process in the Baltic states was believed to require a delimitation from Russia. By representing Russia as a rival, they also represented the structure of the subsystem in terms of rivalry; in a sense, it even became one of rivalry. As a consequence, Baltic security was looked for without and even against Russia, but not with it.

This chapter begins by briefly exploring the patterns of armed conflict in the Baltic Sea region in the 1990s in order to substantiate the claim just made that the structure is indeed one of postrivalry. Because patterns of friendship could not be observed in U.S.-Baltic relations, it would be too ambitious to expect them to exist in Baltic-Russian relations. The absence of violence could, however, have been translated into postrival identities such as pre-friends or nonfriends. Throughout the present chapter and at the end of the next chapter, I discuss whether such a translation could be observed. Following the analysis of patterns of conflict, I analyze and problematize the representations of Russia in the Baltic security discourses over time. Chapter 12 continues along these lines and examines Baltic-Soviet/Russian relations in light of selected representations of historical encounters. The chapter concludes by discussing the implications for the development of friendship in Baltic-Russian relations in terms of nonfriendship and pre-friendship as introduced in chapter 3. In these chapters, the emphasis is on the construction of security and the nation-state and on the role assigned to the Russian Federation in the Baltic states. I largely ignore the construction of security and the nation-state in Russia and the role it assigned to the Baltic states (Sergounin 1997; Herd 1999; Puheloinen 1999; Trenin 2000). The chapters are designed neither as a history of events nor as a chronology of Baltic-Russian relations (Stranga 1996, 1997).

The first part of the analytical task in search of patterns of pre- or nonfriendship can be done relatively easily. Both role identities require the absence of organized violence over a certain period of time. Chapter 2 defined armed conflict, but some words concerning the conception of time are also germane here. The question of the period of time necessary to allow a relevant estimation as to the absence or presence of organized violence is, of course, an intricate one. To a certain extent, every answer is arbitrary. To make the definition as precise as possible is a prerequisite for its applicability. It is difficult, however, because linking pre- or nonfriendship strictly with, say, five, ten, or fifty years of absence of armed conflict would be mechanical and ignorant of the qualitative dimension of change. Thus, rather than suggesting any random period of time, we need to downplay the importance of the temporal aspect and emphasize instead the qualitative dimension of change in the Baltic Sea region.

At first sight, the roughly one and a half decades since the dissolution of the Soviet Union and the reestablishment of independent statehood in Estonia, Latvia, Lithuania, and Russia seems to be a rather short period of time. Historically speaking, it is indeed the blink of an eye. Even if we took into consideration that the Soviet leadership's conviction to deal with the Baltic issue in a nonviolent way can be traced back to the launching of perestroika in the mid-1980s, no more than five more years would be gained. More important, however, is the qualitative scope of change in the Baltic Sea region as sketched in chapter 1. This dimension can hardly be overestimated because it includes such tremendous changes as the dissolution of the WTO, the evaporation of both the Soviet Union and the German Democratic Republic, the reestablishment of the Baltic states, and the establishment of the Russian Federation, as well as large-scale military restructuring, including massive troop withdrawals, the decay of the Russian armed forces, the buildup of armed forces in the Baltic states, and the enlargement of NATO. It also includes large-scale political, economic, and social transformations in the former German Democratic Republic, Poland, Estonia, Latvia, Lithuania, and Russia. Thus, the scope and quality of the changes that occurred in the Baltic Sea region in the 1990s outweigh by far the rather limited time span.

If we use the definition of armed conflict given in chapter 2, we can say there have been no armed conflicts in the Baltic Sea region, not to mention wars, since the declarations of independence in Estonia, Latvia, and Lithuania in early autumn 1991. Even during the years of turmoil that finally resulted in the restoration of independence, no armed conflicts occurred. Although fifteen of eighteen armed conflicts in Europe between 1989 and 1993 took place in the territories of the former Soviet Union and former Yugoslavia (Wallensteen and Axell 1994, 333), the reappearance in the international system of the independent states of Estonia, Latvia, and Lithuania and all the other qualitative changes in the Baltic Sea region indicated earlier were not the result of and did not result in the use of organized violence. Moreover, neither the aggression of Soviet Ministry of the Interior troops against civilians in Vilnius and Riga in January 1991 nor the violence against

Lithuanian border guards qualify as armed conflict according to the definition used here (see chapter 2, note 6). Nonfriendship and one of the ingredients of prefriendship—the empirical absence of violence—can thus be established. Even modest forms of organized force or the threat of force that did not aim to eliminate or dominate the other, but only to realize narrower interests, occurred only occasionally (Clemens 2001, 212). The main feature of a Lockean kind of anarchy, occasional warfare, not to mention the features of a Hobbesian anarchy, thus cannot be shown empirically in the Baltic Sea region. The absence of warfare is certainly no little feat, but is it sufficient for regarding the Baltic Sea region as a mature Deutschian anarchy?

This part of the analysis is no doubt more complex: it leaves the empirical world and enters the realm of perceptions and interpretations, beliefs and representations. Like security communities, prefriendship requires crossing a threshold beyond which a violent escalation of conflicts is neither expected nor being prepared for. However, although the patterns of security community and prefriendship rule out the preparation for armed conflict with fellow members, contingency planning may be practiced and armed forces may be maintained for general disposition. Thus, the presence or, as in the Baltic case, even the buildup of armed forces in itself does not automatically violate the rules underlying prefriendship. Likewise, it is not necessarily indicative of the absence of expectations of peaceful change among a given group of people. Thus, rather than jumping to conclusions and inferring the nonexistence of prefriendship from the existence of armed forces in both the Baltic states and Russia, this chapter proceeds by analyzing the patterns underlying the representations of Russia and justifying the buildup of armed forces in the Baltic states. As the analysis shows, the substance of these representations evolved during the 1990s from predictable malevolence to unpredictability. Russia was no longer depicted as an enemy and only occasionally as a rival. At the same time, however, Russia's ostensible unpredictability was in itself viewed as a threat to security. The chapter analyzes why the representation of Russia persevered in nonfriendship instead of evolving to prefriendship and argues that the cultivation of Russia as a negative reference point can best be understood as an ingredient of both the construction and the reproduction of nation-states.

As shown in chapter 5, the Baltic popular movements for independence had already emphasized the necessity of armed formations and explained it in terms of operational functions. It may, however, be argued that it was precisely the movements' lack of military capabilities that crucially contributed to their success by strengthening the Soviet authorities' commitment to peaceful change and abstention from the use of force in order, among other things, not to undermine the credibility of the Soviet transition process, the success of which was often linked to nonviolence. The success of the non-military approach, however, is sometimes said, following a peculiar logic, to have required its replacement by a military approach. It is argued, for example, that the events in January and August 1991 "made it clear to most Latvians that their national sovereignty required the establishment of their own military forces" (Viksne 1995, 64). Likewise, in the two months following the confrontation between Soviet Ministry of the Interior troops and civilians in Vilnius and Riga in January 1991, it is said to have become clear that "Latvia needed its military structures to counter possible provocation" (Ozolina 1999, 21). This assessment ignores the possibility that instead of deterring provocation, the establishment of military formations in what at that time were still Soviet republics might have been perceived by the Soviet authorities as an unacceptable provocation. Unarmed resistance was certainly less provocative, but its importance to the reestablishment of independence is still insufficiently recognized. Indeed, as a former Latvian minister of defense has put it, the "significance of the barricades has not yet been fully appreciated, either from the political or [from] the military point of view" (Jundzis 1995, 554). In any case, violent interferences by the Soviet central authorities were considered possible and even likely at that time, and peaceful change was clearly not expected. It was, however, the Soviet Union, not the Russian Republic, that was seen as a threat to Baltic security. In the summer of 1991, for example, the leader of the Lithuanian popular movement Sajudis, Vytautas Landsbergis, reportedly hoped that a democratic Russia might contribute to changes in Soviet politics. Furthermore, Landsbergis is said to have argued that the Russian leadership was much more thorough-going in its solidarity with Lithuania than were the Western states, which made too many concessions to the Soviet leadership (Christophe 1997, 302).

As shown in chapter 1, the Russian Republic's leadership had crucially supported the Baltic popular movements, but immediately after the international recognition of independence, the official representation of Russia in the Baltic states changed from a supporter of to a threat to Baltic independence and security. Some Russian politicians' statements and some political practices no doubt facilitated both the representation of Russia's attitude toward the Baltic states as hostile and the depiction of Russia as a threat to Baltic security and independence. For example, the sympathy for the Baltic case displayed by President Yeltsin in 1990 and 1991 swiftly revealed its partly instrumental character and gave way to an at least rhetorical hardening of the Russian government's position on some aspects of the Baltic issue—for example, border delimitation and representations of Soviet-Baltic history—though without calling Baltic independence into question (Carrafiello and Vertongen 1998, 36).

Moreover, Yeltsin occasionally exhibited a high degree of insensibility to neighboring countries. Ignoring Polish historical experiences, he suggested a supply corridor ("a bit of highway") to the Kaliningrad region via Belarus and Poland, thus causing "corridor collywobbles."[1] He also included the radar station in the Latvian town of Skrunda in an April 1994 decree on permanent Russian military bases in the former Soviet republics and by so doing strengthened some observers' concerns as to the Russian political and military leadership's willingness to leave the radar station (Lachowski 1994, 581). To this must be added the psychological impact of both Russia's refusal, as the Soviet Union's legal successor, at least to apologize for Soviet crimes and its rehabilitation of symbols and controversial figures of the Soviet period (Ilves 2001).

Similarly, both the process of the Russian troop withdrawal and Russian representations thereof gave rise to suspicions as to the seriousness of the commitment to a complete withdrawal. The same sense of suspicion resulted from diverse declarations by high-ranking Russian politicians and military officials indicating a lack of willingness to pull troops out of the Baltic states as quickly as possible. Although the troop withdrawal continued constantly, and although the Russian authorities, by entering into negotiations, had accepted the general principle of troop withdrawal, a plethora of official and semiofficial statements invited skepticism about Russia's readiness and ability to fulfill its international obligations and supported the Baltic delegations' proclivity to think in worst-case scenarios. During the process of troop withdrawal, it was occasionally not entirely clear whether the forces favoring, delaying, or opposing the troop pullout would ulti-

1. *The Economist* 1996, 26.

mately gain the upper hand.[2] A unifying bond between the Baltic states and Russia, which might have resulted from the mutual recognition of the principle of troop withdrawal, could therefore not be constituted.

In addition, in the early 1990s Russian violations of Lithuanian airspace occurred as regularly as did violations of transit regulations through Lithuania. For example, Saulius Girnius reports on 194 air-space violations near the Russian-Lithuanian border in April and May 1994 (1995, 46). At the end of the decade, Russian military maneuvers under such provocative titles as "Comeback" (1998) in close proximity to the Baltic states did not facilitate the development of good-neighborly relations, either. In June 1999, the Russian simulation of a Western conventional attack on the Kaliningrad region was equally unlikely to be seen as a confidence-building measure by adjacent countries. After having "failed" to stop the attack by conventional means, the Russian military "succeeded" in defeating the virtual aggressor by using nuclear weapons (Wallander 2000, 11). At the same time, the exclusion of Russian participants (save for observers) from some military maneuvers in Lithuania, such as Baltic Challenge 98, on the basis of a constitutional amendment indicates lasting distrust of military cooperation; overcoming this distrust is still in its initial stages.

It is perhaps not surprising, then, that Baltic political decision makers in the autumn of 1997 rejected Russian security guarantees after internal debate over their pros and cons (Knudsen 1998, 52). As early as September 1995, a member of the Russian State Duma, Vladimir Lukin, had designated Moscow the "natural guarantor to Baltic security" because Moscow rather than NATO had "granted" Baltic independence (1995, 189). Two years later, these ideas were officially sanctioned on the highest political level when the then prime minister, Victor Chernomyrdin, offered Russian security guarantees to the Baltic states in exchange for their remaining outside of military al-

2. According to Lieven, Russian military commanders decided in autumn 1992 that the troop withdrawal was inevitable (1996, 76). According to Skak, already in January 1992 a high-ranking Russian delegation had committed itself to the troop withdrawal from Lithuania (1996, 200). Baev claims that the Russian military leadership had by early 1993 "most probably concluded that rapid withdrawal from the Baltic was inevitable" (1996, 162). Knudsen's speculation that Russia "later seemed almost to regret this commendable move" (1998, 15)—that is, to withdraw its troops from the Baltic states—is not supported by evidence.

liances. This connection—which, according to the Lithuanian Ministry of National Defense, made the proposal "a priori unacceptable" (Ministry of National Defense of the Republic of Lithuania 1999, pt. I, IV, 3)—was later removed when Yeltsin, on 23 October 1997, offered Russian security guarantees to the Baltic states that could possibly be multilateralized (Baranovsky 1998, 122). According to Yeltsin's declaration,

> Russia has already announced that it guarantees security to the Baltic countries. To elaborate on this initiative, we offer these guarantees as a unilateral commitment on Russia's part . . . in view of international law, to be determined between the Russian Federation and each Baltic country, or between the Russian Federation and all three Baltic countries.
>
> We have committed ourselves to making Baltic security guarantees an international priority by including the United States, Germany, France, and other Western countries in an agreement. Additionally, we do not exclude the idea of establishing a space of regional stability and security that includes the Nordic countries.
>
> International agreements based on a complex of regional, economic, human, and ecological measures could be developed into some kind of regional security and stability pact.[3]

This declaration, welcomed by the U.S. State Department,[4] was not at all appreciated in the Baltic states. The Baltic presidents rejected the offer in a joint declaration stating that one-lateral security guarantees and regional security arrangements were not in accordance with the "spirit of the new Europe."[5] The Russian declaration must be seen in light of the Russian leadership's dissatisfaction with the prospects of the Baltic states entering NATO, but it resulted in a strong rhetorical confirmation of Baltic NATO aspirations. From a Baltic point of view, two aspects of the declaration were especially noteworthy. First, from a security perspective, participation in talks on any kind of proposals that could cast a doubt on Baltic determination to become a member of NATO as soon as possible was to be categorically rejected. Likewise, every suggestion resembling an alternative to NATO membership was to be declined. Second, from the point of view of constructing

3. Reprinted in the *Baltic Times* 1997, 8.

4. *U.S. Information and Texts* 1997, 16.

5. *Frankfurter Allgemeine Zeitung* 1997, 9.

nation-states, Russian security guarantees could not be accepted, either. At that time, the Baltic states and their national identities were being constructed with Russia as a negative reference point. By definition, security guarantees offered by the Other cannot be accepted because then the Other would no longer be the Other and lose its identity-shaping capability as a negative reference point. It is therefore of only secondary importance in the present context whether the Baltic leadership had "adequately read" Yeltsin's proposal or not before rejecting it, as reportedly suggested by the Russian deputy foreign minister Avdeyev (see Main 1998, 185). The proposal had to be rejected no matter what.

Latvian historian Aivars Stranga comments on the plethora of proposals concerning Russia's future policies toward the Baltic states published throughout the 1990s by saying that it "is hard to make any final sense of these various contradictory forecasts and guesses, which have become excessive in Russia and which reflect the confusion that exists in that country's strategic thinking in the post-Soviet period" (1997, 186).[6] This may be so, but it is more important in the present context that commentators on Russia's policy toward the Baltic states could easily choose from a variety of statements and proposals whatever they needed to "prove" their interpretations. It was indeed possible to "prove" almost any claim by referring to one or another Russian source (and by forgetting to refer to others). As a consequence, Stranga recommends that commentators "devote serious attention to Russia's *actual* (not rhetorical) approach to the entire post-Soviet territory and the integration processes which are or are not occurring therein" (1997, 186, emphasis in original). Russia's actual policy toward the Baltic states in the 1990s was characterized by at least four things. First, hardly any anti-Baltic statement by Russian politicians led to effective *and* enduring anti-Baltic politics; rhetoric was translated into politics only occasionally. Second, most anti-Baltic statements by Russian politicians were derivative of domestic power struggles in Russia and were meant primarily for domestic consumption. For example, the translation of Vladimir Zhirinovsky's invective into considerable electoral success in December 1993 certainly was an important event and may be seen as one of the reasons for Lithuania's application for NATO membership in early 1994. In general, however, anti-

6. For different frameworks and foreign policy schools within the Russian foreign policy discourse, see Sergounin 2000.

Baltic rhetoric was not rewarded with electoral success. For example, then foreign minister Andrey Kozyrev's nationalistic turn and his attempts "to prove that he was a more genuine nationalist than Zhirinovsky himself" (Baev 1997, 182) were directed at the domestic audience but failed to attract a major part of the electorate; neither did his condemnation of the Estonian Law on Aliens as "apartheid" and "ethnic cleansing" in 1993.

Third, Russian politics toward the Baltic states was mostly reactive rather than active. As Anatol Lieven has pointed out, the official Russian policy "responded to things that the Balts have done or said, for instance on citizenship, military withdrawal, language laws, border claims, property questions, the status of Kaliningrad and so on" (1996, 176). Even the Russian leadership's vociferous and rhetorically aggressive response to the Baltic states' aspirations to join NATO was "a reaction to NATO expansion, viewed by Moscow as illegitimate and threatening" (Lieven 1996, 176). Although many observers within and outside the Baltic states jumped to conclusions as to Moscow's ostensible malevolence and hostility to the Baltic states, the Russian leadership factually displayed, within certain limits, "a deference to Baltic concerns, which is unique in the pattern of Russia's relations with its ex-Soviet neighbors" (Lukic and Lynch 1996, 364). Since 1997, Russia's policy in the Baltic Sea region has indeed been one of "constructive engagement" (Herd 1999, 201), combining soft and hard security initiatives and displaying, behind an occasionally aggressive rhetorical facade, an ability to surprise the neighbors with disarmament initiatives in northeastern Europe and with regional security proposals (Lachowski 2000, 264–65). It is perhaps precisely the constructivity of Russia's dealings with the Baltic Sea region that puzzles many observers—in particular those observers who believe in the need for an antagonistic Russia, the main function of which often seems to be its representation as the Other, which by virtue of its otherness helps stabilize the self.

Fourth, although the Gorbachev/Shevardnadze team in the perestroika-glasnost period was seriously interested in the creation of what they called a Common European House based on the CSCE model—among other things in order to improve the international background for Soviet internal and economic reform (Hirsch 1989, 403–533); although Gorbachev saw the Soviet Union as a European state and believed in a "shared cultural heritage" (Thomas 2001, 229), on the basis of which he aspired to reform the Soviet economic and political order; although the Russian leadership in the

Yeltsin period insisted on Russia's Europeanness and on locating Russia within European borders (Neumann 1997); although President Vladimir Putin has on several occasions reaffirmed Russia's desire to be treated as a part of Europe (Daalder and Goldgeier 2001, 83); and although Western governments are making major efforts to integrate Russia into cooperation networks in northeastern Europe, some observers in the Baltic states have seen things in a different light. I indicated this difference in chapter 8 in connection with the cultural turn in the writing of national security documents, but it can be traced back to the early 1990s. For example, Phillip Petersen quoted the then vice president of the Lithuanian Supreme Council, Ceslovas Stankevicius, as saying in November 1991 that he " 'cannot imagine that Russia could be brought into Europe.' In his opinion, 'Russia is not a European country, but an Asian country in terms of mentality.' He argued that 'European Russia up to the Ural Mountains is the European part of Asia' " (1992, 41). A former chief officer of the Lithuanian Ministry of European Affairs states that "after the reestablishment of independence [Lithuanian foreign policy] was based on the 'realistic' premise that after the Cold War ideas about the 'common security system of Europe' were mere rhetoric of the transition period, a guise for the enduring disagreements between the West and Russia, and that after a certain time new lines of division in Europe would be drawn" (Maniokas 1998, 28).

If the appearance of "new lines of division in Europe" was taken for granted, then trying to locate oneself on the right—that is, Western—side of the future border naturally followed rather than actively trying to prevent the new line of division from materializing in the first place. Agreement and communication with Russia then appeared to be issues of secondary importance because, in this interpretation, Russia was doomed to disappear behind the future border again anyway. Instead, representing oneself as "Western" as possible became a major issue. The cultivation of Russia as both the Other that has nothing in common with "us" and as a threat both to oneself in particular and to "Europe" in general may even have been seen as a promising way to anchor oneself firmly in the West.

Different readings of "Europe" may also be traced in the debate over a regionalization of security in the Baltic Sea area. Despite promising beginnings, the hopes expressed in the mid-1990s that "the Baltic Sea region provides a new and challenging setting for the regionalization of security, and for further demilitarization" (Albrecht 1994, 28) have to some extent remained unfulfilled. Obviously, the governments of the Baltic states did not aim to demilitarize, but rather to build up armed forces. Were they interested in a regionalization of security? According to the official narratives, they were not: security was said to be "indivisible": "there is no 'regional security' at all" (Ministry of National Defense of the Republic of Lithuania 1999, pt. 1.III.3). Likewise, it was said that there "can be no separate security for any separate region in Europe for the security of Europe is a whole" (Tiido 2000). Or, in the words of a Lithuanian minister of foreign affairs: "Confidence and stability around the Baltic Sea cannot be tackled separately from the integration processes. There is no special Baltic security. There is one common security of European nations" (cited in Marsh 1998, 155). Indeed, the Baltic security establishment consistently anathematized a regionalization of security on the grounds of the indivisibility of security in Europe. Ironically, President Putin seemed to see things in a similar light. For example, he argued in September 2001 that rather than expanding NATO, "it would be more correct to create a unified security architecture in Europe which would not create any new dividing lines. . . . And, I repeat once more, [it is necessary] to create a unified system of security, equal for all European states" (Putin 2001). However, whereas the Baltic governments understood the indivisibility of European security in terms of Baltic membership in and Russia's exclusion from NATO, the Russian government would have liked to include Russia firmly in European security structures. Thus, whereas the one understanding of Europe supported the notion of a limited NATO enlargement, the other suggested either NATO's replacement or Russia's inclusion in NATO (which would practically amount to the same thing).

If we take further the lines of thought being built in this chapter, however, it may be argued that even the actual Russian approach toward the Baltic states is a matter of only secondary importance. Rather, what is of interest is the representation of Russia on the part of the political actors in the Baltic states. As Lewis Coser elaborated many years ago, "terminology often provides a clue to orientation" (1956, 22), and the terminological representation of Russia was fairly independent of both the actual and the rhetorical Russian approaches. In fact, threats "may or may not exist in objective reality, but the group must feel that they do" (1956, 104), an assessment that can be applied to the supposed threats emanating from Russian policy toward the Baltic states. One way to make the in-group feel the "Russian threat" was by referring continuously to Russia in terms of a threat. Indeed, "if men define a threat as real, although there may be little or nothing in reality to justify this belief, the threat is real in its consequences—and among these consequences is the increase of group cohesion" (1956, 107). Thus, according to Alexander Wendt, "whether or not states really are existential threats to each other is in one sense not relevant, since once a logic of enmity gets started states will behave in ways that make them existential threats, and thus the behavior itself becomes part of the problem" (1999, 263).

Official and semiofficial statements immediately after independence clearly show that the threat perceptions were directed toward Russia and that peaceful change was not expected. For example, in December 1992, Estonia, in the words of the minister of defense Hain Rebas was confronted with "dangers from the East." Likewise, Colonel Ants Laaneots saw the threat to Estonia emanating from the East (both cited in Vares and Haab 1993, 302, 291). The Speaker of the Estonian Parliament, Ülo Nugis, declared that "NATO and only NATO can provide us with sufficient security guarantees against Russia" (cited in Haab 1995, 51). In February 1993, Lithuanian minister of national defense Audrius Butkevicius saw indirect rather than direct threats emerging from both the increasing instability in what used to be the Soviet Union and "growing tendencies of authoritarianism and nationalism in the territory of the former Soviet Union, above all in Russia," that could possibly involve Lithuania (1993b, 8). In Peeter Vares's assessment, as early as August 1992 "Baltic politicians, dizzy over the ease with which independence was regained, sought to resolve Baltic affairs to the exclusive advantage of [the] Baltic countries, often disregarding political realities as well as international public opinion. . . . Persisting in an anti-Russian vein, the Baltic community, both justifiably and often irrationally, began to challenge everything Russian, adopting the West as the main direction of their international communications" (1995, 55).

Baltic threat perceptions exhibited a peculiar persistence throughout the decade. According to Atis Lejins, in July 1995 Latvian minister of foreign affairs Valdis Birkavs believed that should Russia's transition to democracy fail, "there certainly will be a strong temptation to attack the Baltics since they are the number one scapegoat for all that went wrong in the USSR" (1996, 50). In January 1996, the commander of the Latvian armed forces, Juris Dalbins, spoke about the danger of a military intervention if Latvia fell into the political, economic, and military sphere of influence of its largest neighbor (1996, 8). In 1997, a professor at the Estonian Defense Academy, Rein Helme, identified as an alleged risk coming from Russia the "rebirth of Russian imperialism, the process of which has begun" (1997, 108). On 4 May 1996, Estonian minister of foreign affairs Siim Kallas was quoted in *The Economist* (4 May 1996) as accusing Russia of a "psychology of revanchism" and of planning "the external consumption of neighboring states." According to Eitvydas Bajarunas, the most acute security policy challenge lay in managing the relations with Russia, the more so because Russia, according to the author, was reluctant to accept Baltic independence. Baltic security, Bajarunas claimed, "will always be in jeopardy as long as Russia is hostile and authoritarian." Lithuania thus faced "the risk of renewed Russian expansionism and Moscow's meddling in Lithuania's internal affairs" (1995, 13–14).

Girts Valdis Kristovskis, Latvian minister of defense at the time, saw Russia as still adhering to "old ambitions and historic nostalgia." Likewise, according to Latvian president Vaira Vike-Freiberga, Russia still had "nostalgia for the Soviet Empire."[7] And her Estonian counterpart, Lennart Meri, in a speech honoring the eighty-second anniversary of the Republic of Estonia on 24 February 2000, added that "we see with regret the rebirth of Russian chauvinism and readiness to sacrifice basic human values in the name of power" (2000b). According to Gintaras Tamulaitis, Lithuania, in developing military structures, assumed that it would "face no military threat from the West: Lithuania regards the West as a guarantor of its security, a natural and trustworthy partner" (1994, 13). This statement rendered it unnecessary to state explicitly who was seen as a threat to Lithuanian security. Likewise, Kristovskis emphasized that Latvia has "friendly neighbors to the north (Estonia) and south (Lithuania), but she also shares several hundred kilometers of border with Russia and Byelorussia" (2000). Considering a Russian occupation of the Baltic states "clearly the most dangerous, but not the most likely, threat facing Lithuania," Albert Zaccor identified a plethora of threats to Lithuanian security emanating from a presupposed overall "Russian threat," from which the buildup of national armed forces could then be justified. He included in his threat list criminal activities, environmental and nuclear hazards, smuggling, and even official corruption, but he failed to

---

7. For Kristovskis, see *RFE/RL Baltic States Report* 2000e; for Vike-Freiberga, see *RFE/RL Baltic States Report* 2000f.

make clear why national armed forces should be seen as the most appropriate remedy against such threats (1995b, 12). Indeed, elsewhere he acknowledged that "many of these 'defense tasks' may seem more suitable for a nationwide police force than a national army. The nature of these threats, in fact, has shaped Lithuania's armed forces into a hybrid of border, police and regular troops" (1994, 202).

The discourse on security in the first years following independence was governed by representations of Russia as a direct or an indirect threat to Baltic security, but later the patterns of argumentation changed. To be sure, Russia was still seen as a threat to national security, yet it was no longer the representation of Russia as predictably malevolent that guided this interpretation. Rather, Russia was depicted as unpredictable and unstable. This unpredictability and instability were in themselves seen as a threat to security, and this threat perception was incorporated in the rhetoric of fear. Examples abound. Aare Raid asserted that "the main threat to the national security of Estonia remains the unpredictability of the development of the democratization process in Russia" (1996, 11). According to Atis Lejins, "what threatens Baltic independence is the same as what the West fears—the unpredictability of the future of Russia and its inability to overcome its imperial past" (1996, 45). Migle Budryte specified that "the major external risks for Lithuania's security today are connected with instability on the territory of Russia and the Commonwealth of Independent States (CIS), which is characterized by inter-regional, ethnic-religious, territorial and/or social conflicts" (2000, 128). Finally, Vaira Vike-Freiberga declared that "Russia is extremely unpredictable. . . . The simple fact of unpredictability scares me" (2000a, 200).

If "domestic developments in Russia—the future of its still fragile and incomplete democracy—are still worrisome" (Bengtsson (2000, 357) and are the major uncertainty factor with respect to stable peace in the Baltic Sea region, then the issue of expectations of peaceful change among groups of people who do not yet seem to share the same set of values and identities, familiar from the Cold War period and developed in chapter 3, reenters the analytical and political scene with full force. However, proclaiming Russian unpredictability and representing this ostensible unpredictability in terms of a threat may have the self-fulfilling effect that Russia "will 'respond' and in this way actually become as dangerous to the group as [the group] accused it of being in the first place" (Coser 1956, 106). The articulation of

"Russian unpredictability" nevertheless gained admission to a number of official documents, laws, and reports on security policy in the Baltic states.[8] Russia was permanently represented as either a direct or an indirect threat to Baltic security, although the military ingredient of the "Russian threat" was increasingly downplayed. To be sure, the patterns of argumentation were flexible. Either Russia's actual or potential strength or current weakness, either predictable Russian malevolence or its unpredictability, either Russian capabilities or, if the current capabilities were regarded as insufficient, potential future capabilities or even presupposed intentions were depicted as threatening the Baltic states. The representation of Russia as a threat to Baltic security and independence had made itself fairly independent of Russia's factual Baltic politics. This situation is what should be expected given Robert Jervis's findings that "states that have been expansionist under one set of circumstances or leaders are likely to be seen as posing a continuing threat. The state's aggressiveness will be seen as rooted in factors such as geography and national character that change slowly, if at all. . . . [So] when one country thinks that another is its enemy, the perception of hostility is usually more central than other aspects of the image; it is used to explain much of the other's behavior and is in turn often linked to prior variables such as the other's domestic system or geographical position." Jervis hypothesizes that "when the other acts with restraint, . . .the actor would be more likely to change his view of the other's strength than of its intentions" (1976, 275, 299). In addition, "even if the other state now supports the status quo, it may become dissatisfied later" (1978, 168). This reasoning to some extent explains the stability of the representations of Russia among the political elites in the Baltic states.

However, the persistence in representing Russia as a threat to security cannot be explained exclusively in the terms suggested by Jervis. Rather,

8. See the National Defense Concept of the Republic of Latvia, II. Characterization of the Security Situation of the Republic of Latvia, approved 6 June 1999 (the concept is currently not available in English; the Latvian text, Latvijas Republikas Valsts Aizsardzibas Koncepciza, is available at: http://ppd.mk.gov.lv/ui/DocumentContent.aspx?ID=930. See also Republic of Lithuania, Basics of National Security, pt. 2, chap. 9, first sec., approved 19 December 1996, amended as of 4 June 1998 (see chap. 4, fn. 7, and the appendix, fnn. 1 and 2); Ministry of Foreign Affairs of the Republic of Estonia, National Security Concept, 2.4.3: Relations with the Russian Federation, approved 6 March 2001 (see appendix, fn. 4).

cultivating Russia as a threat to Baltic independence and security also had a function in domestic politics. It served as a vehicle with which to strengthen the Baltic nation-states by disassociating them from everything Soviet, post-Soviet, and Russian. It may even be suggested that representing Russia as a threat to Baltic independence and security during the 1990s had primarily the internal function of constructing and strengthening the collective Estonian, Latvian, and Lithuanian self.

This policy, however, consolidated—and, arguably, in part created in the first place—tensions along ethnic lines in the population, especially in Latvia and Estonia. The cultivation of Russia as a threat was naturally addressed mainly to ethnic Balts (and to the Western states). Non-Balts in the Baltic states, although exhibiting only limited longing for the Soviet Union and identification with the Russian Federation, could hardly be expected to subscribe to the official definitional strategy of declaring Russia a direct or indirect threat. It was furthermore a blow in the face of those ethnic Russians living in the Baltic states who had supported the popular movements and their striving after independence and who suddenly found themselves in the dock. As chapter 7 has shown, the cultivation of Russia as an external threat to the national security of the Baltic states was occasionally extended to include the Russians living in Lithuania, Latvia, and Estonia, who were represented as an internal threat, as saboteurs, "fifth columns," or, in general, people who were working against the independence of the Baltic states, possibly on the basis of political manipulation by "Moscow."

Furthermore, by deepening and in part creating tensions along ethnic lines, Latvian and Estonian political decision makers helped their Russian counterparts to adhere to the image of the two states as hostile to both the Russian Federation and the ethnic Russians residing in the Baltic states. By exaggerating the degree of discrimination faced by the ethnic Russians in Estonia and Latvia and by threatening to implement countermeasures, Russian decision makers for their part helped Baltic decision makers to adhere to their anti-Russian stance. Baltic decision makers, in turn, then presented the cultivation of ostensible Baltic hostility on the part of Russian decision makers as "evidence" of Russia's ostensible hostility toward the Baltic states, against which they felt armed forces were needed. Russian decision makers again in their turn saw the buildup of Baltic armed forces and the aspirations to enter NATO as a "confirmation" of Baltic hostility to Russia that required Russian countermeasures. And so on. It was a vicious circle of mutually intensifying allegations and accusations mainly on the rhetorical level that had little to do with the factual Russian policies toward the Baltic states and the Baltic policies toward both the Russian Federation and the ethnic Russians residing in the Baltic states. Rather, it had much to do with domestic power struggles and nation-state-building processes in both the Baltic states and the Russian Federation.

The rhetoric of fear clearly had a function in domestic politics. Satisfied with the restoration of independence, many people in the Baltic states in the early 1990s seem to have accepted ineffective social and economic policies, widespread political incompetence, and the deficiencies of the reform policy in order not to disturb the internal situation, which they perceived as fragile. Such interference could have endangered at least stability, if not independence. It could also have increased the Baltic states' dependence on other countries, which, directly after the reestablishment of independence, was not seen as a particularly attractive future scenario. The rulers then used the citizens' self-restraint from criticizing them to stabilize their positions and interpreted it as tacit agreement with their political course. Janis Broks and associates have observed that "this method for maintaining political legitimacy" required "maintain[ing] the belief among the Latvian population that the country's independence is somehow threatened. This can only be accomplished if the idea that there are enemies to Latvian independence is actualized, and this process, in turn, has a deleterious effect on Latvia's relationship with Russia, as well as the state's relationship with ethnic Russians within Latvia" (1996–97, 109). In Lithuania, Vytautas Landsbergis is said to have cultivated threat scenarios and an almost permanent state of emergency from the autumn of 1991 onward in order to immunize himself against any form of political criticism and to divert attention away from the deficiencies of the ruling conservative party's policies. By massively mobilizing threat scenarios, Landsbergis arguably aimed to silence domestic opposition by equating criticism of domestic policies with a threat to national security (Christophe 1997, 301–3).[9]

9. According to one observer, Mr. Landsbergis's obsession with de-Russification can even be traced back to the level of his speech. His "once elegant Russian had become gradually more accented; now it seemed almost painfully halting, or pointedly tentative, as if he was carefully shedding any traces of the Soviet past." See Paul Quinn-Judge, "Do Svidaniya," *Boston Globe Magazine* (4 Oct. 1992), 43, as cited in Evangelista 1996, 118.

According to a widely held reading of Baltic history, World War II came to an end in the Baltic states only with their independence, the dissolution of the Soviet Union, and the withdrawal of the Russian troops. In November 1990, for example, Lennart Meri declared that "in the Baltic countries, the situation of the Second World War persists" (1991, 109). Reflecting on the relationship between war and subjectivity, Jervis argues that "because of the dramatic and pervasive nature of a war and its consequences, the experiences associated with it—the diplomacy that preceded it, the methods of fighting it, the alliances that were formed, and the way the war was terminated—will deeply influence the perceptual predispositions of most citizens" (1976, 266). Michael Shapiro, referring to the Clausewitzian understanding of war, describes the relationship between war and collective subjectivity as follows:

> [War] creates the conditions for the production, maintenance, and reproduction of the virtuous self, a way (for men) to achieve an ideal form of subjectivity as individuals and for the state to achieve its ideal form of collective subjectivity. . . . Indeed, the process of fixing stories of past violent encounters plays a role in shaping the spaces and events that constitute the basis for being a "people." Those histories that manage to attain a level of dominance and stability create the imaginative boundaries that contain a people; they exert an influence on the self-interpretations and modes of inclusion and exclusion of the people who embrace them. (1997, 54, 138)

In light of these interpretations, it may be argued that the reading of the Soviet occupation of the Baltic states until 1991 as an extension of World War II served as a kind of substitute, imagined war promoting the maintenance of a collective Estonian, Latvian, and Lithuanian self in the absence of both a "real" war and an "own" state. With independence, an "own" state entered the stage, but the imagined war that could have been used to strengthen collective subjectivity came to an end. Representing Others—and especially Russia, the legal successor of the Soviet Union—as a threat and representing oneself as being in a constant state of danger in which security, independence, and national survival were permanently at risk served as a substitute, imagined war with which the state's collective subjectivity and the citizens' identification with "their" state could be sustained and strengthened.

Therefore, "fighting" imagined wars and cultivating states of emergency and siege are, like war, to be seen in the Clausewitzian sense as practices with which individual and collective subjects are "producing and reproducing themselves" (Shapiro 1997, 56). Constructing (the impression of) a collective self is a never-ending story; "insofar as there is a 'national identity,' it is an ongoing project rather than a fact" (Shapiro 1997, 141). From this, it follows that the collective self has to be continuously confirmed and reconfirmed by way of, among other things, confirming and reconfirming others as the Other in such things as narratives of past violence. Thus, the cultivation of the Other is part of the permanent reproduction of the state. Cultivating a state of emergency is a process, the function of which appears to be as much internal as external: "What the subject/nation represents as a hostile object of an aggressive aim is in part a stand-in for an inward aim; its antagonistic status is produced by the drive for inner coherence, an attempt to assemble harmoniously those elements of the self or the order that defy this coherence" (Shapiro 1997, 59).

If we take this line of thought further, even Baltic participation in the NATO-led military intervention in Kosovo—that is, the replacement of an imagined war by a real one—may be said to have had an internal function because "war retains a significant dimension of both individual and collective identity affirmation" (Shapiro 1997, 77). Here, three aspects of this replacement are important. First, Baltic participation in the war was aimed to recommend the Baltic states as a reliable partner of the West and to increase their prospects of being invited to join NATO. Second, and in contradiction to the first point, supporting NATO in its war in Kosovo was said to have increased the likelihood that NATO might defend the Baltic states against Russia if need be, even before Baltic NATO membership. In a BBC interview on 6 November 2000, for example, the Latvian president reportedly said that "Latvia could expect help from NATO if Russia threatened her country even before it became a member because NATO intervened to help Kosovo, which also is not a member of the alliance."[10] Third, the Baltic states confirmed the role of Russia as a negative reference point by participating in a

---

10. *RFE/RL Baltic States Report* 2000f. Paul Goble, however, has pointed out that this logic may be fallacious: "A claim of a Kosovo precedent is likely to force NATO itself to reiterate that Kosovo is not the precedent it hopes for, thus leaving Latvia and her neighbors in a less desirable position than they were in before such a claim was made" (2000a, 19). Moreover, if NATO were to come to Latvia's aid irrespective of membership, then costly preparations for member-

war that was rejected categorically by the Russian leadership. Collective subjectivity in the Baltic states was furthered by the reading that "our" military strengthens European security, whereas "theirs" is a threat to it. The representation of Russia as the Other, as the non-West, was fostered by the Baltic states' rather uncritical support of NATO, which ostensibly defended and manifested "Western" values in Kosovo.

Based on the lines of thought suggested in this chapter, the development of the Baltic Peacekeeping Battalion (BaltBat) may also be presented in a new light. The erasure in June 1998 of the word *peacekeeping* from the battalion's title in order to make possible its participation in so-called NATO-led peace-enforcement operations considerably expanded the battalion's original function. In the original Memorandum of Understanding, these functions were defined in terms of "carry[ing] out peacekeeping operations mandated by the United Nations (UN) or by the Conference on Security and Cooperation in Europe (CSCE)."[11] Parallel to the expansion of NATO's missions as exemplified in the Kosovo war, a clash between BaltBat's activities and Russia's perceived interests, which some observers had already expected in connection with the Baltic participation in the NATO Kosovo Force (KFOR) and Stabilization Force (SFOR) (Johnson 1999), could not be excluded. Furthermore, the initially positive reaction by representatives of the Russian Federation to the establishment of the Baltic *Peacekeeping* Battalion (Knudsen and Neumann 1995, 25) may have given way to a more critical stance, which Baltic decision makers in their turn may have interpreted as "evidence" of Russia's ostensible malevolence and hostility to the Baltic states. This change of mind, in turn again, may have facilitated the depiction of Russia as the Other and may indeed have been welcomed in the Baltic states as a means to legitimate resource allocation to the military and to strengthen collective subjectivity at the expense of Baltic-Russian relations. With the official dissolution of BaltBat on 23 September 2003 and the termination of all BaltBat-related Memoranda of Understanding and other legal

ship would not be necessary any longer. Domestic support for measures seen as prerequisites for NATO membership may thus have withered away.

11. Memorandum of Understanding Concerning Cooperation on the Formation of a Baltic Peacekeeping Battalion as of 11 September 1994, reprinted in Danish Ministry of Defense 1995. For the erasure of the term *peacekeeping* from the battalion's title, see Ministry of Foreign Affairs of the Republic of Estonia 1999.

agreements,[12] this escalation scenario also ended. The story of BaltBat is nevertheless important even in retrospect because it is indicative of the thought patterns underlying military cooperation by the Baltic states (discussed more in chapter 13). Furthermore, according to a spokesman of the Lithuanian armed forces, Baltic servicemen, while participating in BaltBat, "have learned very well the military interaction procedures of NATO states."[13] Here, the establishment and operation of the battalion is linked to Baltic membership in NATO in a manner that disregards the battalion's function as formulated in the original Memorandum of Understanding. Similar to the retrospective reformulation of the U.S. NEI aims as discussed in chapter 10, the story of the BaltBat is reduced to being a path to Baltic membership in NATO, which, as the original Memorandum of Understanding shows, it was not.

Lewis Coser, referring to Georg Simmel's work, states that conflict serves, among others things, "to establish and maintain the identity and boundary lines of societies and groups. Conflict with other groups contributes to the establishment and reaffirmation of the identity of the group and maintains its boundaries against the surrounding social world" (1956, 38). The politics of negative identification, emphasizing Russia's role as a negative reference point for the Baltic nation-states and seeing Russia as the "antagonist of [Baltic] identity, everything that these nations did not want to be, and wanted to dissociate from" (Medvedev 2000, 15), was during the 1990s—and to a lesser extent still is—an element of the construction of Baltic nation-states according to a modernist script based on, among other things, sovereign units that are clearly separate from one another. The construction of security to a large extent also unfolded by way of demarcation from Russia. It required the cultivation of differences between the Self and both the "internal Other" (Noreen and Sjöstedt 2004, 746) and the external Other. But it also necessitated overemphasizing the level of coherence and neglecting the degree of conflict within the Self—and to some extent also within the internal and external Other: the more coherent the Other appeared to be, the

12. Estonian Ministry of Defense, Public Relations Department, personal communication, 16 June 2004.

13. See *Lithuania Presents* at http://www.tm-lt.lt/main.php?action=main::article.show&article_id=426. Thanks to Erik Mannik for additional information on the dissolution of BaltBat.

more threatening it could be represented. Contrariwise, it would have been more difficult to represent as a threat an Other consisting of atomized individuals (even though the ostensible lack of order in this case was also seen as threatening). Representations of Russia therefore addressed at least three different groups of people, which, however, were in themselves anything but coherent. First, they addressed the Self, aiming at cohesion and order. Second, they addressed the internal Other, intending emulation of and adaptation to the Self's view and, consequently, detaching the Self from that part of the internal Other that refused to imitate the Self and thus to confirm it. Third, they addressed the external Other and aimed to dissociate and reproduce it as a negative reference point, a designation that the Other often "confirmed" by refusing to strengthen the Self's self-perception by means of emulation.

12

It Is All about History— On the Surface at Least

THE CURRENT BALTIC-RUSSIAN RELATIONS started on 24 December 1991, the day when the Russian Federation became the legal successor to the Soviet Union.[1] However, the new beginning was difficult for a variety of reasons, among which the perseverance of the Soviet-Baltic history and the lack of differentiation between the Soviet Union and the Russian Federation in representations of "Russia" figure prominently. Indeed, history matters (Wendt 1995, 77), but its bearing on the cultural attitudes in the present cannot be explained solely in terms of history as that which is not anymore. What is needed in addition is the analysis of representations of historical encounters. These representations may facilitate the evolution of friendship patterns in interstate relations by, for example, differentiating between Baltic-Soviet and Baltic-Russian relations. However, historical representations may also contribute to the extension of the past into the present in such a manner that, rather than learning from the past, actors become prisoners of the past, "insensitive to incoming information" (Jervis 1976, 218).

If the past was dominated by violent encounters, the development of friendship becomes difficult: actors, having "overlearned from traumatic events" (Jervis 1976, 218), may fail to acknowledge that history does not necessarily preordain "our" current behavior toward "them" and "their" current behavior toward "us." They may fail to see that the conceptions of "us" and "them" and the meaning assigned to both concepts may change over time, thereby helping to replace memories of violent encounters with expectations of peaceful encounters as the most important belief underlying mutual relations. Such social concepts as "us" and "them" are rooted in

---

1. The title of this chapter is borrowed from Rankin 2001, 137.

cognition and may therefore be subject to change. As social concepts, they have to be "done" in the sense that "we draw upon typified behavioral patterns to reproduce meaning in a particular setting or situation and, in the same act, reaffirm or modify its sense" (McSweeney 1999, 165).

In representations of Russia, there can indeed often be observed reproductions of typified patterns of representation that reaffirm rather than modify meaning. These standardized representations can in part be explained as reflections of both the intricacies of Baltic-Russian, then Baltic-Soviet, then again Baltic-Russian relations and the peculiarities of the Soviet system, seemingly favoring Russians before the members of the titular peoples of other republics. Baltic-Russian and Baltic-Soviet histories are indeed difficult to disentangle, and there is said to be a "close association in the minds of Baltic decision-makers of Soviet and Russian imperial traditions and ambitions" (Herd 1997a, 94). The task to differentiate is made even more difficult by depictions of the current situation in the Baltic states as one of a simultaneous "struggle for both our past and our future" (Meri 2000b).

The troop withdrawal issue, in which case Soviet troops actually became Russian troops, may illustrate the confusion resulting from a lack of distinction between Russia and the Soviet Union. The Russian army's temporary presence in Estonia, Latvia, and Lithuania was frequently seen through the prism of the historical experience with the Soviet army. Regarding the Baltic-Russian negotiations on the limited Russian use of the radar station in Skrunda, collective memories of 1939 were invoked (Dreifelds 1996, 172–73), in which case demanding mutual assistance agreements and army bases in the territories of the Baltic states had indeed preceded the Soviet occupation.[2] The Russian government's offer in 1997 to guarantee the security of the Baltic states was "not well received in the three Baltic capitals" and was turned down with reference to the notorious 1939 Molotov-Ribbentrop Pact, which had paved the way for the occupation of the Baltic states by Soviet, then German, then again Soviet forces (Rotfeld 1998, 147). Likewise, in some scholarly writings there can be observed a mix-up of Rus-

---

2. In accordance with the Pacts of Defense and Mutual Assistance with Estonia (28 September 1939), Latvia (5 October 1939), and Lithuania (10 October 1939), the Soviet Union set up naval, air force, and coastal artillery bases in the Baltic states and stationed 30,000 troops in Latvia, 25,000 in Estonia, and 20,000 in Lithuania (Misiunas and Taagepera 1993, 15–16).

sia with the Soviet Union. It is said, for example, that the Soviet occupation of the Baltic states in 1940 was "thoroughly prepared by Russia" (Haab 1995, 38) and that in January 1991 "Russia's military threats to Latvian independence became not only a future fantasy, but a reality" (Ozolina 1998, 139). It is but a small step, then, to represent the Soviet Union as "Russia disguised as the Soviet Union" (Järve 1996, 226) or to declare that although Estonia, Latvia, and Lithuania restored nation-states in 1991, "the Russians did actually lose their own state: the USSR" (Viksne 1995, 74).

Language constructs rather than reflects reality. The use of a specific language is indicative of, among other things, the reality that those engaged in a speech act would like to naturalize and the meaning they would like to assign to historical events. These assignments are never uncontested, and counterassignments may always be made. As Pavel Baev points out, for example, rather than posing a threat to the independence of the Baltic states, "the Russian leadership came out decisively [in January 1991] on the side of the Baltic independence movements and was actually their main ally in the struggle against the Soviet center" (1996, 160–61). John Dunlop, referring to Carl Linden's work, argues that the Soviet Union was not a Russian empire and that the Russians in the Soviet Union could not legitimately be called an imperial people. Although Soviet realpolitik often carried with it a certain preference given to Russians, Soviet ideology was essentially not the Russian nation-state's continuation, but its replacement. Accordingly, on average Russians did not live better than the titular peoples of the other Soviet republics (1993, 45–46).

According to Rogers Brubaker's analysis of the Soviet practice of "institutionalized multinationality," the

repression of political nationalism was compatible with the pervasive institutionalization of nationhood and nationality as fundamental social categories. Nationalists' complaints—and Stalin's murderous policies— notwithstanding, the [Soviet] regime had no systematic policy of "nation-destroying." It might have abolished national republics and ethnoterritorial federalism; it might have abolished the legal category of personal nationality; it might have ruthlessly Russified the Soviet educational system; it might have forcibly uprooted peripheral elites, and prevented them from making careers in "their own" republics. It did none of the above. (1996, 37)

Tomas Venclova specifies that in Lithuania Russification ceased to be a regime priority after the Stalin era (1995, 347). Abundant means and methods to suppress and even annihilate the Lithuanian nation and language were still available to the authorities in Moscow, but they did not use them. Graham Smith observes only a few signs of cultural assimilation among Latvians during the Brezhnev era and afterward. To be sure, knowledge of the Russian language was a necessary precondition for entering higher and specialized education, but "urbanized Latvians did not need to assimilate into the Russian language and culture in order to gain position and status in the republic" (1996, 154). Latvians were not excluded from administration, economic management, and the professions.

At the same time, however, change of habitation from the Russian republic to the Baltic republics continued and after a short decline in the early 1980s increased again from the mid-1980s. Many Russian-speaking migrants' reluctance to learn the Latvian language helped produce the impression of a threat to the Latvian language, the more so because the central Soviet authorities in the late 1970s restarted to accentuate the teaching of the Russian language in schools and universities at the expense of the Latvian language. As a consequence, in June 1988 the Latvian Writers' Union represented the demographic situation and the suppression of the Latvian language as a "threat of Latvian ethnic annihilation." As late as 1979, however, the titular nationality had exhibited a remarkable power of resistance against language assimilation: 97.4 percent had declared Latvian as their native language (Muiznieks 1993, 192). This resistance against assimilation is important because language, according to Anssi Paasi, is "the medium through which discursive stories about us (and them) are produced and reproduced" (1996, 91). In Vieda Skultans's words, using a Baltic metaphor, language is "the amber which preserves relics of past feelings and ideas" (1998, 24). This assessment is not meant to minimize the oppression, censorship, assimilation pressure, and control exhibited by the Communist Party *nomenclatura* in Moscow, Tallinn, Riga, and Vilnius, but the results left much to be desired from the *nomenclatura*'s point of view. Furthermore, trying to overcome national diversities "through a sustained progress of nations" rather than through their assimilation strengthened national, republicwide identities and was supported by investments in culture and education (Baev 1997, 175). According to Romuald Misiunas and Rein Taagepera in what is still the best survey of Soviet-Baltic history,

Cultural assimilation in the Soviet Union may have been overestimated by many earlier observers. Correctly noting the utterly limited political and economic autonomy of its national republics, such observers underestimated the boost to national identity supplied by the mere existence of those republics, not to mention the very real cultural autonomy that existed in many of them. The Baltic nations made full use of their restricted opportunities, and by so doing they appreciably extended their historical depth as modern nations, compared to the situation as it was in 1940.

By 1980, the Baltic nations could look back to six rather than two decades of native-language universities and republic-level administration. In the 40-year perspective, the use of Russian in those fields had advanced; but in the perspective of 80 years, the overall picture was still one of a massive shift from Russian to the national languages. The very territorial units called Estonia, Latvia and Lithuania, as applied to the ethnolinguistic areas, were only two decades old in 1940. They too had now tripled in age. (1993, 273)

Brubaker aptly summarizes Misiunas and Taagepera's findings by saying that the Baltic nations emerged from decades of Soviet assimilation pressure "much more firmly established and consolidated as nations than they had been in 1940" (1996, 38).

All the same, because personal and collective experiences and memories of experiences do not necessarily have to coincide with the results of scholarly analysis, the attitudes toward Russia among major portions of the Baltic peoples are often said to be characterized by "a strong *feeling of insecurity* vis-à-vis Russia" or even a "fear of Russia." These feelings appear to be "part of the personal affective repertoire . . . formed on the basis of personal experiences, pre-existing knowledge and available information" (Carrafiello and Vertongen 1998, 32–33, emphasis in original). For instance, many ethnic Latvians are said to "associate the brutality of communism in its Stalinist guise with Russians" (Muiznieks 1993, 184). This association, in turn, seems to have resulted in internalized, stereotyped, and dichotomizing personal images of, on the one hand, Russia and "the Russians," and, on the other, Estonia and "the Estonians," Latvia and "the Latvians," Lithuania and "the Lithuanians." According to Vieda Skultans's poignant analysis of illness narratives in post-Soviet Latvia, for example, love of solitude, of one's home, of the homeland, love of work and self-control are all consciously used to symbolize the boundaries between Latvians and non-Latvians. Conversely, non-Latvians are distinguished by the absence of these traits. . . . For example, Latvians' love of quiet and solitude are contrasted with Russian love of noise and crowds. References are made to the Latvian tradition of living in isolated farmsteads and contrasted with the Russian tradition of villages. The ideal situation for a house is thought to be one where no sign of human habitation is visible. The Latvian habit of speaking quietly and calmly is contrasted with Russian habits of loud talking and shouting. Latvians pride themselves on being self-controlled, whereas Russians are thought to lack discipline and self-control: they swear and drink. Latvians characterize themselves by their extreme attachment to locality and to the idea of a Latvian homeland. Russians are perceived as rootless, ready to move to where the going is good. (1998, 128–29)

These contrasting images are neither a Latvian speciality nor an exclusive trademark of the post–Cold War era. Referring to others as "rootless" is an ingredient of the standard European repertoire of representing the Other, be it a German, a Briton, or a Russian. Only in the case of Russia, however, this feature is said to have been "a stable feature of discourse throughout the 19th century and beyond" (Neumann 1997, 166).

As the previous chapter has shown, the politics of negative identification in the Baltic states throughout the 1990s used Russia as a negative reference point against which the Baltic nation-states were constructed. This construction made use of and cultivated typified patterns of representation, including reproductions in both histories of literature and anthologies of literature of stereotyped ostensible national characteristics that, as Skultans writes with respect to Latvia, "undoubtedly influence[d] the Latvian image of Russians" (1998, 160).

Likewise, in general, representations and memories of past violent encounters without doubt influence most citizens' perceptual predispositions (Jervis 1976, 266). For this reason, I take a closer look in this chapter at some recent representations of the anti-Soviet resistance movements of the late 1940s and early 1950s in what then were Baltic Soviet republics and at their relationship with the current security policies. The chapter is not intended as an original contribution to the research on the history of the resistance movements. It is not meant as a critique of the movements, either. With respect to the historical resistance movements, I make no claims to originality and offer no political or moral judgment. My discussion is exploratory only, sketching possible future research areas rather than trying to give conclusive answers. Furthermore, although I focus on the Forest Brothers (defined later), I do not claim here that the representations of the Forest Brothers are the only or even the most important legitimacy provider for the current security policies in the Baltic states; they are one among others.

Past violent encounters as well as memories and representations of them are often more important to the definition of the in-group and its demarcation from out-groups than are other forms of encounters. Under the circumstances of war, the distinction between in-group and out-groups can be both established relatively straightforwardly and defended relatively easily. According to Georg Simmel, in war situations "groups . . . are not tolerant." They do not accept and cannot afford "beyond a definitely limited degree" individual deviations from the basic principles on which they are resting (cited in Coser 1956, 95). War, however, is the exception rather than the rule. As a means with which to define a community, it is unavailable most of the time. Histories and memories of war, however, are usually available. Memories of past violent encounters are indeed a powerful and durable vehicle with which to define and strengthen communities, national and otherwise. The analysis of "the promotion by states of particular interpretations of the past" (Halliday 2000, 65), with which interests are articulated, identities are confirmed, and political ends meet, has been suggested as an ingredient of a future research agenda. The analysis of interpretations of violent encounters is especially salient here because to "locate a war in a people's memoryscape is, among other things, to engage in a politics of interpretation," including the politics of identifying a "people" in the first place: "The process of fixing stories of past violent encounters plays a role in shaping the spaces and events that constitute the basis for being a 'people.' Those histories that manage to attain a level of dominance and stability create the imaginative boundaries that contain a people; they exert an influence on the self-interpretations and modes of inclusion and exclusion of the people who embrace them" (Shapiro 1997, 138).

Furthermore, with respect to questions of war and security, access to information is severely limited for both the scholarly community and the general public. This lack of information renders difficult the renegotiation of memories in the light of historical data and documents, and it facilitates the construction of what may be called an official national memory, integrated into or even forming the basis of the idea of the state. Incorporated into the idea of the state, a certain reading of the past, although changeable over time, tends to become nonnegotiable at any given moment. The rightfulness and moral legitimacy of "deviating" views—that is, views not in accordance with the official view of the past—are and, from the perspective of the polity, have to be called into question. Through their very existence, alternative views on the past challenge the officially maintained historical foundation on which a given nation-state rests.

Such imagined communities as nation-states in fact rely heavily on the "cultivation of the past, of a particular past" that has to be seen as "a key area of cultural reproduction, of moral regulation and of identity maintenance" (Schöpflin 1999, 9), binding the citizens together on the basis of collectively shaped and shared propositions that are held to be true. The nationalizing strategies of newly independent states or of states that have regained their independence can indeed be expected to rely heavily on references to the past because "all beginnings contain an element of recollection" (Connerton 1989, 6). To this must be added the powerful function of a state-sponsored culture of didactic memory, binding for and teaching the nation, in the form of mnemonic practices, *lieux de mémoire*, school textbooks, monuments, national commemoration days, historical museums, commemoration stamps, and so on. Indeed, "stressing certain elements of the national past" is "a powerful device of shaping people's identities and influencing their future action" (Lagerspetz 1999, 24), and "control of a society's memory largely conditions the hierarchy of power" (Connerton 1989, 1). Representations of the past and their incorporation into collective memory are components of cultural hegemony in the sense that they not only express the interests of a dominant social group, but also are accepted "as 'normal reality' or 'commonsense' by those in practice subordinated to it" (R. Williams 1976, 145).

Representations of the anti-Soviet movement in the Baltic Soviet republics of the 1940s and 1950s, the so-called Forest Brothers, occupy a very prominent position in recent approaches to the Communist past and, by implication, to the post-Communist present. Because "conceptualizations . . . are best developed in the context of specific historical episodes" (Shapiro 2004, 35), the following section presents a general overview of the resistance movement, directed mainly to those readers unfamiliar with the history of the Forest Brothers. Even though the name was also used in the aftermath of the 1905 revolution in Russia and in connection with the resistance to the German occupation of the Baltic states from 1941 to 1944, the collective name "Forest Brothers" primarily designates a group of people in the Baltic Soviet republics who in the second half of the 1940s and the first half of the 1950s retreated into the forest to fight Soviet armed formations. The group consisted of individuals with rather divergent backgrounds and reasons for withdrawing into the woods. There were those who had fought in the German army, people wanting to avoid Soviet draft and deportation, Red Army deserters, people suffering from Soviet land redistribution and farm collectivization, men and women with religious motives (mainly in Lithuania), Baltic national patriots, as well as adventurers and looters, although the latter were seemingly in relatively small numbers because the stakes were too high (Misiunas and Taagepera 1993, 84; Gaskaite-Zemaitiene 1999, 27–28; Laar 1999, 214).

Thousands of men and women withdrew into the woods motivated by, among other things, memories of the Soviet rule in 1940 and 1941; Soviet resumption of the practice of deportation in February 1945; expectations of a military confrontation between the Western states and the Soviet Union after the defeat of Germany, leading to the liberation of the Baltic states; and social structures of support established during the years of independence that made guerrilla warfare appear sustainable (Vardys and Sedaitis 1997, 81–82). They fought Soviet armed formations either actively or passively, offensively or defensively, individually, in small groups or in larger military structures organized as follows: "In 1945, the partisans had organized into divisions and regiments, but later mostly in companies or platoons or assault groups with their staffs and commanders. The partisans had their written regulations, battle orders for divisions, regiments and companies, their pseudonyms and sometimes identity cards or other means of recognition. The partisans gave and signed solemn oaths, organized and carried out their own system of information and signaling" (Strods 1997, 156).

The Lithuanian partisan movement strove for countrywide organization and command. From 1945 to 1947, it was organized into seven and then nine regions, only two of which integrated themselves successfully under a single command, although a common underground command structure was discussed in 1947 (Vardys and Sedaitis 1997, 82; Gaskaite-Zemaitiene 1999, 32). They were later fused into three military regions, and between

1946 and 1951 a central organization was restored from the bottom up (Mis-iunas and Taagepera 1993, 87). Establishing and maintaining contacts with the West as well as keeping the urban intelligence and the population at large informed by means of leaflets and even printed journals were the par-tisans' other aims at that time. In Latvia, the formation of four partisan divi-sions led to temporary control over several districts. In general, however, the partisans were lacking coordination and common leadership. With the struggle's advancing duration, sources of supply, shelter, and information became scarce (Strods 1999, 154–55). The partisan warfare in Estonia—like its Latvian and Lithuanian counterparts afraid of Soviet infiltration (and rightly so)—seems to have been more individual. After the destruction in late 1944 of the partisans' organizational core, the movement was lacking in-teraction and structure. It operated primarily in independent small groups, although some organizational framework was established and existed until early 1949 (Taagepera 1993, 80; Laar 1999, 222–29).

The resistance is said to have hampered the Sovietization of the coun-tryside and helped prevent Lithuania from being Sovietized to the extent ex-perienced by Estonia and Latvia (Lieven 1994, 95; Dreifelds 1996, 43). The movements in all three republics are often referred to as an embodiment of the willingness to resist Soviet occupation and to strive for the reestablish-ment of independent states. Militarily, however, the movements failed; "none of the hopes of the fighting countries were fulfilled" at that time (Kuodyte 1999, 82).[3] In fact, the movements were suicide commands: "Life in the forest was dangerous—most perished after an average of only two years" (Vardys and Sedaitis 1997, 81). Armed insurgency provoked strong and merciless countermeasures, including the involvement of "fake guerril-las," special provocation groups, and assault agents. "Victory over the guer-rilla forces became the first priority of the reestablished Soviet government"

3. Here, the emphasis is on "at that time" and on "militarily." The reindependence of the Baltic states in 1991 indeed followed from principles directly opposite to those applied by the resistance movements of the 1940s and 1950s—namely, nonviolence and dedication to peaceful change. It may be argued, however, that the resistance movements succeeded in the long term in the sense that they became an important identity provider and legitimacy provider with re-gard to the claim to independence. Furthermore, the fact that these movements occupy a prominent position today within the body of scholarly and popular literature on the history of the Baltic nations may be seen as a retrospective success.

(Vardys and Sedaitis 1997, 81). To this end, the Soviet army and secret police forces, local extermination squads, armed groups of regional or city committee activists, and a network of secret informers performed antiguerrilla actions and extensive "cleaning-up" and combing operations in the forests in 1944 and 1945. Later, the territory of the Baltic republics was divided into districts, antipartisan groups were placed in thousands of locations, and a comprehensive network of secret agents and special armed groups employing special methods was made use of (Strods 1997, 157).

Outnumbered and without foreign assistance, the partisans were bound to fail. However, their military failure at that time does not prevent their war from serving as a positive reference point today. This function can in part be explained in terms of the meaning assigned to the forest, which is an integral component of representations of landscape in general and of the landscape-memory-history nexus in Baltic narratives in particular. Landscapes are culture before they are nature, "constructs of the imagination projected onto wood and rock" (Schama 1995, 61). In Latvia, for example, "certain rocks, lakes and rivers are linked to ideas of national destiny" (Skultans 1998, 49). Landscapes are both material (physical environment) and cognizant (mental constructions resulting from human observation and reflection), but the borders between the real and the imagined, the material and the constructed, are blurred in forest wars: the enemies may be hidden invisibly behind every tree; they are virtually everywhere. Even if the enemies are not physically present de facto, they *might* be there; hence they *are* there: they are present even if factually absent; there is no escape from their virtual or factual omnipresence. Moreover, especially in winter, the forests—idyllized by Polish romantic poet and playwright Adam Mickiewicz as, in Simon Schama's words, "naturally fortified shelters, where the Polish-Lithuanian nation had begun and to which, harried on all sides, it would finally retreat"—become "a forest prison" (1995, 61). Whereas American revolutionaries in the 1790s made sure that the trees chosen to represent the revolution symbolically "should cast no tyrannical shade; daylight emphasized the natural equality of those who encircled them" (Daniels 1988, 57), in forest wars it is precisely the tyrannical shade that determines the natural equality of those people involved.

The Baltic forests have for centuries been battlefields. These battles, in turn, have provided "important traditions" (Zaccor 1995a, 14) and identities built on military traditions. This is not to suggest that military traditions are the only or even the most important ingredient of the national identity in the Baltic states. Neither is it meant to indicate that the Baltic states are the only states that incorporate military traditions in the respective national identity. All the same, the meaning assigned to the military over time as an identity provider should not be ignored, either in the Baltic states or elsewhere. For example, the Estonian style of partisan activity after World War II has been compared with the "Estonian and Lithuanian styles of resistance to Germans in the 1200s" (Taagepera 1993, 80). Lithuanian post–World War II partisans, by refusing to cut their hair "until Lithuania was free again" (Lieven 1994, 88), visually resembled their medieval predecessors. The current military reportedly refer to the military traditions of the fifteenth- and sixteenth-century Lithuanian empire as "a source of military pride and tradition" (Zaccor 1994, 203). Here, we encounter one of the most powerful self-interpretations of communities: "the images of themselves as continuously existing" (Connerton 1989, 12).

The Baltic forests have also for a long time been places of refuge and shelter. For example, the aspirations of the Russian imperial government under Czar Nicholas II to Russify the Baltic provinces completely and the punitive military measures undertaken in 1905 in reaction to unrest and strikes, the destruction of property, and the violence against landowners made many Latvian elementary schoolteachers seek refuge in the forests. The name "Forest Brothers" in fact goes back to 1905. During and after World War II, it was used in all three Baltic languages to designate the partisan movement. As Skultans writes with respect to Latvia, "forests provided a physical refuge for many in the post-war decade. Their importance for personal memory, however, lies as much in their contribution to moral as to physical survival. Stories of forest life provide a moral structure to lives and enable others to achieve a sense of agency" (1998, 83).

How, then, is the forest war depicted in individual narratives and memories of the anti-Soviet resistance? To begin with, the memories even within such a relatively small and seemingly coherent group as the resistance movement are ambivalent and contradictory, reflecting that it is perhaps more appropriate to speak about resistance *movements*, in the plural.[4] A unit-

4. Connerton reports on different groups' memories of the Great War and concludes that "it is possible to imagine that the members of two quite different groups may participate in the same event, even so catastrophic and all-engulfing an event as a great war, but still these two

ing factor, however, is that the narratives of memories describe what is often perceived as indescribable. Forest experience exceeds both human imagination and language. What has often been observed with respect to Holocaust survivors is also true of forest survivors: they "often acknowledge the poverty of speech, how words echo hollowly against their remembered pasts" (Linden 1993, 18). Resort to ready-made images, often derived from literature, is frequently required to verbalize such experience. In fact, individual memory is never pure, but always mediated through a "prefabricated discourse," including standardized perceptions of landscapes (rocks, rivers, forests) in historical narratives, primary school textbooks, and popular folk songs (Skultans 1998, 48–49).

In her selection of accounts of forest life in Latvia based on more than fifty letters received from and more than one hundred interviews conducted with people suffering from neurasthenia, Vieda Skultans identifies two interrelated themes—namely, "pastoral and adventure themes" and "the bleak recollection of uncertainty, fear and famine" (1998, 84). The former theme reflects the transfiguration and idyllization of the forest life in school textbooks as well as the mythological superelevation of it in nationalistic-romantic literature, whereas the latter theme captures the harsh realities of the forest life. As a result, "narratives of forest life blend personal memories of hardship with literary images of the forest" (84). Personal memories are adapted to and made bearable through romanticized literary treatments of the forest. Contrariwise, the city, in particular Riga, serves as a negative reference point. Riga is associated with the central prison and with militia men causing fear through their visibility. The Forest Brothers thus replaced what they perceived as the "insecurity of civilian life" (Misiunas and Taagepera 1993, 84) with what they considered to be the security of and in the forest.

Memories of the forest, however, are ambivalent. One former Forest Sister remembers never having been afraid in the forest (Skultans 1998, 87), but another Forest Sister, having been in the same group of partisans as the first one, has a quite different recollection: "You know, people lived in continual fear, the enemy was all around. As we read in the newspapers now, it had to be a terrible life where the enemy is behind every bush. All the time you

---

groups may be to such a degree incommensurable that their subsequent memories of that event, the memories they pass on to their children, can scarcely be said to refer to the 'same' event" (1989, 20).

have to be so tense" (87). "How shall I put it to you?" she asks. "On the nervous system, you feel that someone is following you the whole time" (87). Two Forest Brothers add: "There was a bitterness felt by everyone, particularly during that autumn when everyone was living in hope alone. There was enemy all around, there was no hope" (87). "There is no front line here," they say. "The enemy can be all around" (87).

Even worse, it is not in *some* forest, but in *our* forest, in this case the Latvian forest, that fear is experienced (Skultans 1998, 88). This fear is seen as profoundly disturbing and incompatible with the notion of the forest as a shelter, as home, as being to some extent more humane than human beings themselves. Besides fear, randomness—in particular "the arbitrariness of a system which claims to be logical" (104)—and betrayal are prominent motives among the forest war memories collected by Skultans: betrayal on the part of the Soviet forces; betrayal by some fellow Latvians who refused to assist the partisans and even informed on them to the Soviet authorities; betrayal of hope by the British and other Western powers who refused to come militarily to the aid of Latvia. At the end of the partisan war, the U.S. government even considered promising to engage on the side of the Soviet Union should Germany attack (Trachtenberg 1998, 137). Here again, recollections are ambivalent and emphasize at the same time both the generosity of local people and the potential danger they posed to the partisans. Childhood memories of order and calmness are confronted with moral chaos; "the beauty and goodness of nature [linked with] the innate goodness and wisdom of childhood stands in contrast to the adult world of lies and treachery" (Skultans 1998, 93); the protectiveness of the forests is juxtaposed with the harsh realities of forest life; the "spiritual unity with the forest and its animals" is opposed to human unreliability (97). Encounters with human beings in the forests were indeed ambivalent: trust and suspicion, reliability and untrustworthiness, solidarity and betrayal were part of the forest life and are part of forest memories, impervious to simplification, objectification, and simple truths. Likewise, the reasons for withdrawing into the woods—and the reasons for not doing so—were as manifold as the resisters' objectives.

How the past relates to the present depends essentially on acts of representation. Here, it is especially important whether the past is represented in terms of historicity and truths (in the plural) or in terms of actuality and the one single, nonnegotiable, and obligatory truth. In light of what has been said here about representations of the past as the ingredients of newly independent states' nationalizing strategies, the latter form of representation can be expected, wherein a collective memory is appealed to and constructed. Like nationalizing strategies, collective memory

> simplifies; sees events from a single, committed perspective; is impatient with ambiguities of any kind; reduces events to mythic archetypes. Historical consciousness, by its nature, focuses on the *historicity* of events—that they took place then and not now, that they grew out of circumstances different from those that now obtain. Memory, by contrast, has no sense of the passage of time; it denies the "pastness" of its objects and insists on their continuing presence. Typically a collective memory, at least a significant collective memory, is understood to express some eternal or essential truth about the group—usually tragic. A memory, once established, comes to define that eternal truth, and, along with it, an eternal identity, for the members of the group. (Novick 2001, 4, emphasis in original)

Thus, nationalizing strategies and collective memory often join hands. Both have a problematical relationship to historical consciousness. Whereas history is at all times the problematical, incomplete, and relative reconstruction of that which is not anymore, collective memory is an actual phenomenon, is shaped in the light of present concerns, knows only the absolute, is capable of all kinds of omissions, cuts, and projections, and is interested only in that with which it can be confirmed (Nora 1998, 13–14). Representations of the past as components of nationalizing strategies therefore usually appeal to memory rather than to history.

The preceding discussion offers some avenues for future research, links with the overall questions of the book, and implications for friendship patterns. A good starting point for discussion of these elements is the interpretation of the resistance movement in official representations. It is important to note here that the movement is not depicted as something past. Rather, it is linked to the current security policies in the sense that it serves as a legitimacy provider for these very policies. This function is not surprising because what animates appeals to the past is, among other things, "uncertainty about whether the past really is past, over and concluded, or whether it continues, albeit in different forms, perhaps" (Said 1994, 1). This uncertainty and the need to "base our particular experiences on the prior context in order to ensure that they are intelligible at all" (Connerton 1989, 6) are reflected, for example, in chapter 7 of Lithuania's Basics of National Security. Here, the Lithuanian security policy is said to be based on, among other things, "the Nation's experience of a decade-long, postwar partisan struggle effort against the troops and the occupational regime of the Soviet Union." Likewise, the voluntary territorial defense forces in Lithuania reportedly prepare themselves "to carry on the kind of warfare waged by their spiritual predecessors, the rebels who fought Soviet troops from 1944 [to 1953]" (Zaccor 1994, 211). The commander of the Estonian defense forces is said to have explained in September 2000 the priority given to territorial defense by referring to, among other things, Estonia's guerrilla warfare in the 1940s and 1950s (Huang 2001, 30). By stating that defense shall be prepared for on the basis of the partisan struggle (and, in the case of Lithuania, on the basis of the partisan struggle against the occupation regime of the Soviet Union, not against the occupation regime of Germany between 1941 and 1944), techniques of future warfare are explicitly built upon those techniques of past warfare, and past enemies are implicitly represented as (at least potential) future enemies.

The past, however, has often been a rather unreliable guide to the present. "Often the actor would perceive more accurately had he not undergone the earlier experience" (Jervis 1976, 220). Increase in knowledge on past events does not necessarily improve the quality of decision making. Governments are inclined to prepare for the previous war rather than for the next war, and only if the next war resembles the previous one will learning from the previous war improve decision making. Likewise, while constructing peace treaties, decision makers are often sufficiently aware of the sources of previous wars, but they frequently fail to register "important new issues already appearing on the diplomatic horizon" (Holsti 1991b, 344). Turning Santayana's maxim on its head, Jervis pointedly states that "those who remember the past are condemned to make the opposite mistakes" (1976, 275). As Jervis acknowledges, this assessment is an overgeneralization, and a more appropriate observation seems to be Tzvetan Todorov's (1982) message that even if we know the past, we still do not know what to do today and tomorrow. To this can be added what Pierre Nora calls the acceleration of history, resulting in a profound "uncertain[ty] as to what form the future will take" (2002).[5] This uncertainty concerns decision makers and society alike. It results in what may be called a crisis of memory in the sense that, first, the policies built on and appealing to memory appear to be as unreliable as memory itself. Second, in a period of accelerated history, memory seems to be an even poorer guide to present and future policies than it usually is. Third, policies based on memory are likely to focus on those aspects that resemble the past rather than on those aspects that deviate from the past. In other words, they tend to prioritize continuity rather than change. This resistance to change supports the intrinsically conservative character of security policy, which tends to respond to change only slowly.

If the past can be said to be only a poor guide to present and future security policies, it can nevertheless function as a powerful legitimacy provider for these policies. The historical armed resistance movements of the 1940s and 1950s failed militarily, whereas the popular movements of the 1980s succeeded in reestablishing independence in 1991 by completely nonviolent means. As shown in earlier chapters, however, directly after independence, the Baltic states replaced nonviolent and civilian security conceptions with military security strategies. The acceptance of this change by the population, which was very critical of the military at that time, may have been facilitated by the official celebration of the earlier armed resistance movements as an integral element of Baltic history and by the narrative integration of the movements' endorsement into representations of this history. This history, in turn, is said to be a means that "sustains the Lithuanian people in their daily work and creative endeavors" (Adamkus 1999, 4). It seems indeed to be insufficient to explain references to the violent past in recent national policy documents exclusively in terms of technical-operational issues. In addition, these references are to be seen as a legitimacy provider for the basic lines of thought underlying the current security policies—that is, the buildup of military capabilities and military integration.

Collective memory is said to depend on repetitive acts of communication between individuals, either directly or indirectly as transmitted

---

5. "Acceleration of history" means that change is the most permanent feature of the modern world. As a result, we can no longer rely on the past to prepare, order, and organize the present and the future: we "are utterly uncertain as to what form the future will take. . . . We do not know what our descendants will need to know about ourselves in order to understand their own lives" (Nora 2002).

through institutions (Connerton 1989, 39). The substance of these acts of communication is frequently nonnegotiable. It is ritualistic in that it "does not employ forms of communication which have propositional force," and the language used is both "performative" and "formalized," allowing neither variation nor dissent (Connerton 1989, 57–58). Thus, in order to evaluate the meaning of the Forest Brothers for the constitution of national communities in the Baltic states in the 1990s, the acts and the substance of communications referring to the resistance movements have to be studied thoroughly.

In Lithuania, for example, the Forest Brothers' fight is depicted as the most important way of approaching the Communist past. One of the first measures undertaken by the new conservative majority in the Lithuanian Parliament after the elections of 1996 was to acknowledge officially the partisans' military ranks—an important pecuniary measure, but an equally important symbolic means to underline the legitimacy of the resistance and to strengthen its position within the national community. Numerous television broadcasts, articles in journals, and autobiographies have been dedicated to the Forest Brothers, thus keeping them alive in public memory. One academic journal exclusively publishes documents and articles on the Lithuanian resistance (Tauber 1997, 15).

Expressed in the form of a tragic story, the Forest Brothers became fixed in a stereotype that was memorialized in local monuments all over Lithuania and, to a lesser extent, in Estonia and Latvia. The stereotypization of historical experiences is, however, problematical because a memory fixed in a stereotype "install[s] itself in the place of the raw memory and grow[s] at its expense" (Levi 1988, 24). It is important to note that this characteristic "is not a peculiarity of Holocaust survivors' memories [such as Levi's]; it is in the nature of memory" (Novick 2001, 351). In addition, the transmission of knowledge between individuals across generations is an important issue. Vieda Skultans's work on memory of the forest life (1998, 83–101) is an important starting point. A next step would be an investigation into how subsequent generations remember, based on the narratives delivered by their parents and grandparents, the forest war that they themselves have not experienced.

As stated earlier, memories of war and narratives of war memories are important glue that holds communities, national and otherwise, together. It is an identity provider, defining the in-group and distinguishing it from outgroups. The Manichaean element of the partisan war—it was either yes or no, life or death, friend or foe, "no one must stay neutral" (Misiunas and Taagepera 1993, 89)—historically had consequences for those people who wished to stay neutral and were thus at least suspected of, if not liquidated for, siding with the enemy. It may be expected, then, that memories of the partisan war help keep alive both a Manichaean worldview and memories narratively constructed in terms of "we" versus "them." Thus, the identity of the in-group is reaffirmed, but the boundaries against the surrounding social world are reaffirmed as well (Coser 1956, 38). From the in-group, constructed around the memory of the partisan war, are excluded those people who remember the war in a different way or not at all, such as those who settled in the Baltic republics only after the epoch of the partisan war. This cleavage may block communication between the in-group and out-groups, pose barriers, and inhibit the development of mutual understanding and trust. At the same time, it may foster and solidify the respective worldviews in both groups.

Without communication, however, "[s]elf-perpetuating vicious cycles in conflict" (M. Deutsch 1987, 40) may follow. Others are no longer persons with whom one disagrees on a specific issue; rather, they become ontological Others. Psychologists thus emphasize that memories associated with past anxieties are inclined to result in cognitive closures in the sense that people lose both their potential for thinking in alternatives and their ability to perceive conflict as a constructive experience (1987, 40). This argument is not just theoretical, but also has diverse practical consequences in the Baltic states' situation, six of which I briefly mention here. First, in an essay contest for youth organized by a right-wing Latvian publishing house in the spring and summer of 2001, the tradition of the Forest Brothers was invoked in some essays (Rislakki 2001, C2). The book that resulted from the contest (*Nevienam Mes Latviju Nedodam* [We Do Not Give Latvia to Anyone])—"a pretty boring read . . . filled with sentimental nationalist poetry, inane comparisons of the Soviet Union to the EU, and suggestions of how best to promote the 'voluntary repatriation' of the Russian minority to Russia" (Muiznieks 2002, 88)—was harshly condemned by Russia, and both the Council of Europe and the OSCE investigated, prodded by Russia.[6] Second, in Lithuania, the 1993 mutiny of some Voluntary Service of National De-

---

6. The book was condemned by both the prime minister and the president (Muiznieks 2002).

fense members found its expression in their abandonment of their posts and withdrawal into the woods (S. Girnius 1993, 44–47). Third, "retreating into the woods," in addition to its historical connotations, came to be seen as synonymous with a general disapproval of the political circumstances (Kerner 1995, 25). Fourth, several incidents involving the Estonian home guard in the early 1990s were openly justified as being a continuation of the armed resistance of the 1940s and 1950s against the Soviet occupation of Estonia.[7] Fifth, issues of remembering and forgetting continue to "shap[e] the strategic alignments of the future" (Gong 2001, 45). Finally, cultivating the Forest "Brothers" has a political effect relating to gender as well. Subsuming female resistance under the gendered designation "Forest 'Brothers' " tends to ignore the existence of Forest "Sisters" and helps to ascribe an inferior status to resistance by women, quite regardless of whether it is a self-designation or a name imposed by outsiders. It contributes to the maintenance of dominance by men and the marginalization of women in the sphere of security.

Conflict "festers underneath the surface and has many indirect effects" (M. Deutsch 1987, 39). If temporarily suppressed, it is likely to reappear or be displaced onto other issues, thus destabilizing the relationship between groups even more. In the Baltic societies, there still coexist those people who withdrew into the woods and those people who made them withdraw (and their descendants); those people who lost their lands through confiscation while retreating into the forest and those people to whom the confiscated land was given; those people who depict the antipartisan war as a legitimate fight against "bandits" and "fascists" and those people who see the partisan war as a legitimate fight against "occupants" and "colonists." This coexistence has not resulted in open conflict, but it has made communication between different groups of people difficult because each adheres to its specific interpretation of the past.

Conceiving of memory in a nondominative way, however—that is, as a space where multiple norms, values, and identities meet and potentially interact with one another in a positive manner—challenges the very nature of memory. It requires the willingness both to historicize memories and to learn to live with difference, including different views on the past. In Nick Couldry's assessment, "each person carries with them an individual history

7. See *Estonian Review* 1992.

of reflection which cannot be reduced to shared cultural patterns. Partly pure accident, and partly structured, this history is a trace of that person's perceiving, absorbing, interacting, reflecting, retelling, reflecting again, and so on, a sequence endured by that person alone" (2000, 51). A nondominative memory would need to acknowledge each person's individual bag of experience and to accept the simultaneous existence of different memories or, in other words, of different ways of seeing "a neutral set of historical events" (Schöpflin 1999, 11), which are in themselves equally "true." While accepting the basic assumptions upon which the relationship is founded, different groups of people may use conflicts over the past as a means with which to resolve tensions and, by so doing, to stabilize and integrate their relationship with one another (Coser 1956, 80).

What, finally, do the discussion of the Forest Brothers here and the analysis presented in the preceding chapter tell us about Baltic-Russian relations in light of the concept of nonfriendship and prefriendship? As noted earlier, during the 1990s the representations of Russia in the Baltic states were unfolding in terms of "security," but the issue was frequently one of constructing nation-states and legitimizing the modes of domination built upon them. Boris Yeltsin's engagement in January 1991 in support of the Baltic popular movements' course of independence could have been the starting point for the development of good neighborly relations and eventually friendship between the Baltic states and the Russian Federation. For a variety of reasons, it was not. Rather, role positions—for Estonia, Latvia, and Lithuania, on the one hand, and Russia, on the other—stagnated at the level of nonfriendship at best; the role structure was one of a preparatory stage for a Deutschian anarchy. Mutual representations on the basis of dependable expectations of peaceful change could not be observed. The absence of armed conflict was not translated into expectations of peaceful change; nonfriendship did not evolve into prefriendship; mutual representations continued to unfold along a security "logic" that defined others as threats to one's own security, independence, and eventually survival. In Baltic representations of Russia, unpredictability and instability replaced predictable malevolence and hostility, but both characterizations were still equated with a threat to the Baltic states.

The representation of Russia as hostile was facilitated by various (not only speech) acts by leading Russian politicians, but it ignored at least four things. First, almost no anti-Baltic statement was translated into an enduring anti-Baltic policy; anti-Baltic policies were usually of short duration, applied in a rather unsystematic manner. Second, many anti-Baltic statements were directed at a domestic Russian audience, but regularly failed to attract the electorate (with Vladimir Zhirinovsky being an important exception, but even in his case it may be argued that he was not elected primarily because of his anti-Baltic statements). Third, Russia's de facto policies toward the Baltic states were reactive rather than active and predictable rather than unpredictable (in language only did these policies exhibit several ingredients of rhetorical rivalry and even enmity). From 1997 onward, Russia's security policy in the Baltic Sea region was constructive rather than destructive, innovative rather than conservative.

Finally, whereas one of the main objectives of the Russian political leadership during the 1990s was locating Russia within "Europe," some Baltic political decision makers seem to have been convinced of the futility of such an endeavor from the beginning. They took a future division in Europe for granted and tried to locate the Baltic states on the Western side of the taken-for-granted future divide. For some Baltic leaders, trying to prevent the new line of division from materializing seems to have been a question of only minor importance. Rather, they were occupied with cultivating the image of Russia as the Other, the otherness of which they then used to consolidate the collective self. Representing Russia in terms of peaceful change could only be seen as obstructive to strengthening collective subjectivity. Conflict with Russia was cultivated in order to contribute to "the establishment and reaffirmation of the identity of the group and [maintain] its boundaries against the surrounding world" (Coser 1956, 38). As a result, the Baltic nation-states were constructed in part at the price of alienation from Russia.

Chapters 11 and 12, as well as earlier chapters, explored the preconditions for and consequences of using Russia as a negative reference point. Cultivating Russia as the Other was built on patterns of argumentation referring to geography. Geography was elevated to a natural law to which political decision makers could only respond as they must. It thus was used to locate political decision making outside the sphere of open and democratic discussion, to pretend the alternativelessness of decisions, and to prevent the democratization of foreign and national policy decision making. Thus, the necessarily social construction of meaning assigned to geography was ignored.

Applying to the Baltic states an interpretation of war as contributing to the consolidation of the self also helped shed new light on both the participation of military contingents of the Baltic states in the NATO-led military intervention in Kosovo and the redesignation of the Baltic Peacekeeping Battalion as the Baltic Battalion. Cultivating the Other also required specific representations of the Baltic-Soviet and the Baltic-Russian histories, dichotomizing the Baltic states and the Russian Federation as well as equating Russia with the Soviet Union. These representations integrated collective memories into a nonnegotiable set of officially approved historical "facts." This process was an indispensable precondition for strengthening imagined collective subjectivity, but it also served as a powerful mode of domination by excluding alternatives and those people adhering to them from the collective self. It neglected the collective self's lack of coherence as well as the internal and external Other's not-so-otherness. In fact, it had to ignore the latter because it perceived the establishment and reaffirmation of the in-group's identity as being dependent on the maintenance of boundaries against the surrounding world. And the issue was indeed one of boundaries in the sense of definite lines of separation. It was thus the opposite of "partial identification" with one another, seen by Deutsch and his team (1957, 36) as a precondition for a sense of community, which in turn is a precondition for security communities.

Both words, *partial* and *identification,* are equally important here: *identification* because it neglects the notion of a complete separateness between "us" and "them"; *partial* because it emphasizes that each person carries with him or her a bag of experience that is an individual property and cannot be reduced to patterns shared and common with others. The use and construction of a nonnegotiable collective memory perpetuated the Self-Other contrast within the Baltic states as well as between them and the Russian Federation, instead of defining "us" and "them" in terms of different degrees of we-ness and instead of identifying "them" as "less us" and "us" as "less them."[8] It inevitably led to representations of Russia that were in fact a parody of Russia (as indicated in the epigraph at the beginning of this book).

It has also been argued that the interpretation of both the Soviet occupation of the Baltic states until 1991 and the years following reindependence as a kind of war situation—the former as an extension of World War II, the lat-

---

8. According to Ole Wæver, the EU defines various neighbors not as "anti-Europe," but as "less Europe," thus breaking with the habit of Otherization (1998, 100). In Baltic-Russian relations, this habit still prevails.

ter as a substitute war or an imagined state of siege—contributed to the lack of differentiation between the Soviet Union and the Russian Federation. This interpretation also resulted in a strong conformity pressure, narrowing the tolerated scope of individual departures from the official way to see and represent things. Refusing to accept the incoherence, disorder, and otherness within the individual and collective self, however, ultimately leads to representing the nonimitative—internal as well as external—Other as a threat (Shapiro 1997, 72). Cultivating imagined war situations and seeking to exploit them in order to increase internal cohesion furthermore prevented conflict over the past within the collective self and fostered the modes of domination built upon a specific reading of the past. Struggles between the collective self and the Other over the past, however, were promoted because they had the same effects—imagining internal order and strengthening power relations within the self—but inevitably resulted in seeing the Other as a threat to the imagined domestic order and, ultimately, to survival.

Finally, cultivating the internal and external Other consolidated—and arguably in part created—tensions along ethnic lines in Estonia and Latvia by discriminating, at least rhetorically, against the Russian speakers as internal enemies. By intensifying tensions along ethnic lines, some Baltic political decision makers participated in a cycle of mutually reinforcing allegations and speech acts that then could be interpreted and represented as a "confirmation" of the preformulated views. Rhetorical escalation led to a strengthening of prejudgments rather than to their interrogation and reworking. By equating criticism of effective politics with a threat to national security, some decision makers aimed both to divert attention away from the deficiencies of domestic politics and to silence the domestic opposition and the public, the political passivity of which may be explained in terms of self-censorship in order not to undermine domestic stability.

As Trevor Barnes and Derek Gregory have noted, "words have the ability not only to represent but also to create worlds, to offer possibilities, to produce action" (1997, 4). Words, however, also have the ability to prevent all this. Although chapter 10 shows that words have indeed triggered political action, with respect to Baltic-Russian relations they have rather had the opposite effect: they helped prevent the development of bilateral relations based on good neighborliness and froze them in nonfriendship. In the case of Baltic-Russian relations, language has been used to cultivate otherness. Indeed, words rather than actions are indicative of the mutual cultivation of otherness. All this exchange is advertised in terms of "security," but references to "security" with its high-policy aura of urgency, importance, and survival have served mainly as a pretext with which to disguise other aims—namely, the solidification of the nation-states, the modes of domination built upon them, and the unexamined, taken-for-granted acceptance of both. It furthermore has had a function in the domestic power struggle, albeit arguably a diminishing one.

Representing bilateral relations in terms of (a threat to) "security" and equating security with the military may be seen as the conditions of impossibility for the development of role identities based on prefriendship, not to mention friendship. From the point of view of the constructors of the nation-states, the development of (pre)friendship among the Baltic states and Russia was hardly desirable during the 1990s: Others cannot be friends or even prefriends. Others, however, were needed to help construct the nation-states by serving as the negative mirror image. Thus, as long as the nation-states were seen to be on shaky ground or as long as the representation of the nation-state as fragile was considered a useful tool with which to legitimize specific modes of domination and to realize specific interests or both, the development of prefriendship role positions was as unlikely as were mutual representations in terms of prefriendship. Further, focusing on security and seeing security in terms of independence and survival prevented alternative views on security from entering the political agenda in more than rhetorical fashion. To be sure, security documents defined security increasingly in a comprehensive way (indeed, in too comprehensive a manner, which was characterized in chapter 8 as nonmanageable). The sincerity of this redefinition, however, may be called into question because it was not reflected in the representation of Russia or in the security-political consequences derived from it. It may thus be speculated at this point that redefining security in the Baltic states in light of a comprehensive reading of security was primarily a rhetorical adaptation to NATO's self-representation throughout the 1990s as a political rather than a military organization. Be that as it may, a comprehensive understanding of security was granted admission to documents and speeches. This change indicates the transition from a Lockean to a Deutschian kind of anarchy because the possibility of occasional warfare during a Lockean kind of anarchy makes security elites equate security with the military, prevents alternative views from entering the stage, and monopolizes the security business by excluding nonstate and societal actors. In the next chapter, I analyze to what extent the current redefinition of security is only a surface adaptation—in other words, how much of the effective security policy is still dominated by a military reading of security.

In the light of Baltic decision makers' negative attitude toward Russia and Russians, it is surprising that the Latvian awarding committee decided in February 2000 to grant former Russian president Boris Yeltsin the highest state award, the Order of the Three Stars, "for his role in supporting the restoration of Latvia's independence." It is not surprising, however, that Yeltsin refused the award.[9]

9. *RFE/RL Baltic States Report* 2000a and 2000b.

13

Security Options

IN THIS CHAPTER, *security options* refers to the paths to security effectively chosen by the decision makers in the Baltic states. It is not meant to naturalize these policies: security policy is always a choice between different alternatives and includes those options that the decision makers decided not to prioritize. Rather than discussing the financial, operational, and technical aspects of Baltic security policy, this chapter searches for the patterns of argumentation on the basis of which the key decisions regarding security policy were made in the early and mid-1990s: to build up national armed forces and to apply for membership in NATO. Thus, the chapter discusses the patterns of argumentation underlying the basic security political decisions rather than the technical and operational aspects of security policy.

The chapter sketches the evolution of the Baltic states' security policy in the 1990s without retrospectively rationalizing these policies: with respect to NATO membership, it was not a linear process from application to negotiations to membership. Likewise, the buildup of national armed forces was a complicated and ambivalent process rather than a smooth and undisturbed process. In the first part, the chapter discusses the buildup of armed forces in the Baltic states; in the second, it examines the issue of Baltic NATO membership. One of the most humiliating and violent aspects of the Soviet mode of domination, the *dedovshchina* system, serves as the chapter's starting point.

In soldiers' parlance, *dedovshchina* meant a specific mode of rule and subordination in the Soviet army—namely, the almost unlimited use of violence exerted by second-year service conscripts ("grandpas," "elders") on first-year conscripts ("youngsters," "sonnies"). Regardless of the military leadership's condemnation of it, *dedovshchina* remained a constant manifestation of violence within the Soviet armed forces. It aimed to destroy individual moral standards and to socialize the recruits according to the norms and values prevalent in the armed forces (subordination, obedience, discipline, acceptance of authority and hierarchy, and so on). This system of socialization assumed alarming proportions, especially in the 1980s. According to independent experts, from 1986 to 1991 physical deformation, suicides, desertion, and mental illness resulting from *dedovshchina* matched the officially disclosed figures of Soviet casualties during the ten years of war in Afghanistan. Within the armed forces, it was difficult to escape from the system: once recruits became "elders," they were committed to behave according to the role model. Former victims thus became perpetrators, thereby continuing the process of resocialization from a civilian to a perverted military code of behavior (Lewada 1993, 128–36).

As a result of universal conscription in the USSR, *dedovshchina* concerned millions of young men, year after year. During the perestroika years, it became a main issue of criticism. Public campaigns by, for example, the Committee of Soldiers' Mothers or in Latvia the League of Women (Karklins 1994, 71–74) contributed to the armed forces' loss of reputation, which was already considerable as a result of, among other things, the devastating performance in Afghanistan. An antiharassment mass movement, triggered by a July 1988 article in *Komsomolskaja Pravda* on cases of extreme harassment in the armed forces, started in the Baltic republics. The recruits had to serve outside of their home republic, and *dedovshchina* was increasingly being replaced by *gruppovshchina*, "hazing based on ethnicity and national affiliation" (S. Meyer 1991–92, 22). *Gruppovshchina* resulted, among other things, from the increasing activities of the draftees in nationalist political organizations. Draftees from the Baltic republics were massively engaged in these organizations and therefore became an object of *gruppovshchina*. As a consequence, Baltic dissidents demanded, among other things, that recruits from the Baltic republics should complete their military service in their home republics. The Soviet military leadership rejected this claim, arguing that the present system had helped defeat Germany in World War II and that other systems were bound to fail.

In the late 1980s, the Baltic popular movements pushed their demands further and called for a demilitarization of the whole Scandinavian-Baltic area (Haab 1994, 154). These calls occasionally included ideas on east-west transborder disarmament. For example, the call for abolishing the military airfield in Tartu, Estonia, was accompanied by a proposal to abolish a similarly located west European airfield. Some proposals, however, were aimed primarily at Soviet rather than overall demilitarization. Some leaders of the Estonian Popular Front in October 1988 demanded the recall of the three hundred Estonian officers serving in the Soviet armed forces and the revival of the Estonian Rifles Corps (Clemens 1991, 100–101). Both the growing number of deserters from the Soviet army, encouraged by the nationalist political elites in the republics, and competitive legislation on conscription passed by the republics' Supreme Soviets increasingly challenged the Soviet Union's claim to rightful rule in the Baltic republics and as such were directed at the core of the Soviet power structure. According to the 1949 Geneva Convention, citizens of an occupied country may not be forced to serve in the occupation army. For the Soviet Union, acknowledging the legitimacy of the claim not to serve in the Soviet army therefore would have meant subscribing to the reading of occupation rather than voluntary accession of the Baltic states to the Soviet Union. In chapter 5, I described both Soviet reactions and the Baltic popular movements' position on the creation of national armed forces and need not repeat that discussion here.

Many observers have noted that after independence the Baltic armed forces had to be built up from scratch (once the decision had been made to build up armed forces in the first place). Many of them took the creation of national armed forces for granted, because of either their operational or their symbolic function.[1] Only very few seem to have opted for what has been called the "Icelandic solution"—that is, no national armed forces, but international security guarantees (Vitkus 1997, 69). Even this alternative, however, was not detached from equating security with the military because the issue was one of military security guarantees. Likewise, the option of neutrality was discussed only in the form of armed neutrality and rejected on historical and financial grounds (Vares 1993, 20; Nekrasas 1996, 69; Miniotaite 1998, 187; Ozolina 1998, 158; Vike-Freiberga 2000b). Conceptions demanding the demilitarization of the Baltic-Scandinavian area also soon disappeared from the agenda.

Why national armed forces? Throughout the 1990s, the question "Do Es-

---

1. Some authors, however, have called into question the need for armed forces comprising army, navy, and air force. Others have opted for a specialization of the Baltic armed forces in order to complement NATO forces (see Ilves 2000). Such "role specialization . . . is about making virtue out of necessity: scarce financial, technical and human resources have to be channeled towards viable objectives" (Missiroli 2004, 123).

tonia, Latvia, and Lithuania need full armed forces?" was answered in the affirmative for a variety of reasons. First, although the Baltic states would have little to put up against a full-scale military aggression, national armed forces might be able to defend these countries against less serious external threats such as civil unrest spilling over from Russia (Hillingsø and Dalsjö 1999). Second, the existence of regular armed forces might facilitate Baltic integration into Western security structures by providing rather than consuming security. Third, national armed forces, together with other states' armed forces, could "work together for common security" (Linkevicius 1995, 102). Fourth, according to then Latvian minister of foreign affairs Valdis Birkavs, the lack of a defense strategy might provoke others by "displaying weakness and vulnerability" (cited in Simm 1995, 12). Fifth, armed forces could be used to fight internal threats such as political instability, organized crime, or major ecological disasters (Zaccor 1994, 202). It is ironic, however, that even the commander of the Latvian Armed Forces, Juris Dalbins, acknowledged in November 1994 that "at the moment there is no perceived need for an army so it is difficult to persuade the people that Latvia needs armed forces" (cited in Simm 1995, 14).

Moreover, the armed forces were widely seen as one of the symbolic and functional cores of the classical nation-state and, as such, an important ingredient of the reconstructed Baltic nation-states. Furthermore, as Dana Eyre and Mark Suchman have argued, highly technological militaries can to some extent be understood in their symbolic linkage with "modernity, efficacy, and independence." Militaries are symbolically connected with "sovereign status as a nation, with modernization, and with social legitimacy" (1996, 86–87). This suggestion fits neatly into the tie between the modern nation-state and the military, referred to in earlier chapters.

Indeed, it does not necessarily follow from the functional value of weapons systems that they are acquired primarily or even exclusively *because of* their functional value. The armed forces of the Baltic states indeed resembled for several years the general pattern developed by Eyre and Suchman primarily in respect to the so-called Third World: they were armed forces, the "actual military utility [of which was] exceptionally open to question" (1996, 82). This resemblance supports the argument suggested by the authors that modern militaries are at least as much a cultural phenomenon as they are a military phenomenon. Thus, the transformation of Audrius Butkevicius from an advocate of civil resistance and psychological defense to a supporter of conventional armed forces appears to be consistent. This transformation coincided with Lithuania's evolution from a Soviet republic to a sovereign nation-state and with Butkevicius's personal evolution from general director for national defense to minister of national defense. In this latter position, he reportedly represented the creation of conventional armed forces as "simply necessary" (Kerner, Stopinski, and Weiland 1993, 31). Phillip Petersen's observation on this transformation is revealing: "the Lithuanian Defense Ministry was originally called the Society Security Department. Indicative of the Department's commitment to non-violent resistance, its Director General, Audrius Butkevicius, established a Center for the Research of Nonviolent Resistance inside the Department. After independence was achieved and the government's policy switched to *armed* defense of the country's independence, the Center was moved outside the government and renamed as the Center for Nonviolent Action" (1992, 36, emphasis in original).[2]

Eyre and Suchman argue that "the world political and social system builds modern nation-states, which in turn build modern militaries and procure modern weaponry" (1996, 82). In this structural argumentation, the motivation and interests of the people who actually make the decision to build up armed forces somehow disappear behind the logic of action determined by structure. Furthermore, it does not follow from the symbolic value of weapons systems that they are procured primarily or even exclusively *because of* that value.

For example, the Lithuanian minister of national defense Ceslovas Stankevicius stated unmistakably that "we are building strong and credible armed forces for the defense of the country, able to deter a potential aggressor or inflict on him maximum damage in case of invasion" (1999, 80). Accordingly, the Lithuanian government is said to have "purchased anti-tank weapons from Sweden in order to help address a formidable potential tank threat" (Butrimas 1999, 47). Likewise, maintaining stocks of antipersonnel

2. It should be noted, however, that the concept of civilian defense, which had been institutionalized as early as 20 February 1991 by the establishment of the Commission for Psychological Defense and Civil Resistance in the Department of Defense, is said to have been accorded an "important role in nearly all conceptions of national security proposed in 1992–1995" (Miniotaite 1997, 205).

mines and retaining the option to employ them do not serve as a symbol of modernity; they have a clear operational function. As the Lithuanian commander in chief Jonas Kronkaitis explained several months *after* Lithuania had signed the Mine Ban Treaty, the state retained the option to employ them to protect antitank minefields "in case of a serious crisis" (1999, 117). This interpretation violates to some extent Lithuania's aspired self-image as a modern, Western country because antipersonnel mines are increasingly ostracized in the Western states. This latter position was indeed acknowledged by Lithuanian vice minister of foreign affairs Rokas Bernotas in May 1999 when he said that the Mine Ban Treaty "is a victory of humanitarian consideration over military doctrine" (cited in Human Rights Watch 2000, 770),[3] but it was not translated into military thinking at that time. Of course, both interpretations—one focusing on modern weaponry's symbolic value, the other emphasizing its functional value—are not mutually exclusive, but rather mutually supportive in the sense that combined with each other, they powerfully support the buildup of national armed forces and effectively marginalize alternatives.

However, as a result of, among other things, the Soviet army's practice in the Baltic republics, the predictable futility of armed resistance in case of a Russian military aggression, and the widespread reluctance to spend several months in the armed forces when the social and economic fabric of the society was about to collapse and each person had to lay the foundations of his or her livelihood, the armed forces were not popular among considerable parts of the population of the Baltic states throughout the 1990s. Trust in the armed forces was limited, especially in Latvia and Lithuania (Rose 1997, 30), and would-be conscripts strongly resisted entering the armed forces. This resistance manifested itself in high numbers of deserters or young men simply not showing up when inducted into military service. For example, in the autumn of 1992 in Lithuania, 11,517 out of 17,034 draft notices were avoided; the following spring, 18,000 draft notices were issued, and 8,000 draftees avoided service (Vitas 1996, 78). According to the Latvian defense

---

3. Among the Baltic Sea states, Estonia, Finland, Latvia, and Russia have not yet signed the Mine Ban Treaty. The nonsignatories' policy seems to be one of passing the buck to one another—that is, justifying one's own nonsignature by the nonsignature of other Baltic Sea littoral states.

minister, in 1996 approximately 78 percent of potential recruits found reason to avoid the draft.[4] In Estonia, in 1997 only 50 percent of all men ready for conscription are said to have been conscripted in reality, with one-quarter of all conscripts ignoring the call-up (Gießmann 1997–98, 154).

The deputy chief of staff of the Lithuanian Voluntary Service of National Defense described those conscripts who did show up as "conscientious, mentally slow, or possess[ing] no wealthy and influential family or friends." A member of the Lithuanian Parliament's National Security Committee phrased it more succinctly: "Only people with empty pockets serve in the military" (both cited in Vitas 1996, 79–80). The actual realization of the principle of conscription was unexpectedly difficult and required not only persuasion but also executive power and enforcement capabilities. The Baltic enforcement agencies, however, could not make use of the Soviet conscription system, which, as a system based on unionwide and extraterritorial service and dependent on local law-enforcement agencies, had collapsed already in the Soviet Union's final stage, when central military instructions were ignored by republican authorities (Allison 1993, 8).

Like the buildup of the armed forces in general, the realization of a new conscription system based on new legislation took time. Unfortunately, the Baltic national armed forces in their early stages adopted from the Soviet army many of its worst features. They suffered from what a former national security adviser to the president of Latvia has called a "post-Communist mindset," key elements of which included:

> the persistent avoidance of difficult decisions, deliberate short-sightedness in planning; the frantic avoidance of administrative systems, structures and documents that compel accountability and delimit maneuverability; decision-making without regard to the feasibility of implementation; a preference for unnecessary crisis management; the creation of an illusion of being "in control of the situation"; a chronic avoidance of coordination and the exchange of information; limited understanding of the value of public relations; administrative rigidity at the cost of results; process-oriented rather than result-oriented management; in short, collective irresponsibility. (Zalkalns 2002, 37–38)

---

4. *Baltic Times* 1996, 5.

The Baltic armed forces thus displayed a "decidedly Soviet flavor" (Zaccor 1994, 204–5) and were seen by some observers as "a continuation of the warped Soviet ethical system and the abuses of the Soviet military" (Vitas 1996, 77) rather than as a substantial deviation from it. For example, in the autumn of 1997, carelessness resulted in death by drowning of fourteen participants in a peacekeeping exercise in Estonia, thus virtually eliminating a complete platoon of the Estonian peacekeeping contingent (Gießmann 1997–98, 153). Resulting mainly from the violation of military discipline in the Latvian armed forces in the early 1990s, twenty-one soldiers perished within twenty-one months. Disrespect for the soldiers' lives and physical integrity was said to be even worse than in the Soviet army (Kerner 1994, 33). In 1997, the CBSS commissioner on democratic institutions and human rights, including the rights of persons belonging to minorities, reportedly stated that conscripts in the Latvian army did not enjoy the same rights as the average Latvian citizen. In particular, they suffered from an inadequate complaint procedure (Kahar 1997a, 8). In the summer of 2001, Latvian officials were still concerned about ongoing harassment in the Latvian armed forces.[5] Furthermore, "an excess of presidential whim that the parliament was unable to control" resulted in chaos in Estonia's military leadership, with five individuals serving as commander or acting commander of the armed forces between 1999 and 2001 (Huang 2001, 35). All this did not increase the attractiveness of serving in the armed forces. It furthermore undermined the claim that the armed forces of the Baltic states were indeed completely different from the Soviet army.

Regardless of all these problems, the military was frequently represented as an institution that could help strengthen and "unify the country" (Zaccor 1994, 203). Conscription armies were depicted as "a positive element in establishing social harmony and lessening social and ethnic tensions" (Ozolina 1996, 51). The army was seen as the place where "we will learn to love our independent country and its values" (Meri 2000b). The demand that "pacifist thinking" in the Baltic states be "countered resolutely" was a corollary to this view (Skujins 1996, 645), underlined by the rhetorical discrimination against objectors, or, as then Estonian president Lennart Meri put it, "all the shirkers and cowards, whose 'unique intellect' somehow

5. *City Paper: The Baltic States* 2001, 7.

prevents them from defending their home" (2000b). They were banished from the collective self, and the quality of their characters and bodies was called into question: "Loose morals, and a loose body to go with [them]" (Meri 2000b). Thus, rather than living with difference, including different views on the armed forces, the state rhetorically otherized and banned deviators from the collective self.

This mindset explains Latvia's preoccupation from April 1996 to March 1997 with passing a law on military conscription that aimed to increase the participation in and reduce the number of legal discharges from the military service (Bebriss 1996a, 1996b; Kahar 1996a, 1996b), which found its continuation in Estonia in early 2000 (Sindrich 2000; Huang 2001, 36). Likewise, pre–military service training, including courses on state defense in secondary schools, began to be emphasized. For example, in 1997 Tallinn Pedagogical University graduated fifty-eight national defense teachers. National defense instruction began to be included as an elective in the curriculum of public and vocational schools (Ministry of Defense of the Republic of Estonia 1999, 35). In Latvia, courses on state defense were introduced as an optional subject in secondary schools in 1997 and are now offered in nineteen schools (Ministry of Defense of the Republic of Latvia n.d.). The Lithuanian Basics of National Security state that, among other things, "instruction in the means of resistance and training in the skills of resistance shall be a constituent part of compulsory school education program" (pt. 1, chap. 7, fourth sec.), although teaching military subjects at school is prohibited according to the Law on Education. In March 1997, President Algirdas Brazauskas's support of military training in secondary schools met with a negative response because it was reminiscent of the Soviet education system (Vitkus 1997, 72–73). The Estonian Ministry of Defense's statement that "in the future, national defense instruction should be mandatory in all institutions of learning" (1999, 35) was the opposite of the proposals submitted by popular-front representatives to the authorities in Moscow in 1989, which included the reduction or abolishment of military training in schools (Holloway 1989–90, 18). Special military training of university students is also indicative of the role assigned to the armed forces in creating or imagining internal order. Objectives of this training are, among other things, "to facilitate patriotic education and training of youths for state defense purposes" (Ministry of Defense of the Republic of Latvia n.d.).

All these cases may be explained in functional and operational terms, but they also have both a symbolic dimension and a function in the imagination and construction of the domestic order. This argument may also be applied to the voluntary territorial defense forces and especially to their youth and (in Estonia) women's organizations. In Lithuania and Latvia, the voluntary defense forces are frequently seen as an embodiment of historical continuity: the Lithuanian National Defense Volunteer Forces and the Latvian Zemessardze in a more spiritual manner,[6] the Estonian Defense League (Kaitseliit) explicitly by having reenacted its 1931 statute and revitalized the interwar youth and women's organizations. These unarmed organizations are said to have been "recreated on the principle of historical continuity and aim at strengthening patriotic feeling among the population" (Haab 1995, 42). The Defense League oversees the Women's Home Defense with more than five hundred members; the Homedaughters, a voluntary organization for girls between eight and eighteen years with one thousand members; and the Young Eagles, a voluntary organization for boys between eight and eighteen years with two thousand members. The reestablishment of the Homedaughters disregarded the popular movements' celebration in 1989 of the end of preinduction military training for girls as a step toward the "final elimination of militarism from the education process" (Holloway 1989–90, 19). In Latvia, the Jaunsargi (Young Guards) is a voluntary youth organization of the National Guard for boys and girls between twelve and eighteen years. The Jaunsargi is divided into two groups: one, based on the Jaunsargi Training Program, for boys and girls between twelve and fourteen years; the other, based on the State Defense Training Program, for boys and girls between fifteen and eighteen years. Participation in the organization is officially said to contribute to useful leisure activities as well as to preparations for the conscript service, a potential military career, and studies in the National Defense Academy. Jaunsargi groups are assigned to a brigade or battalion of Zemessardze (Ministry of Defense of the Republic of Latvia n.d.).

Likewise, following from the meaning assigned to armed formations, voluntary paramilitary organizations such as the Riflemen's Union in Lithuania are officially promoted as a connecting link both between the armed forces and society as well as between the current Lithuanian state

---

6. Reportedly, the National Defense Volunteer Forces "[saw] itself as basically an organized guerilla force, preparing itself to carry on the kind of warfare waged by their spiritual predecessors, the rebels who fought Soviet troops from 1944 [to 1953]" (Zaccor 1994, 211).

and the interwar state. As Peeter Vares and Mare Haab report, the Riflemen's Union was reestablished in 1989 as a "remodeled heir of the powerful
pre-war Sauliai-Rifles" and was composed of "extremely radically
minded," armed, and uniformed members. Directly after its reestablishment, the union was largely free from state control and funded by "sympa-
thizers in the West." In 1992, union members, seeing themselves as both the
guardians of independence and the representatives of the "true" Lithuanian
values, even "warned that their organization would take steps to prevent
any acts of 'betrayal' by the government formed by the Lithuanian Democratic Labor Party" (1993, 304).[7] Regardless of its nationalist orientation, the
Riflemen's Union is officially given an important role in linking armed
forces and society:

> The Sauliai (Riflemen) Union has a significant role in informing society
> about the missions and activities of the Armed Forces. This is a civilian, vol-
> untary para-military organization that helps to raise national conscious-
> ness and develop state defense activity. . . . The Sauliai Union is active in
> preparing for civil self-defense and armed land defense. They help to pre-
> pare citizens for universal and armed resistance in the event of war threats
> or occupation, thereby strengthening the state's mobilization reserve. The
> Union also helps the police, it establishes subdivisions of young Sauliai,
> and organizes various social activities. (Ministry of National Defense of the
> Republic of Lithuania 1999, pt. 3.II)

During the 1990s, the armed formations in the Baltic states frequently
were a part of the problem rather than a part of the solution; they were a
source of insecurity rather than one of security; they competed rather than
cooperated with one another.[8] This feature may in part be explained by

---

7. In a June 1991 interview, one of the Riflemen's Union leaders said that "of course, if they
[a left-wing government] were properly elected they would be allowed to take power. But we
would watch them very carefully; and if we saw that they were betraying Lithuania, we would
have to act to save the nation" (cited in Lieven 1994, 74).

8. For "military personnel engaging in criminal acts," see Vitas 1996, 76–77; for the
Lithuanian army's "inability to guard even its own facilities adequately," see S. Girnius 1993,
43; for officers' being suspicious of the principle of civilian control of the military, see Zaccor
1994, 205; for "indiscipline and the persistence of the Soviet-style abuse" in the Latvian armed
forces, see Zaccor 1995c, 13.

scarce resources, making competition and struggle over funds inevitable. It may also be explained by a lack of clear-cut political decisions as to whether the regular armed forces, the volunteer defense forces, the border guards, the police, or other armed formations should be prioritized. The buildup of some armed formations followed from decision makers' personal ambitions and political motivations. In Lithuania, for example, Vytautas Landsbergis is said to have initiated the Parliamentary Guard as a "presidential body-guard" (Lieven 1994, 328).

The problems emanating from the armed forces may also be seen as a corollary of the self-image of parts of the armed formations as the true representatives of national values, a self-image strengthened by the functions officially assigned to these formations. In particular, the volunteer defense formations fancied themselves in the image of true guardians of independence. This self-image was not totally unfounded because they emerged in Latvia and Lithuania from the gathering of volunteers at such strategic places as parliamentary buildings and television and radio facilities in January 1991. This self-understanding, however, also led to distrust of the other armed formations, among which especially the regular armed forces reciprocated by displaying contempt for what they considered undisciplined hobby or part-time soldiers. Again, this latter position was not completely unfounded.[9] Lack of political, civilian, and often even military control of both regular and volunteer forces led to an explosive and incalculable situation. The newly created regular armed forces were subordinated to the ministries of defense, whereas the volunteer forces initially reported to the supreme councils and subsequently to the heads of state and were subordinated to the defense ministries only later. The situation was one of parallel structures of armed formations, the control over which was rendered difficult by insufficient legislation (Kerner et al. 1998, 127).

In addition, until August 1993 Russian armed forces in considerable number were still stationed in Lithuania, and the last Russian soldiers did not leave Estonia and Latvia (with few exceptions) until August 1994. Thus,

9. "Since anyone could join [the volunteer forces], just about anyone did. The territorial forces attracted criminals and hooligans along with decent, patriotic citizens" (Zaccor 1995c, 12). Furthermore, "initially a military background or competence was not needed to obtain leading positions: preference was given to people who belonged to the political parties in power" (Vares and Haab 1993, 300).

three types of armed formations were temporarily present at the same time: regular armed forces, volunteer forces, and Russian armed forces. For a short period in May 1991, the buildup of another type of armed force loomed in Lithuania when some activists of the Polish minority promoted the creation of the Polish Autonomous Territorial Region within Lithuania, including its own flag, assembly, police force, and even army (Lieven 1994, 169; Norkus 1998, 147).

In Estonia, tensions existed between the Defense League and the National Defense Initiative, which was launched secretly in the late 1980s and in the summer of 1992 organized the Läänemaa Voluntary Jaeger Company. After the first democratic elections in Estonia in October 1992 and the subsequent refusal to give leading representatives of the company posts within the official state and military structures, the company remained an armed formation outside the legal state military structures. It was not integrated into the structure of the national army until 8 March 1993. Tensions between the company and the General Staff continued and culminated in July 1993 when the General Staff deployed the company as an additional national military unit in the city of Paldiski, then still largely under the control of the Russian military. As Peeter Vares reports, "the Company withdrew from the city and the commander of the unit refused to disarm his fighters as demanded by the Government and the General Staff. The military unit had placed itself outside the jurisdiction of military, state and civil control and the situation was analogous to that of mutiny" (1997, 37). Although the company renewed its declaration of subordination to the General Staff, relations remained strained and were resolved only in 1994 when the company was dissolved.

A governmental commission investigating the case came to the conclusion that the company's activities had threatened the independence of Estonia and that they could have resulted in an involvement of the Russian troops stationed in Paldiski. Furthermore, the existence of alternative military structures with an alternative ideology was seen as a threat to the integrity of the state military structures (Vares 1997, 37–38). Civilian and military control over the Läänemaa Voluntary Jaeger Company was largely nonexistent, both before and during the period of its subordination to the state military structure. The company's disobedience was left "politically and militarily open-ended" (Vares 1997, 38), which some observers saw as indicative of "benevolent passivity" on the part of "some superior authorities" (Kerner 1994, 34).

Tensions existed also between the Estonian Defense League and the Homeguard (Kodukaitse). The establishment of the Homeguard by the reform Communist leadership in 1990 as a counterbalance force against Soviet military forces and militia aimed to unite the supporters of Estonian independence. The ultranationalist Defense League, however, called into question the legitimacy of the government and called upon members of the Homeguard not to defend the government in case of an attack by the Inter-Front. Until its dissolution in March 1996, the Homeguard was a voluntary reserve organization under the jurisdiction of the Ministry of the Interior, mainly assisting the police force. Its relations with the Defense League remained strained; its creation by the reform Communist government was often used as an argument against the Homeguard, which was suspected of having a left-wing political orientation (Vares 1997, 21). Because the Defense League was initially composed of ultranationalists, it is not surprising that the relations between it and the regular armed forces (Kaitsevägi) were also far from smooth. This was because the Kaitsevägi to a certain extent recruited former Soviet armed forces personnel,[10] which in the eyes of the volunteer organization made them untrustworthy and compromised. Many officers of the General Staffs, in particular, were seen with suspicion because of the "wait and see strategy" they ostensibly had adopted in January 1991 (Zaccor 1995c, 12).

In Lithuania, the tensions between the volunteer forces and the regular armed forces paralleled the tensions in Estonia. Because the Voluntary Service of National Defense emerged from the gathering of volunteers at strategic locations in January 1991, many of its members are said to believe that "the Lithuanian nation is indebted to them" rather than to the regular armed forces, which were created only later (Vitas 1996, 81).[11] Furthermore, because of the volunteer forces' anti-Communist self-understanding and right-wing political orientation, both the LDDP's victory in the parliamentary elections

10. In order to avoid becoming too dependent on former Soviet military personnel, the Estonian armed forces used some members of the Estonian National Corps, disbanded by the Soviet army in the late 1950s, and officers of the Estonian interwar army. This procedure naturally aggravated the age structure of the armed forces (Jones 1998, 4).

11. According to the 5 May 1998 and 14 January 1999 Law on the Organization of the National Defense System and Military Service, the Voluntary Service of National Defense was reorganized into the National Defense Volunteer Forces.

and Algirdas Brazauskas's victory in the presidential elections in 1992 were viewed with suspicion. This suspicion increased when the Voluntary Service's share in the Defense Ministry's budget decreased in 1993 and when Brazauskas, on 29 September 1993, nominated a 1967 graduate of the Leningrad Military Academy who was also a long-time Communist Party member and former Soviet military advisor in Ethiopia to be the commander in chief of Lithuania's armed forces.

Like in Estonia, Lithuanian regulars looked down on the reservists as part-time soldiers, and reservists reciprocated by distrusting many of the former Soviet officers in the regular forces. Like in Latvia, the volunteer forces in Lithuania are said to have displayed a greater eagerness to adapt themselves to Western military conceptions than were the regular armed forces (Clemmesen 1998, 243). In addition to the rivalries between the regular armed forces and the volunteers, the Zemessardze in Latvia developed conflictual relationships with both the regular police forces and the Security Service, one resulting from overlapping competencies and ethnic tensions between the predominantly Russian-speaking police forces and the predominantly ethnic Latvian Zemessardze, the other emanating from a lack of legislation regulating the Security Service's responsibilities and from competition between the two formations (Kerner 1994, 33; Lieven 1994, 326; Dreifelds 1996, 106).

To sum up this discussion so far, it may be said that the Baltic armed formations were given, in addition to their operational function, an important function as to both the construction of the national identity and the imagination of domestic order and historical continuity. The latter function is indicative of the conception of the state underlying the security policies in the Baltic states—a conception that links the security of the state closely with the military. Even those security policy options that were rejected (such as neutrality) were conceived of only in military terms and rejected mainly on operational and financial grounds. The buildup of armed forces is frequently depicted as being simply necessary or natural, and alternatives are often rejected, without much discussion, as "unrealistic" (Bajarunas 1995, 11), "questionable" (Lydeka 1995, 115), or "too expensive" (Nekrasas 1996, 69). From the state-military nexus followed the implementation of some of the procedures that had been criticized ten years earlier with respect to the Soviet army: preinduction training for girls and courses in national defense at secondary schools. Rivalries among the armed formations were a corollary

of the volunteer organizations' self-understanding as the guardian of independence and the true representative of the national values. These rivalries
were facilitated by the functions officially assigned to the armed formations,
but initially not clearly regulated by law. Strongly politicized armed elements were temporarily a source of insecurity rather than a source of security, a part of the security problem rather than a part of the solution. This
situation alone explains why NATO officials were far from enthusiastic
when in 1993 the Lithuanian president inquired about the possibility of
Lithuania joining the alliance.

Throughout the 1990s, the relations between NATO and the Baltic states
were contradictory. NATO, especially after the adoption of the new Alliance
Strategic Concept in 1999, was advertising itself as mainly a political organization, whereas the Baltic governments were aspiring to enter NATO as a
military organization. NATO emphasized its function and responsibility for
security in Europe as a whole (and even beyond Europe), whereas the Baltic
governments saw the alliance primarily in its capability to defend its members collectively. Although the Baltic governments were adapting themselves to NATO parlance as part of their preaccession strategy, their
understanding of "Europe" deviated from NATO's. After having entered
the Partnership for Peace (PfP) in 1994 only reluctantly, the Baltic governments represented PfP as a direct path to membership in NATO. In contrast,
NATO saw PfP as an alternative to its own rapid enlargement. By inviting
Poland, Hungary, and the Czech Republic to membership, NATO extended
security guarantees to those central and east European states that were often
said to be the least threatened. Yet it refused to give security guarantees to
those states that, when the whole discussion on NATO enlargement got
started in the mid-1990s, regarded themselves as the most threatened. The
Baltic governments wanted to anchor Estonia, Latvia, and Lithuania irreversibly and as tightly as possible in Western military organizations, but
these organizations' representatives seem to have preferred a more open
and flexible approach to the Baltic states. Finally, Baltic decision makers
wanted to enter NATO as a means with which to increase security *against*
Russia, but NATO was looking for security *with* Russia.

Hans Binnendijk and Richard Kugler suggested that NATO should
think more about its own strategic aims and less about "the political merits
of individual countries" (1999, 128), a proposal displeasing to Baltic decision
makers. For example, they said, "admitting the Baltic states is not out of the

question, but NATO should do so only if its strategic purposes are served by such a move" (135). Binnendijk and Kugler also argued that NATO should work on solutions that, if certain measures were fully pursued, could make Baltic NATO membership "become less important because these countries [would] be secure without it" (135).[12]

The ways to secure the Baltic states outside NATO were, however, as disputed as the ways to make them become NATO members. For example, Zbigniew Brzezinski argued that full membership in the WEU would suffice because it would make the Baltic states the "automatic, though informal, beneficiaries of the NATO umbrella" (1995, 27). Ronald Asmus and Robert Nurick, however, argued that this was precisely the reason why EU membership for (one of) the Baltic states should be decoupled from WEU membership: "Any attempt to blackmail or shame the West into giving a security guarantee which it is not yet prepared to give will only backfire" (1996b, 134). As regards the possible paths to NATO membership, both group and individual approaches were recommended. Andrew Winner, for example, argued that "it will be much, much easier for Washington to gather the domestic political support necessary to bolster NATO membership for all three Baltic states than for one at a time or for one and not the others" (1998, 56). The Independent Task Force U.S. Policy toward Northeastern Europe (1999), sponsored by the Council on Foreign Relations, disagreed and recommended differentiating between the Baltic states and admitting them into NATO individually rather than as a group.

The debate over Baltic NATO membership since 1994, with its reiterating, almost ritualistic declarations of intentions ("open door" and so on), showed that the "fundamental flaw in NATO's expansion policy" had not been remedied: "Its declared purpose is to enhance stability in Europe as a whole, but the obvious risk is that, while reassuring some nations, it would intensify the sense of insecurity in others and create a Baltic front of friction between the West and Russia" (Jakobson 1996, 54).

---

12. These measures include "helping them build strong governments, viable economies, and better military forces. . . , encouraging them to develop security ties with their Nordic neighbors, the United States, and other European powers . . . [and] bringing them into the European Union." These suggestions are very close to those made by Asmus and Nurick (1996b) and are indicative of the lack of development in conceptual thinking about Baltic security in the 1990s.

Furthermore, without a coherent and credible strategy toward the Baltic states, NATO enlargement was seen as a "prime candidate for a crisis between the West and Russia." If mishandled, the Baltic issue was said to have the potential to "poison the West's attempt to develop co-operative relations with Russia" (Asmus and Nurick 1996a, 1).[13] Moreover, the conceptualization of NATO's policy toward the Baltic states did not seem to make substantial progress during the 1990s. For example, Ivo Daalder and James Goldgeier wrote in the spring of 2001 that "while those Baltic nations that meet NATO criteria should be allowed to enter the alliance, NATO should take steps to reassure Moscow that the alliance poses no threat to Russian security" (2001, 83). This position was basically the very same strategy that many observers had recommended since 1994 but that had failed to both convince the Russian leadership of NATO's benevolence and make them accept NATO membership for the Baltic states (Black 2000). It furthermore failed to convince the Baltic leaders of its ingenuity.

It follows from the contradictory recommendations and the lack of progress in conceptualizing Baltic security that no matter how the Baltic decision makers responded to the various suggestions, their reactions could be turned against them. If they prioritized military cooperation in the framework of BaltBat or other international cooperation schemes, this cooperation might have been applauded, but neglecting to build up national armed forces might have been criticized. Conversely, if they prioritized the buildup of national armed forces, this buildup could have been applauded, but the lack of military cooperation might have been criticized. Trying to do both at the same time could have led to a perception of both as insufficient. The struggle over increasing the defense budget to 2 percent of GDP revealed how difficult it still was at the end of the 1990s for Baltic decision makers to increase military expenditures. In 2000, even BaltBat, a prestige project that would not have been possible without considerable foreign financial and practical assistance in the first place, was still said to be *very* dependent upon external support" (Simon 2000, 4, emphasis in original).

Scarce financial resources certainly required decisions on priorities, but every decision could be turned against the Baltic states because it could always be said that the "wrong" issue was given preference. The Baltic states'

---

13. I am grateful to an anonymous source in the German Ministry of Foreign Affairs for providing me with the unofficial, prepublication version of the article.

participation in the PfP program, regarded by Baltic decision makers as a precondition for NATO membership, was even criticized as a "somewhat mixed blessing for the Baltic states" because some PfP activities, having a rather symbolic value, were said to undermine the development of both self-defense capacities and the capability to cooperate militarily in NATO Article 5 operations (Clemmesen 1998, 247).[14] In October 1996, the U.S. secretary of defense, William Perry, referred precisely to the lack of capability to take on NATO's Article 5 responsibility to justify the exclusion of the Baltic states from accession talks with NATO (Kozaryn 1996). The Baltic states nevertheless were very active in international military cooperation (Möller and Wellmann 2001, 101–14), not least because, as Christopher Jones has put it, they wanted to be "seen-to-be-doing security" (1998, 12)—that is, they eagerly wanted to direct NATO's attention to their capabilities—and because, in the words of an Estonian senior defense politician cited by Jones, cooperation with NATO members would "make it harder to sell us out" (13).

However, no matter how intensively the Baltic states participated in international military cooperation and quite regardless of the praise they earned for so doing, their contribution remained negligible from the NATO point of view and hardly sufficient for changing their basic predicament with respect to their longed-for NATO membership: "simply put" (perhaps too simply), "they do not have the votes" (Asmus and Nurick 1996b, 123). Baltic decision makers seem to have been clearly aware of this lack of support. How, then, did the debate over NATO membership of the Baltic states evolve during the 1990s?

According to a standard representation of the relations between the Baltic states and NATO in the 1990s, "beginning with the re-instatement of independence in 1991, the goal of joining NATO has enjoyed the broad support of Estonia's political parties and the population alike, regardless of changes of government and in the composition of the Parliament" (Ministry

---

14. PfP inhibited the development of a self-defense capability because the means invested in international programs could not be spent for self-defense purposes. It inhibited the development of the capability to cooperate militarily in NATO Article 5 operations because the purpose of the program was not to prepare partners for NATO membership. Accordingly, "it was realized that in order to increase the quality of experience and to make a full use of different exercises and activities the number of different PfP activities and events should have been better prioritized" (Ministry of Foreign Affairs of the Republic of Lithuania 1999).

of Foreign Affairs of the Republic of Estonia 2001a, pt. 2.1). Why, then, did the Estonian Parliament wait until May 1996 to approve the goal of NATO membership? Why did Estonia never officially apply for membership? Why did only 20 and 28 percent of those people asked in a poll commissioned by the Estonian Ministry of Defense in February 2001 support joining NATO "definitely" and "rather," respectively, with 16 percent "rather against" and another 16 percent "definitely against" NATO membership, and another 20 percent still not sure (Ministry of Foreign Affairs of the Republic of Estonia 2001b)?[15] Some scholars agree with the official view, which is cleansed of ambiguities and ignorant of the process character of developing and implementing security policy. It was said, for example, that EU and NATO membership "has been pursued with equal force by each successive national government since 1991" (Lejins 1998, 9). Why, then, did the Defense and Foreign Ministries wait until 1998 to produce a joint paper on Latvia's policy toward NATO (Ozolina 1999, 24)? Calling the official representation into question does not mean, of course, that everyone was opposed to NATO membership, but it means that not everyone was in favor of it in 1991. It also means that the policy orientations of the Baltic security establishment changed to some extent during the 1990s. At the end of the decade, EU and NATO membership was indeed pursued by the national governments of the Baltic states and supported by most opposition parties, although, curiously enough, major parts of the population were still showing a somewhat distanced attitude to NATO membership.[16]

Retrospectively rationalizing political decisions includes active, purposive forgetting and submitting to oblivion all those statements and events at variance with current policies. It means representing one's own policy over

---

15. The ministry refers to the poll under the title "63% of Estonians Support Joining NATO." The title shows how "broad support" is being constructed, namely by "leav[ing] aside people with no clear position on this matter" and by ignoring the non-Estonian part of the population.

16. In the summer of 1999, 39.3 percent of Latvian residents said they would vote against membership in NATO, and 16 percent were undecided (*Baltic Times* 1999, 2). The Lithuanian government decided in February 2001 to spend more money on informing the public about NATO membership because 35 percent of the population was still against it (*Newsfile Lithuania* 2001). At the same time, the *RFE/RL Baltic States Report* (2001a) reported on a December 2000 poll according to which only 22.3 percent of the Lithuanian respondents were against NATO membership.

time as logical and consistent, implying that only that which is consistent can also be logical. The impression of a specific security policy as natural is imprinted on a society's collective memory, and alternatives are effectively marginalized. Indeed, making the security of the Baltic states dependent on NATO membership, even representing Baltic NATO membership as "a question of survival" (Graube 1999, 66), inevitably supported the equation of security with military security and marginalized civilian approaches to security. After all, what distinguishes NATO in essence from other international organizations is the quality and quantity of its military capabilities. Furthermore, it is precisely NATO's military capability that, on the basis of a conventional understanding of security, made it appear superior to, for example, the EU as a means with which ultimately to guarantee Baltic security. For example, according to a Latvian foreign minister, "while EU membership will indirectly strengthen our security, especially economic security, only NATO membership can fully ensure that democratic values and stability are defended" (Berzins 1999, 56).

It may be, however, that EU membership gives a "sense of belonging" that in an almost Deutschian sense is "an effective guarantee of security."[17] From most Baltic political decision makers' point of view, however, EU membership was never considered as a sufficient path to security. Governments of many NATO candidate countries viewed with suspicion the development of the European Security and Defense Policy, for example, because they feared that it could, among other things, be developed into an alternative to NATO enlargement (Missiroli 2004, 122). Furthermore, linking NATO membership with survival supports the notion that security policy, "overflow[ing] the normal political logic of weighing issues against each other" (Buzan, Wæver, and de Wilde 1998, 24), is legitimately detached from open and public discussion. As the Conservative Lithuanian member of Parliament Rasa Jukneviciene reportedly put it, the "Lithuanian nation is too small to afford the luxury of disagreeing on the security of its state and citizens" (cited in Tracevskis 2000b, 6). Any debate over defense and security is indeed explicitly unwelcome. Responding to an appeal made by the president, policymakers in Estonia are said to have reached consensus that "defense policy is simply non-negotiable" (Meri 2000a).

In the field of security policy, then, democracy is interrupted. Apart

17. Romano Prodi, personal communication, 27 April 2000.

from being profoundly undemocratic, this procedure also contradicts the very nature of security, which lies precisely in the absence of a generally agreed definition of it and consequently in the simultaneity of different views on and paths to it. Thus, within this procedure, NATO membership and military paths to security become "natural," "logical," and "taken for granted" and not just various paths to security among possible others; they become sedimented in collective knowledge and memory and thereby become a powerful legitimacy provider for current and future policies.

What changes in the policy orientations of the Baltic security establishment can then be observed during the 1990s? To begin with, it has been argued that the Estonian security policy changed from an emphasis on establishing national guards and territorial defense to approaching NATO. According to Peer Lange, this change was a result mainly of "Russian misbehavior in the extremely reluctant and controversial handling of the withdrawal of Russian troops" (1995, 135). Although chapter 5 of the present book suggests a different interpretation, the security policy orientation of Estonia (and of Latvia and Lithuania as well) indeed changed, and it did so in reaction to perceived Russian policies (discussed more fully later). The claim that "Estonian decision-makers have never approached the question of NATO membership from an anti-Russian perspective" (Klaar 1997, 19) is obviously beyond belief: some officials made unmistakably clear that the aim of NATO membership was indeed to secure the Baltic states against Russia (see chapter 11). Having realized that emphasizing the dangers to Baltic security ostensibly emanating from Russia would not increase NATO's eagerness to give security guarantees to the Baltic states, these states pursued the issue of NATO membership differently, starting with what is considered Lithuania's application for membership in January 1994.

This application for NATO membership gave new momentum to the debate over NATO's future (a momentum that NATO soon absorbed in the PfP program). Initially, some Baltic decision makers' position on NATO was somewhat ambivalent, as was their position on the West in general: "There are a good many radical nationalist ideologists in the contemporary Baltic [states] who cordially despise the modern West, a contempt increased by what they see as the West's 'cowardice' and 'treachery' in failing to do more to save the Balts from Soviet conquest in the 1940s, and to support them after 1989, while the Balts were sacrificing themselves to save Europe from a new Soviet attack" (Lieven 1994, 71). Having realized independence from the Soviet Union only recently, some decision makers were inclined to maintain a certain distance also from west European institutions. That membership in the EU meant getting "from one Union into another" was more than a bon mot. It showed some mental reserve toward west European organizations, which made the Estonian president reportedly feel obliged in May 1993 to "reassure his co-nationals that integrating into Europe does not mean 'dissolving' into Europe" (D. Smith 2001, note 89). In Lithuania, preserving sovereignty not only from Russian but also from Western influence, especially in its cultural features, became an issue in influential nationalist-conservative circles. Here, the issue was depicted as one of protecting the Lithuanian nation, seen as an embodiment of "true" European values—values ostensibly lost in the West a long time past. In 1991, for example, Minister of Foreign Affairs Algirdas Saudargas is said to have told a group of stunned Western diplomats that "Western democracy is a 'sham,' and . . . true democracy stems from the heart of a nation" (Lieven 1994, 391). The position on the West thus was ambivalent: on the one hand, both a return to the West and a clear separation from Russia was aspired to; on the other hand, the Lithuanian nation was considered more Western than the current western Europe, which was seen as having sold its soul for the benefit of material gains.

Vytautas Landsbergis, for example, is said to have seen the return of Lithuania to the West as a chance of renewing the European family on the basis of the Lithuanian nation's experience in its struggle with the Soviet Union. In his view, this struggle resulted in both a spiritual maturity that made even tanks fall back and a feeling of moral unassailability—especially because it contrasted sharply with the West's relations to the Soviet Union, which had ostensibly been based on inferior material motives (Christophe 1997, 308). According to Landsbergis, the peoples in neither Russia nor other countries could be grateful to the West for its ambivalence toward the Soviet Union and Russia, an ambivalence that had ignored moral principles and European values such as freedom. In an argumentative about-face, however, Landsbergis, forgetting his own reservations toward the West, then depicted Lithuania as the "outpost of Western civilization" (1997). The military flavor of this metaphor shows that Western civilization's encounter with non-Western civilizations was not thought of as a peaceful and mutually beneficial one. This understanding, in turn, fit conveniently into the representation of Russia as a cultural threat along the lines of thought that predicted a clash rather than a dialogue of civilizations.

Regarding Lithuania's military integration policy, it has been emphasized that an alliance policy replaced a policy of neutrality not before 1993 (Miniotaite 1998, 175). But even then, equidistance from the East and the West was stressed in several policy manifestations. This position was to some extent an expression of pragmatic policies vis-à-vis Russia. In October 1992, the LDDP was elected as a result of, among other things, the weak economic performance of its conservative predecessor. Many people simply hoped for improved living conditions, which were to be achieved on the basis of good relations with Russia, at the time Lithuania's most important trading partner. It thus seemed advisable on economic grounds to consolidate relations with Russia rather than to undermine them by too energetically heading west. At the same time, the LDDP carried with it an ideological heritage that made some of its representatives legitimize their foreign policy preferences by pointing to the danger of economic peripheralization resulting from an exclusive orientation toward the West's developed markets. In a peculiar alliance with national-conservative circles, some also warned against a loss of cultural autonomy resulting from an unconditional opening to Western mass culture and its ostensibly leveling influence (Christophe 1997, 312). Thus, in 1993 the minister of national defense depicted protecting Lithuania from being dominated by both Eastern *and* Western powers as the main objective of Lithuania's external policy (Butkevicius 1993a, 171). Or, as he stated elsewhere (in of all places an official NATO publication) even more explicitly: "Only the status of an independent state will enable us to protect the interests of our people. That is why the main objective of our policy is to seek a balance in our relations with East and West, preserving a maximum of independence. This is also true of the economic, cultural and military influences" (Butkevicius 1993b, 8).

Western military advisers are said to have encountered some initial resistance in Lithuania (Zalkalns 2002, 45). The idea of a "union with NATO," formulated by Lithuanian foreign policy decision makers in late 1991, did not mean security guarantees, but rather security assistance, for example with respect to the Russian troop withdrawal (Petersen 1992, 33). Consequently, Landsbergis, then in his capacity as the chairman of the Supreme Council, requested in a letter of 18 December 1991 that NATO "examine the possibility of sending observers to Lithuania to observe the actions of [the Russian] forces up to and during the period of withdrawal" (Ozolins 1994, 62).[18] The Constitutional Act of 8 June 1992 prohibits Lithuania from joining any new political, military, economic, and any other state alliances or commonwealths formed on the basis of the former Soviet Union, but alignment with the West did not immediately follow from this Constitutional Act. In the autumn of 1992, Landsbergis again asked NATO to send observers to monitor the Russian troop withdrawal. Before then, Lithuania's and in particular Sajudis's position on security was one of prioritizing national ways to security by developing military capabilities:

> At the outset [1989], two of *Sajudis'* factions favored formal neutrality as the orientation of security policy, whereas Landsbergis merely wanted "autonomy". . . . In the following year, Lithuania went as far as offering Moscow that it would become a member of the Warsaw Pact in exchange for state sovereignty. When Landsbergis visited NATO headquarters in Brussels in the autumn of 1992 he expressed his belief in the fruitfulness of deepening cooperation between NATO and the Baltic states, but stressed that the real safeguard against the emerging security vacuum in the Baltic states would be the national military build-up in the Baltic states. . . . He expressed his appreciation for NATO, but did not raise the issue of membership. (Skak 1996, 210)

Landsbergis's successor as head of state, however, did raise this issue. In a letter to the then NATO secretary-general, Manfred Wörner, Brazauskas carefully avoided emphasizing Lithuania's need for protection from Russia. Rather, he stated that Lithuania "seeks to contribute to the security of the North Atlantic area by joining [NATO]." To this end, he invited a NATO fact-finding mission to Lithuania. In the letter, Brazauskas expressed his conviction that "the results of that mission will affirm the progress that the Republic of Lithuania is making toward meeting the requirements necessary for the enhancement of security in the North Atlantic area" (cited in Skak 1996, 209–10). However, whereas a June 1993 Draft National Security Concept saw *cooperation* with NATO as one of Lithuania's aims in the

18. In April 1992, NATO made monitoring the withdrawal dependent on agreements determining concrete withdrawal timetables. It thus expressed its willingness to become involved in Baltic-Russian affairs, but only subsequent to conflict resolution.

process of integration into global structures,[19] in late 1993 *integration* into NATO became the dominant issue.

From then on, most political decision makers in the Baltic states consistently blocked regional approaches to security, especially those approaches that were not just constructed in preparation for NATO membership. Frequently referring to the ostensible indivisibility of security in Europe, the Baltic security establishment was throughout the 1990s obviously afraid that security arrangements specifically tailored to the Baltic Sea region could be developed further as an alternative to Baltic membership in NATO or at least could postpone it. Promoting the idea of the indivisibility of European security resulted in the paradoxical situation that the more promising regional arrangements appeared to be, the more strongly they had to be rejected (Tiido 2000). Likewise, the Baltic governments regarded subregional security arrangements, especially those arrangements limited to the three Baltic states, as a means to an end, not as an end in itself. The ultimate aim was membership in NATO, but in the meantime trilateral military cooperation served mainly as a means to prevent regional security arrangements that would encompass the whole Baltic Sea region, as suggested in some scholarly writings (Knudsen and Neumann 1995). The military cooperation among the three Baltic states was often depicted as an expression of having learned from the conspicuous lack of military cooperation in the interwar period. Rather than substantially increasing the defense capabilities, however, such cooperation aimed primarily to achieve interoperability with NATO and to underline the ambition of the Baltic states to become providers of European security rather than merely to be consumers of it. BaltBat is a case in point. Its relevance to national and regional security was always limited. Because of its dependence on external financial assistance, it has even been described as "the best demonstration of the failure of multilateral efforts" (Heinemann-Grüder 2002, 23).[20] But it was an important symbol of the

19. Draft National Security Concept of the Republic of Lithuania, drafted by an ad hoc working group headed by G. Kirkilas, 23 June 1993.

20. BaltBat was launched in 1994 as a trilateral battalion, consisting of three infantry companies, for participation in international peacekeeping operations under a UN or OSCE mandate. The command was distributed on a rotating basis. I discussed some substantial changes in BaltBat's character in chapter 11. The BaltBat project was officially closed in September 2003. The Baltic Naval Squadron (Baltron) was initiated on 18 October 1996 and has been operational

Baltic states' willingness and ability to cooperate militarily, and the decision makers saw it as an equally important means with which they could adapt to NATO standards. This can also be said about Polish-Lithuanian military cooperation in the framework of the Lithuanian-Polish Peacekeeping Battalion (LitPolBat), which had an important preaccession function.[21] It is indicative of the limited character of the regional military cooperation that even BALTSEA, regardless of its all-encompassing acronym, does not include representatives of the Russian Federation. From the point of view of security community building, then, the regionalization of security as it unfolded in the Baltic Sea region during the 1990s was a suboptimal strategy that increased social interaction and communication between only some, but not all Baltic Sea states.

As stated earlier, the Baltic political decision makers' security policy orientation changed in reaction to Russian policies. This assessment may appear to be inconsistent with the analysis presented in chapter 7, which emphasized the cooperative Russian attitude to the Baltic states in the most important issue negotiated at that time, the troop withdrawal. However, although the troop pullout indeed unfolded slowly but surely, its completion did not pave the way for the development of good neighborly relations. In addition to the politics of representation as discussed in chapter 11, this situation was the result mainly of Russian hardliners' intransigence and great-power manners (which, in turn, contributed to the representations of Russia as discussed in chapter 11). Although the Lithuanian government exhibited

---

since August 1998. Besides effective cooperation between the Baltic navies, the squadron is being used for mine sweeping and countering, search and rescue operations, and international peace operations. The Baltic Regional Airspace Surveillance Network (BALTNET), established in late 1996 following a U.S. sponsored feasibility study (Regional Airspace Initiative), is a Baltic airspace surveillance and communications network, operational since May 2000 and integrated into NATO's Integrated Air Defense System on 7 April 2004. The Baltic Defense College was established in 1999 and provides staff-level officer training for officers from the Baltic states. And, finally, since 1997 BALTSEA serves as a forum for coordinating international military assistance for the Baltic states.

21. LitPolBat was inaugurated on 14 April 1999. The first joint exercises, however, had already been held in September 1998, and the decision to establish the battalion goes back to an intergovernmental agreement that came into force on 3 December 1997 (Möller and Wellmann 2001, 113–14).

considerable readiness to make concessions, the Russian leadership adhered to uncompromising positions regarding four components of the troop pullout: military and civilian transit to Kaliningrad; most-favored nation status for Lithuania; border demarcation; and territorial integrity, which at that time was frequently violated by the Russian air force. The conflicts were often and quite arbitrarily linked with one another, thus making a resolution even more difficult. The Russian intransigence showed that the Brazauskas government's obliging attitude toward Russia did not totally pay off. Brazauskas's policy faced difficulties mainly because it "coincided with the strengthening of Russian hawks" (Skak 1996, 283).

Furthermore, even though the troop withdrawal was perceived and represented as the most important issue in Baltic-Russian relations at that time, it was not the only one. The considerable electoral success of Vladimir Zhirinovsky's totally inappropriately named Liberal-Democratic Party of Russia in December 1993 was profoundly disturbing. It seemed to show that Baltic independence could not yet be considered irreversible and that the patterns of interaction in the Baltic Sea region could indeed be reversed toward enmity.[22] Sustainable independence was seen to require military guarantees from the Western states. At the same time, however, the hardening of the Russian position toward the Baltic states and Zhirinovsky's success deterred NATO from extending security guarantees to the Baltic states. NATO evaded the issue by initiating the PfP program—according to Meri, a "nice but empty scent-bottle."[23]

With respect to timing, two more things may be said. The completion of the Russian troop withdrawal in autumn 1993 facilitated the decision to

22. In a September 1992 interview in a Lithuanian newspaper, Zhirinosvky said that "the territories of the Baltic republics are native Russian lands. I'll destroy you. I'll bury waste in Smolensk *oblast*, along the border. I'll move the Semopalatinsk test site to your area. You Lithuanians will die from diseases and radiation. The Russians and Poles I'll evacuate . . . Soon there will be no Lithuanians, Estonians, and Latvians in the Baltic. I'll act the way Hitler did in 1932" (cited in Simonsen 1996, 182). Zhirinovsky's Liberal-Democratic Party of Russia won 22.8 percent of the vote in the 1993 Duma elections.

23. Cited in *Osteuropa-Archiv* 1994, A314. According to Korbonski, in eastern Europe PfP was "either received with contempt as a meaningless gesture and a poor substitute for full NATO membership, or seen as still another reaffirmation of Washington's traditional lack of interest in the area of Europe between Germany and the former Soviet Union" (1995, 130).

apply for NATO membership and made the application's success, however unlikely it may have seemed at that time, more likely.[24] The Polish push for NATO membership, expected at the NATO summit on 19 January 1994, furthermore urged Lithuanian decision makers to act (Knudsen and Neumann 1995, 17). If Poland were admitted to NATO and Lithuania were not, Baltic decision makers feared, Lithuania would (again) be located on the wrong side of the European divide, at least located in what was perceived as a gray zone between NATO and Russia or, worse, in the Russian sphere of influence, the legitimacy of which NATO would have seemed implicitly to approve by excluding Lithuania from membership. At the same time, should Poland be invited to NATO, but Lithuania should not, Poland's NATO membership could be used to help Lithuania gain membership on the basis of a newly invented strategic partnership. To this end, Poland had to be elevated from "the greatest threat to Lithuania" to a "strategic partner,"[25] and Lithuania's wish to enter NATO needed to be expressed unequivocally (T. Snyder 1995, 331–32). Furthermore, it had to be presented in such a manner that it did not emphasize the ostensible threat to Baltic security posed by Russia as Baltic decision makers consistently perceived it at that time.

It is a reasonable assumption that the Baltic strategy toward NATO at this time followed the notion that an invitation to NATO would hardly follow from representations of Russia as a threat to Baltic and European security. Representing the Baltic states as threatened by Russia and as a bulwark against (often presupposed) Russian aggressiveness would rather be counterproductive, for at least two reasons. The prospects of becoming entangled in a conflict between the Baltic states and Russia would rather induce NATO to avoid too direct and ambitious an engagement in the Baltic region. Moreover, NATO was representing Russia as a partner, and recommending oneself as a bastion against a partner would neither make sense nor be a

24. The expected failure of the application was one reason for Latvian and Estonian decision makers not to follow the Lithuanian example (*Osteuropa-Archiv* 1994).

25. At the end of 1991, the Lithuanian defense minister reportedly had designated Poland "the greatest threat for Lithuania" (Miniotaite 1998, 168). But in 1997, the presidents of Lithuania and Poland referred to the relationship between the two countries as a "strategic partnership" in their joint declaration of 19 June (Ministry of Foreign Affairs of the Republic of Lithuania 1998).

promising preaccession strategy.[26] Representing Latvia as "a front-line state for Europe" may have been an honest statement, but tactically it was ill advised (Valdis Birkavs, as cited in Main 1998, 182). Good or at least normal relations with Russia came to be seen as a precondition for membership, rather than membership a precondition for the development of normal relations with Russia.[27] The rhetorical adaptation to representations of Russia as unpredictable rather than outright hostile followed, as did advertising the Baltic states in terms of protecting "Europe" from general uncertainty and

26. Noreen and Sjöstedt dismiss what they call "a sophisticated realist balance-reasoning" because "it lacks the support of any empirical evidence" in the accessible sources at the Estonian Ministry of Foreign Affairs (2004, 742). For this strategy to work, however, it is essential that no respective references can be found in the openly accessible official material. Noreen and Sjöstedt emphasize the Estonian officials' genuine identification with the norms and values represented in NATO policy manifestations and with the Western political culture. Such identification shall not be disputed here. However, the authors, although analyzing the Estonian integration process in terms of a process of socialization, learning, and institutionalization of language, also come to the conclusion that the process of socialization "also includes a process of adapting the official Estonian language to the language of the West. A case in point is the language of security. In Western Europe, the United States, and in Scandinavia especially, one can discern a shift in the vocabulary concerning security policy issues as far back as the beginning of the 1990s. Most striking in this respect is that Russia is not characterized as a potential enemy, as was often the case during the Cold War. Instead, security must be maintained not against, but in cooperation with Russia. In addition, the NATO membership issue is not connected to assessments of possible threats from Russia, which is in line with the Westernized security language of the 1990s. Estonian policymakers are reframing security as well as their identity. Within this context, we also discern a tendency to talk about Russia or Russians in more inclusive terms" (747–48). This conclusion strengthens the interpretation offered here. If the Estonian integration process can be understood in terms of a learning process, then the decision makers have certainly learned how to represent the motivations underlying their policies in order to achieve their goals.

27. Some Baltic decision makers may have wished that, from a position of strength as NATO members, they would be able to force Russia to exhibit a more pro-Baltic attitude. Lieven, for example, has argued that "[some Baltic leaders] think that to 'cure' Russia of neo-imperialism and force it to become a Western-style democracy and submit to a junior part in a U.S.-led security structure, it needs to suffer another shattering geopolitical defeat, and that NATO expansion up to Russia's borders would provide such a defeat" (1996, 175). The Independent Task Force U.S. Policy toward Northeastern Europe also recommended this policy, arguing that "Baltic membership in NATO would help cure Russia of its imperial nostalgia" (1999).

unpredictability rather than singling out Russia as a potential source of danger (Saudargas 1999, 78). In itself, however, this strategy was hardly sufficient to solve the contradiction that for internal purposes Baltic decision makers kept representing Russia as the negative reference point, whereas for external consumption they depicted Russia as an at least potential partner. While developing arguments for Baltic NATO membership that aimed at an international audience, Baltic decision makers therefore increasingly sidetracked the Russia issue. They started both to emphasize what the Baltic states had in common with the Western states rather than what distinguished them from Russia and to advertise their strivings after membership in terms of a general (re)integrationist policy. It was a rhetorical adaptation to NATO's anticipated set of expectations and, in that sense, a learning process. It was also an expression of the myth of the return to the West characteristic of Baltic postindependence policy representations.[28]

By emphasizing NATO's character as a community of values (Öövel 1996, 67) rather than a community of defense, as a political organization (Birkavs 1997a, 20) rather than a military organization (respectively as an organization aspired to for political rather than military reasons), as something "new" (Saudargas 1999, 77)—"the new NATO"—Baltic decision makers adapted themselves to official NATO parlance beginning in the mid-1990s. Realizing that their prospects for membership were inversely proportional to both the volume of Russian protests and NATO's tacit acknowledgment of the protests' legitimacy,[29] Baltic decision makers also tried to dedramatize the impact of their NATO strivings on their relations with Russia. They repeated NATO's position that enlargement was directed against no one; it would increase everyone's security, even Russia's by providing for a zone of stability and predictability at Russia's western borders and thus enabling Russia to focus on the internal reform process. However, whereas Baltic decision makers represented NATO enlargement as "the end of a centuries-old

28. For patterns of argumentation directed at the domestic audience, see Christophe 1997 and Kuus 2002.

29. What mattered was not the volume of the protests in itself. NATO's tacit acknowledgment of the Russian protests as legitimate meant that NATO planners knew that if they were in their Russian colleagues' shoes, they, too, would have protested and thought of countermeasures.

era of power politics and spheres of influence" (Vike-Freiberga 2000b), others likely saw it as a continuation of these very politics.[30]

Ultimately, it did not really matter what the enlargement of NATO "really" was: throughout the 1990s, Russian decision makers followed their own logic and perception of NATO enlargement, and their perception certainly deviated from both NATO's self-image and the candidates' perceptions. Fortunately, NATO officials, while implementing enlargement and other policies, were much more cautious than the language they regularly applied would lead one to assume. They ultimately took into consideration the possibility that Russian reactions might not show a belief in NATO's "new" self-image, but rather a profound skepticism of it. NATO representatives seem to have expected their Russian colleagues to behave according to their own logic because, with reversed premises, they themselves would have exhibited the same behavior. They abstained from the position, aptly designated as "nonsense" by Michael MccGwyre, that "the future composition of NATO is solely a matter for its present members" (1997, 158). And again, it is in one sense not relevant whether the expected Russian reactions effectively materialized because the possibility that they could have materialized was in itself a sufficient cause for NATO decision makers to display a cautious enlargement policy. Preventing NATO-Russian relations from collapsing carried the day, so that defense guarantees were not provided to the Baltic states too early. That the Baltic governments were "more concerned with the idea of obtaining NATO membership, rather than [with] what happens the day after they have become members" (Main 1998, 181), is hardly surprising. It is equally clear that NATO representatives were more concerned with what would happen the day after the Baltic states became members rather than with the idea of Baltic membership in the first place.

NATO thus seems to have followed Daalder and Goldgeier's assessment that a "Europe without Russia cannot be peaceful, undivided and democratic" (2001, 83). Throughout the 1990s, NATO prioritized good relations with Russia rather than jeopardizing these very relations by inviting

30. The long-term Russian policy guidelines toward the Baltic states, published on 14 February 1997, described NATO enlargement as "eastward expansion to fill a security vacuum left by the end of the Cold War" and argued that NATO was "merely act[ing] according to old geopolitical considerations" (Herd 1997b, 252).

the Baltic states to membership without a convincing strategy to make the Russian decision makers accept that membership. What was considered "convincing" was to a large extent decided in Moscow, to a lesser extent in Brussels, Washington, and other NATO capitals, and to an even lesser extent in Tallinn, Vilnius, and Riga. In effect, during the 1990s this juggling act amounted to both cautiously implementing NATO enlargement as if Russia had a veto (while strongly denying this scenario in public) and dealing with the Baltic issue in a way that the Baltic states were likely to perceive as a Western punishment for having been occupied by the Soviet Union. This strategy showed that NATO took seriously into consideration the advice that the alliance should be more interested in its own strategic aims and less in the situation of individual candidate states—even more so because it apparently saw good relations with Russia as an effective interim strategy to assure the security of the Baltic states.

In terms of security community building, the opaqueness and ambiguities inherent in NATO's approach to the Baltic states during the 1990s were not the worst of all possible scenarios. After all, in the original writings on security communities, military alliances were found to be a rather poor pathway to integration. The forms of cooperation initiated to postpone NATO enlargement or to cushion its consequences, among other things, may indeed have been more important to security community building in the Baltic Sea region than enlargement itself—by, for example, ensuring among all Baltic Sea states "more information about one another, more attention to that information, more joint operations, and more actual contact" (Deutsch et al. 1957, 164). To be sure, NATO enlargement, if implemented cautiously, is not the condition for the impossibility of a security community in the Baltic Sea region. This is especially so if NATO and Russia intensify their cooperation, for example, on the basis of the agreement to set up a Russian liaison group at NATO Supreme Headquarters Allied Powers Europe or in the NATO-Russia Council, which is said to have been "particularly valuable" in the early 2000s because "it ha[d] the potential to ease any tensions over [NATO's] projected expansion to Russia's western neighbors" (IISS 2003, 88).

When the Baltic states' joined NATO in May 2004, the period of testing as to security community building in the Baltic Sea region entered a new phase (discussed in the next chapter). Neither Poland's nor the Baltic states' membership in the alliance seems to have had severe negative consequences yet. The perspectives for a smooth implementation of Baltic NATO member-ship seem to be brighter than they were a few years ago, when Moscow "ap-peare[d] to have drawn its line in the sand at the Baltic borders" and NATO was said to have to "calculate precisely the risk it would be taking for cross-ing that 'red line' " (Black 2000, 221). President Putin has shown a more re-laxed (or defeatist) attitude to the question of Baltic NATO membership than his predecessor and accepted it even before the "war on terrorism" brought the United States and Russia closer together (Kramer 2002, 748), thus relaxing the attitudes toward NATO enlargement in both Washington and Moscow and reducing it to a question of only secondary importance. However, although NATO enlargement does not seem to have triggered an increase in tensions in the Baltic Sea region, it has not sparked new momen-tum as to security community building, either.

14

A Period of Testing

SECURITY COMMUNITIES allow a period of testing to make sure that there really is a security community. In the original writings, this period is characterized by a process of slowly withdrawing troops from common borders, followed by the demilitarization of the internal relations once integration is assured. Surely, this path is but one way to a security community among many others. According to a reading of security community building as a cognitive project, during the period of testing something else has to happen: the crossing of a mental threshold as a result of a process, in the course of which groups of people develop a sense of community and expect the elimination of war as a social institution. Having crossed this threshold, general agreement is reached about peaceful change. Integration follows from a process of social learning; it includes the possibility of unlearning.

The term *threshold* thus should not lead one to misinterpret integration in terms of "once and for all." Rather, integration may involve a broad space of transition, during which time the threshold may be crossed and recrossed several times. Security community building is not a linear process, and it may take decades or even generations until dependable expectations of peaceful change are finally established and fully internalized. Only then are groups of people integrated. They know that intragroup conflicts will be handled without recourse to violence because this knowledge is socially defined as reality and is individually as well as collectively experienced, remembered, and subsequently expected. Means of force may be maintained for more general disposition, but within the community they are not perceived as threatening any longer; their use in intracommunity affairs has virtually become unthinkable. The security dilemma is ameliorated to the extent that it stops being a dilemma in its literal sense as a choice between alternatives that are equally unfavorable to the actor. Although security communities are not systems of collective security, national security interests are to some extent absorbed in and amalgamated with community security interests. Both community interests and national interests are defined in terms of peaceful change. Although other norms and values are negotiable, peaceful change is not.

Paradoxically, following from the introverted reading of security, security communities do not cease being security communities even if they become *in*security communities for outsiders (for example, through military aggression of one or more members against nonmembers). In the absence of worldwide integration, regional security communities say something about the social relations within the community, but relatively little about the community's relations with outsiders. Security communities consisting of liberal democracies may be inclined to peaceful relations toward other states that are equally seen as liberal democracies. States and societies, however, that do not fall within this frame of representation may perceive as threatening the military means that the members of a security community are permitted to maintain for general disposition. The security dilemma may thus be ameliorated within the community, but it may still govern the relations between members and nonmembers.

Moreover, the apparently growing intolerance of "liberal" democracies toward those states and societies that are based (or are represented as being based) on a different set of values and norms shows that even security communities composed of liberal democracies may present a threat to others. This caveat does not deprive the concept of value, but it shows its limits especially when security communities are equated with one particular form of political and societal organization. It may furthermore be seen as an invitation to think more carefully about the inside-outside *problematique* that neither the original writings on security communities nor the current social constructivist adaptations have managed to address adequately. The writings on security communities sometimes display a self-congratulatory attitude, as if problems are always owing to nonmembers. In chapter 3, a more self-critical attitude is suggested, one that does not absolutize one's own values and that emphasizes compatible rather than common values and, by doing so, appreciates rather than eliminates difference.

In the Baltic Sea region, the appreciation of difference gathered momentum in the mid-1980s with the coming to power of Mikhail Gorbachev and with the subsequent changes in Soviet foreign policy. The period of testing as to security community building was started. The basic lines of thought underlying the new thinking in Soviet foreign policy rendered impossible the use of military force to prevent both the Baltic states from gaining reindependence and the Soviet Union from disintegrating. A place for the Soviet Union in the common European house necessitated dealing with the Baltic issue through negotiation rather than through force. The use of force would have meant success for the enemies of perestroika, but the nonuse of force was more than just instrumental. It reflected substantial ideational changes in the Soviet leadership's understanding of foreign and security policy, in which the nonuse of force was pivotal. The reading of the dissolution of the Soviet Union and the reindependence of the Baltic states in terms of a dedication to peaceful change on the part of the Soviet leadership does not minimize the equally strong dedication to peaceful means on the part of the Baltic popular movements. Neither does it neglect some Soviet violence, displayed in the Baltic region occasionally and elsewhere in the Soviet Union more systematically. It indicates, however, that the Soviet leadership was at that time convinced of both the necessity and the possibility to resolve the Baltic issue without recourse to violence. The Baltic popular movements were skeptical of this change and, in light of Soviet-Baltic history, quite understandably so.

Security community building requires at one point a break with conventional orders of security production in terms of military security, a balance of power, membership of alliances, strict border regimes, and so on. Throughout the 1990s, however, the security policies of Estonia, Latvia, and Lithuania persevered to some extent in these conventional frames of security policy. The patterns of argumentation referring to history, geography, and material capabilities have proven to be powerful means with which to prevent security from being reimagined and rethought. These patterns of argumentation were rehearsed throughout the book and need not be repeated here. From the mid-1990s onward, academics have thought and rethought the concept of a security community and applied it to regional settings, among which the Baltic Sea region has held an eminent place. In chapter 3, I reviewed the recent writings on security communities, with special attention given to Alexander Wendt's friendship pattern and types of anarchy in international politics. The term *friendship* refers to interstate relations based on both nonviolence and mutual assistance and thus combines security communities with collective security. Although collective security has usually been conceived of globally, Wendt constructs it as a regional security pattern, arguing that security subsystems may be relatively autonomous. However, whatever else governs the international relations in the Baltic Sea region, it certainly is not friendship as defined by Wendt. Although within the Baltic Sea region there can be discovered interstate and even multilateral relations based on expectations of peaceful change, the international relations in the whole region are not governed by the logic of a security community, let alone by that of a friendship. In particular, the Russian Federation is conspicuous by its absence, in most scholars' representations, from existing or emerging security communities among the Baltic Sea littoral states.

The exclusion of the Russian Federation may be a result of distrust resulting from historical experiences with the Soviet Union up until the mid-1980s (a recurrent theme throughout the book is the identification of the Russian Federation with the Soviet Union in political and academic representations). However, it may also follow from two ingredients of the recent writings on and the political representations of security communities alluded to already. The first is the narrow reading of security communities in terms of liberal democracy, which obscures parts of the original concept's potentialities—namely, expectations of peaceful change encompassing liberal democracies, on the one hand, and polities perceived as organized according to different principles, on the other. The second is the replacement of compatible values by common values in some social constructivist approaches to security communities.

Apart from the value of "peaceful change" that the members of a security community indeed have to have in common, the compatibility of other norms and values is stressed in this book, for three reasons. First, groups of people may disagree on many major values in terms of which they define their identities and interests and as a reflection of which they make their political decisions. At the same time, they may agree upon peaceful change and nonviolence as the basic principle organizing mutual relations. Second, compatibility of values means taking the multiplicity, complexity, and ambiguities of values seriously rather than streamlining them. It also means accepting different sets of values as equally valuable. Ideally, it may result in a conception of security communities as nondominative and nonhierarchical forms of political and social organization, rejecting the idea that the core of a security community, by virtue of its power, can force fellow members to behave as it sees fit. Third, nonhierarchical and tolerant forms of social interaction within the security community may also lead to outsiders' perception of the security community as benevolent and thus may prevent conventional relations from arising between the members and the nonmembers of a security community.

In order to prevent internal peaceful changes from being accompanied by conventional outside relations, a spatially limited security community requires two more features. First, its external fringes must be understood in terms of frontiers or zones of contact rather than boundaries or strict lines of separation. To have a positive function in security community building, borders must therefore be conceived of as vehicles with which to tie communities together rather than separate them from one another. This is one of the problems resulting from the application of the EU's border regime to the Baltic Sea region and especially to Russia's Kaliningrad region, which promoted separation rather than cooperation and rendered difficult the understanding of borders as meeting places.

Second, to avoid conventional security dynamics between the members and the nonmembers of a security community, the maintenance of troops for general disposition within the community requires a high measure of sensitivity vis-à-vis the nonmembers. If disarmament, starting with withdrawing troops from common borders, is the ultimate proof of a security community's members' sincerity in their commitment to peaceful change, then it should also be given some credit for contributing to the evolution of dependable expectations of peaceful change between members and nonmembers, which ultimately would mean the security community's enlargement. In the Baltic context, the high concentration of Russian military power in the Leningrad MD is an obvious challenge to security community building (IISS 2004, 108). Another challenge is the Baltic states' membership in NATO, to which NATO secretary-general Jaap de Hoop Scheffer reportedly responded by stating that "there were no plans to create Alliance infrastructures on the territory of new member states" (IISS 2004, 100). Furthermore, NATO is said to have reassured Russia that it would not deploy nuclear weapons to the territories of those new members states bordering Russia (IISS 2003, 88) and to have "quietly dismissed" a suggestion by the Lithuanian minister of defense "that NATO should set up military bases in Lithuania or in other Baltic states" (Lachowski and Sjögren 2004, 722). However, it has been noted that "on 29 Mar[ch] [2004], the same day that the new members formally joined NATO, 4 Belgian fighter aircraft were stationed in Lithuania to patrol the airspace of the Baltic states, and 100 NATO troops were stationed at the same site to support the aircraft. Russia objected, but its reaction was not as strong as had been feared" (Lachowski and Sjögren 2004, 722).

From the point of view of security community building, the wisdom of this exercise may be called into question. Instead of implementing policies that are expected to result in strong reactions, a security community would employ a less introverted reading of security and define the community's security in a way that would take into consideration the security interests and perceptions of nonmembers or not-yet members. Security would have to be thought of as an intersubjective process rather than as something that states or clusters of states can perform in isolation. Security would have to be defined with others, rather than independent of or against others. In the second half of the 1990s, NATO, by claiming that its future composition is only of concern to its present members, rhetorically disregarded others' legitimate interest in that composition. Its actual enlargement policy, however, was much more careful than the statement suggests. The actual policy, not the enlargement rhetoric, should serve as an example for NATO's policy after enlargement in order to render a security community in the Baltic Sea region possible.

NATO enlargement seems to have influenced the evolution of a security community in the Baltic Sea region only modestly. This lack of influence is in accordance with the original writings on security communities, which tended to downplay the importance of military integration to security community building. NATO enlargement does not automatically result in a security community, but it does not make a security community impossible, either: a security community is the result of diverse processes of social interaction, which in themselves need not have anything to do with security, let alone with military security.

Chapters 6 and 8 showed the conceptual understanding of security in the reindependent Baltic states—in other words, what issues were referred to as security issues. At the beginning of the 1990s, Baltic decision makers clearly identified security with independence, and they thought sustainable independence required national military capabilities. In retrospect, this understanding is surprising because completely nonviolent popular movements had achieved independence. Nonviolence was the condition for the possibility of the movements' success. Once independence was achieved, however, sustaining it was said to necessitate military means; the nonviolent path to independence was said to require a military change in direction. Prioritizing military means and thinking about security in military terms were perhaps reflections of one important military issue that entered the agenda immediately after independence—namely, the negotiations on the CFE Treaty. After all, the CFE Treaty was also referred to among Western observers as the cornerstone of European security. Aspiring to "return to Europe" and equating "Europe" with "western Europe" made the Baltic states adapt to Western interpretations. Perhaps the fixation on military security came about as a consequence of interpreting the issue of the Russian troop withdrawal as a military issue rather than as a political issue. These explanations are, however, not sufficient without referring to the conception of statehood underlying the independent Baltic states. This conception had—and still has—at least two dimensions. The first helps explain why Baltic decision makers, although being keen on pleasing the West, opted against Baltic membership to the CFE Treaty and, by so doing, made a decision that resulted in what they perceived as a deterioration of Baltic security: the concentration of TLE permitted in the Baltic MD in Russia's Kaliningrad region rather than its distribution among Kaliningrad and the Baltic states. The powerful narrative of state continuity had given the strivings for independence legitimacy, but now made it appear impossible to enter the treaty as Soviet successor states.

The second dimension of Baltic conceptions of statehood is its profound conventionality. Judy Batt has argued that "there is a long-neglected heritage of 'fuzzy state' thinking in [central and eastern Europe] that has been squeezed out by the steamroller logic of the sovereign nation-state." *Fuzzy statehood* is an umbrella term under which we can group "various trends apparent in Europe towards political fragmentation, the resurgence of regional and ethnic minority identities, and the devolution of state power" (n.d.). This heritage, however, was almost completely ignored in the reindependent Baltic states.[1] In its stead, conventional European nation-states were established based on, among other things, the classical state-sovereignty-security-military nexus discussed in chapter 4. This latter conception may perhaps be seen as a corollary of having been sequestered for decades from political developments working toward unification in western Europe.

1. For a possible exception—the 1993 Estonian Law on Cultural Autonomy of National Minorities—see D. Smith 2001.

Many observers have also identified a return to the Europe of the 1930s as a motive underlying Baltic state construction in the early 1990s. The priority assigned to the construction of the state can be observed throughout the 1990s, during which time the construction of security, although rhetorically emphasized, was in fact pursued in the shadow of the construction of the nation-state, which was accompanied by a search for negative reference points, internally and externally. The representations of Russia, discussed in chapter 11, clearly fit into this logic, but also many decision makers were domestically in search of internal enemies in order, among other things, to divert attention away from the deficiencies of the internal reform process. These internal enemies were mostly imagined and existed mainly in the decision makers' thinking in terms of worst-case scenarios, which may be understandable given their perceptions of the fragility of statehood at that time. Importantly, the attempts in Latvia and Estonia to securitize the Russian military retirees failed. The general public refused to see them as a threat to security and, by so doing, served as a desecuritizing corrective against decision makers haunted by a Manichaean view of life.

Constructing the nation-state sooner or later required replacing the ad hoc security policies of the initial postindependence years with policies based on elaborated national security documents. The ad hoc policies cannot, however, be considered a failure. Regardless of their inexperience, the decision makers managed such complex and difficult processes of social interaction as the negotiations on the withdrawal of the Russian troops. They did so regardless of, among other things, a weak material and institutional position resembling quasi-statehood rather than statehood, limited external support, limited Russian readiness for compromise as to selected aspects of the troop withdrawal, and a rhetorical hardening of the Russian position vis-à-vis the Baltic states.

As chapter 8 showed, the writing of national security documents proved to be difficult. The documents retained the narrow focus on military security and enumerated an endless list of internal and external threats and possible threats, risks and possible risks, as well as challenges and possible challenges to national security. As always, such enumerations reveal some degree of analytical helplessness (Albrecht 1986, 34) and prepare the ground for practical inapplicability and governmental overstretch. However, the national security documents of the Baltic states should also be read as political documents that aimed to indicate the degree of progress these states had made in their adaptation to the patterns of representation and practices prevalent in western Europe. The Baltic decision makers obviously wanted to be seen to be approaching Western lines of thinking on security and Western ways of implementing security policies. The incorporation into the documents of a broad reading of security instead of a narrow one focusing on the military should thus not be taken at face value.

Even today the military dimension of security is still given an important role in the overall conceptions of Baltic security. Introducing into the documents a broad reading of security certainly reflected learning to some extent, but it also was an adaptation to the representational modes in the West and especially to NATO's security formula, which throughout the 1990s defined the alliance as a political rather than a military organization. From this followed both the almost formalistic modes of expression with which the wish to enter NATO was promoted in the second half of the 1990s, in language that closely followed the modes of representation established in Brussels, and the conspicuous disappearance of Russia as a threat to Baltic security from both the Estonian national security document, which was the last one to be written, and the policy manifestations in all three Baltic states.

To be sure, documents do not necessarily reflect what decision makers "really" believe, and they do not necessarily preordain subsequent policies. Still, documents are important representations of the elite's self-understanding and vehicles for framing security policies. However, because "the individual text simply does not work as a stable reference point" (Couldry 2000, 77), documents should not be read in isolation; rather, they should be contextualized. For example, adhering to a broad understanding of security in policy manifestations and documents by no means necessitates abandoning the de facto focus on its military aspect. Such a representational strategy may mean adhering to military security conceptions behind the broad public-security smokescreen. In the case of the Baltic states, throughout the 1990s the focus on the buildup of armed forces, membership in a military alliance, and the lack of appreciation for the Clinton administration's foreign and security policy in northeastern Europe show the perseverance of an essentially military understanding of security that is camouflaged in documents but adhered to in practice.

Concepts of security vary over time and across actors. Chapter 10 showed that especially the Clinton administration's conception of security for northeastern Europe differed substantially from both the initially quite simplistic security concepts of the Baltic governments and the equally simplistic role assigned to the U.S. government as an advocate of Baltic NATO membership. The designation of the United States as a "friend" is both conspicuous by its positive emotional attachment and indicative of the importance that Baltic notions of security attached to the United States. Yet according to the historical reading presented in chapter 9, the historical relations between the United States and the Baltic states hardly gave a reason for excessive Baltic optimism as to the degree of amalgamation between U.S. and Baltic interests. The U.S. policy toward the Baltic states always reflected U.S. interests, which were occasionally defined very narrowly indeed.

A third-degree internalization of friendship resulting in U.S. assistance to the Baltic states even in situations that, from a strictly national point of view, might suggest nonassistance could not be expected. Referring to the United States as a "friend" therefore was an attempt to bring moral pressure to bear upon the United States, a procedure that U.S. decision makers had almost invited by representing U.S.-Baltic relations in a very moralistic tone. During the 1990s—that is, before the Baltic states were invited to NATO— referring to the United States as a "friend" was a political-rhetorical strategy, the aim of which was to motivate the U.S. government to act as if it were a "friend" because friends are supposed to help each other even when not formally allied. Cultivating the United States as a "friend" followed from representations of international politics in northeastern Europe in terms of what Wendt calls a Lockean kind of anarchy, characterized by the occasional occurrence of violence (although the other's right to existence is in principle respected).

Chapter 10 explored the evolution of the U.S. foreign policy from a policy toward the Baltic states to a policy toward northeastern Europe (and increasingly toward northwestern Russia) and from a policy emphasizing a military path to security to one focusing on nonmilitary means. Early attempts to deal with the Baltic states were fairly conventional and indicative of the U.S. government's wish not to become entangled in Baltic-Russian affairs too directly. In particular, explicit or implicit security guarantees were avoided. At the same time, some building blocks for Baltic security outside NATO were constructed, and it is difficult to say whether this construction was intended as a temporary or a permanent strategy. If one relies on policy manifestations, one can say it was an interim strategy. This strategy, however, gathered momentum to a degree probably unforeseen even by those

U.S. decision makers who initiated this policy. It developed from a rather simple and unoriginal set of building blocks for Baltic security to a sophisticated policy approach that, however, was unoriginal as well. The U.S. policy toward northeastern Europe is indeed the rare case of a U.S. foreign policy adaptation to a regional environment rather than a superposition on that environment, and it is an equally rare case of a U.S. foreign policy that emphasized learning from the regional actors. By synchronizing especially the NEI with the EU's Northern Dimension and by adapting the U.S. policy to existing patterns of cooperation in Europe's north, the Clinton administration both assigned to other actors, especially the EU, the task of playing the leading role in northeastern Europe and performed a nonprovocative policy that helped ease the remaining tensions in the region.

Although copying the policies that had already been performed for quite a long time by other actors, especially the Nordic states, the Clinton administration's policy toward northern Europe probably contributed to the reimagination of security in the region more than did similar policies pursued by others. Here, especially the Baltic governments became to some extent entangled in their own politics of representation. They simply could not afford to reject policy initiatives by their U.S. "friend" without the risk of undermining the U.S. support for Baltic membership in NATO. All the same, the Baltic governments' enthusiasm for the U.S. policy approach toward Europe's north was lukewarm and obligatory at best, probably because the United States seemed to repeat here what many Baltic decision makers considered the basic deficiency of the EU when it comes to security—namely, the lack of a credible military and defense dimension (Berzins 1999, 56). Indeed, during the Clinton administration, the U.S. policy toward northeastern Europe conceived of security in the region in far more than just military terms. To borrow a term used earlier, it may be called *fuzzy* security in the sense that it envisaged cross-border cooperation and regional linkages among a variety of actors, thus breaking with the strict separation between "us" and "them" that the Baltic governments preferred in their relations with Russia. It aimed to create zones of contact rather than dividing lines and to encourage NGOs and nonstate actors to participate actively in network and security building in the region. Thus, it not only promoted interstate cooperation, but also put emphasis on intersociety cooperation "below and beyond" the nation-state (Chaturvedi 1996). It aspired to establish a sense of community and security among groups of people in northeastern Europe.

Approaches to international and intersocietal relations that go "below and beyond" the nation-state have many similarities with the original concept of a security community. Indeed, bringing the society in was an important, albeit largely unexplored ingredient of the original writings on security communities. These writings did not exclusively refer to interstate relations, but rather to a sense of community among groups of people in general. "Groups of people" may refer to nation-states, but it may also mean social groups within nation-states, noncontiguous groups—what Emanuel Adler and Michael Barnett call "cognitive regions" (1998b, 33)—and, most relevant in the present context, cross-border communities, thus adding to their respective national identity further identities with respect to place, region, values, language, religion, a common cultural heritage, and so on. Identities can be thought of as concentric circles, among which the national identity is but one layer. They can furthermore be conceived of as projects rather than properties; they are flexible and susceptible to change. This change may follow from, for example, cross-border cooperation, which after the Cold War gained new momentum. However, the physical borders of the Cold War period had also resulted in mental borders. Like every cultural change, tearing down mental borders is more difficult and time-consuming than initiating institutional reform—and not just in transition societies (Lauristin 1997, 27).

If applied for a sufficiently long time, the U.S. policy, by promoting cooperation in such traditional low-politics areas as environmental protection, may have helped pave the way for cross-border cooperation in high-politics areas. Even within traditional high-politics areas, low-politics issues may have been linked with high-politics issues, thereby obscuring the once inpenetrable border. Taking this line of thought further, security may also have been made an affair of ordinary citizens by dispossessing security of its aura of secrecy, urgency, and high politics, and by disassociating security from the state and the military. Security and a sense of security may have been linked with citizens' everyday experiences and interests (van Ham 2000). The U.S. policy toward northeastern Europe may have made military security appear unnecessary in the long run as a result of the increasing insignificance of the military in intraregional and intersocietal relations and as a result of a growing sense of security among the citizens stemming from daily cooperation in fields of practical relevance to everyone's everyday lives.

The substance of the U.S. policy toward northeastern Europe was to a large extent to be found in its language and the consistency of the language. Especially the nonarticulation of security was an important ingredient of the U.S. policy, which indicated that northern Europe did not require extraordinary means to deal with problems and conflicts. The nonarticulation of security furthermore contributed to security community building because in some cases "security might be furthered exactly by the downgrading of security concerns" (Wæver 1998, 77). This strategy requires referring to a traditional security issue in nonsecurity and nonurgency terminology; it means dedramatizing an issue by not calling it a security issue. Without any doubt, the U.S. policymakers were aware of both security, conventionally defined, in the Baltic Sea region and the perception of it by the state actors in the region, especially in the Baltic states, with its emphasis on discrepancies in military capabilities and a lack of security guarantees. However, although the U.S. policy acknowledged the existence of conflicts, it aspired not only to resolve them within the normal rules of the political game prevalent in and among Western democracies, but also to deal with them as political, social, economic, and environmental problems rather than as military problems. Issues were politicized but not securitized. Accordingly, solutions to the problems were sought in the political realm and not in the military realm.

To be sure, the ideas underlying the U.S. projects in Europe's north changed to some extent over time, the focus of the projects narrowed, the initial selectivity evaporated, some of the envisioned synergy effects remained unfulfilled, and duplication loomed (Palosaari and Möller 2004, 265–73). Furthermore, the basic lines of thought underlying the NEI frequently seem to have been either misunderstood or correctly understood but rejected. In connection to security community building, however, the U.S. policy in northern Europe was important precisely because of those ingredients that were most often criticized: its lack of a military dimension, its emphasis on language and cooperation, its focus on cooperation among citizens rather than between states, and its ambition to alter thought patterns, including those on security.

Under the current Bush administration, not much public attention is being devoted to northern Europe, but the U.S. policy toward northeastern Europe has never made it to the front pages, anyway. During the Clinton administration, too, the policy was conspicuous by its pleasant lack of grand gestures, solemn language, and the sensational. Recent reformulations, however, indicate that only modest changes in language may result in considerable changes in substance. The NEI is currently experiencing several alterations in its redesign as the e-PINE. The initiative is now explicitly linked to conventional security issues and related to negatives rather than to positives. The NEI is being retrospectively diminished to a deliberate and target-oriented but also fairly traditional approach to the integration of the Baltic states in NATO. Its emphasis on cooperation with Russia has been replaced by a somewhat vague eight-plus-one approach. Ironically, these changes, although reducing the importance of the U.S. projects in Europe's north with regard to security community building, may increase the U.S. policy's legitimacy among the political decision makers of the Baltic states. After all, Baltic leaders have always been fairly skeptical of the idea of a security community in Europe's north.

The U.S. project in the north of Europe seems to enjoy sufficient domestic support to be continued, albeit in a different form. The lack of emphasis on cooperation with Russia is perhaps the most obvious alteration. However, after September 2001, U.S.-Russian cooperation has evolved in many other forums so that the lack of emphasis on cooperation with Russia in e-PINE does not have to be indicative of an overall lack of U.S. interest in cooperation with Russia. As a strategic partner of the United States in its "war on terrorism," Russia is for the time being deotherized and depicted as a part of "us," although what links the United States and Russia together seems to be primarily a common goal to pursue, without foreign interferences, national interests in their respective "wars on terrorism." The fragility of this representational mode means that should the U.S.-Russian cooperation fail, negative effects on northern Europe cannot be ruled out, the more so because the Baltic governments' support of the U.S. position on the "war on terrorism" has so far been very audible indeed.

Russia's "return to the West," its representation in terms of "us" rather than "them," was prepared for during the 1990s by efforts to integrate the Russian Federation in many political and societal networks in northern Europe. This effort was a basic ingredient of the U.S. policy toward northeastern Europe from its beginning and was opposed to the negative representations of Russia in the Baltic states. From the Baltic point of view, there was too much talk about integrating Russia in the West (considered by many

Baltic observers as bound to fail anyway) and too much consideration of Russian sensitivities. From a Baltic perspective, Russia was the security problem, not its solution. But Russia was more than that. During the 1990s, it served as the negative reference point against which the Baltic nation-states were being constructed. Wrapped in "security" language, the security policies toward Russia were often a part of the nation-state-building projects. These projects required downplaying the Russian Federation's support for the independence of Estonia, Latvia, and Lithuania; the nonviolent process of the Russian troop withdrawal; and the reactive and quite reasonable Russian foreign policy in the Baltic Sea region. It also required identifying Russia with the Soviet Union and eliminating from collective memory any traces of positive representations of the Soviet Union. Gorbachev, for example, is not being remembered as the person who initiated new thinking. Rather, he is depicted as the man who in 1991 "unleash[ed] the military and use[d] tear gas, rifles, and tanks against unarmed people" (Vardys and Sedaitis 1997, 176).

Immediately after the international recognition of the independence of the Baltic states, the representation of Russia in the Baltic states indeed changed from its being a supporter of Baltic independence and security to its being a threat to them. Some aspects of Russian politics and rhetoric at that time certainly facilitated this change. During the 1990s, however, Russia was permanently represented and in part constructed either as a direct or indirect threat to Baltic security. Either Russia's actual or potential strength or its current weakness, its predictable malevolence or its unpredictability, its material capabilities or, if the current capabilities were regarded as insufficient, its potential future capabilities or even its presupposed intentions were seen as threatening the Baltic states. The stability of this representational mode is not totally surprising given the impact of past hostilities on present perceptions and politics. However, cultivating Russia as a threat to Baltic independence had also the function in domestic politics of constructing and strengthening collective Estonian, Latvian, and Lithuanian subjectivity. Tensions along ethnic lines in the Baltic states, especially in Estonia and Latvia, were consolidated or created in the first place; relations with Russia entered a vicious circle of mutually intensifying allegations and accusations; and domestic opposition was largely silent because any kind of opposition could easily be represented as a threat to national security.

Likewise, the interpretation of both the Soviet presence in Estonia,

Latvia, and Lithuania until 1991 as a continuation of World War II and the years following independence as a state of emergency in the presence of a permanent Russian threat served the consolidation of the nation-state. The continuous confirmation and reconfirmation of the self unfolded, among other things, through the equally continuous confirmation and reconfirmation of others as the (ontological) Other in wars, imagined wars, and narratives of past violent encounters. This is also one of the reasons why the Baltic governments participated in the NATO-led intervention in the former Yugoslavia: we-feeling was furthered by participating in a war rejected by the Russian leadership, which thus "confirmed" its Otherness and "rightful" exclusion from the West. This is also one of the reasons why the term *peacekeeping* had to be erased from the title of the Baltic Peacekeeping Battalion and why the scope of the battalion's missions had to be expanded, enabling Baltic participation in NATO-led operations even without UN or OSCE mandate. Parallel to the expansion of NATO's missions as exemplified in the Kosovo war, tensions between the battalion's activities and Russia's perceived interests could not be excluded. Such a development could then have been easily represented as a confirmation of Russia's Otherness as well as its malevolence and hostility to the Baltic states.

Chapter 12 presented a case study on the relationship between representations of the past and present policies. The chapter explored the use of representations of the anti-Soviet resistance movement, the Forest Brothers, as a legitimacy provider for the current security policies in the Baltic states. It was not claimed that these representations are the only or even the most important legitimacy provider for the current policies; rather, they are one legitimacy provider among others. It was argued that the representations of the resistance movement follow to a large extent rather simplified representations of the past that have more to do with the perceived requirements of the present than with an adequate representation of the past. If constructed by and incorporated into national historiography, collective memory, narratively constructed in the form of a tragic story, frequently becomes nonnegotiable. Deviating views can be adhered to only at the price of exclusion from the collective self. The level of coherence within the self is overemphasized, just as is the level of coherence within the Other and consequently the difference between the self and the Other. Representations of the past are inclined to see past enemies as future enemies rather than to assign past enmities to history. Security communities have a history, but it is equally important to

acknowledge that they also have a memory and that this memory, by per-
petuating representations of past enmities, is often an obstacle to security
community building.

Deutsch and his team saw partial identification between self and another as a precondition for a sense of community, which, in turn, they saw as
a precondition for security communities. To repeat what has already been
said in chapter 12, the words *partial* and *identification* are equally important
here: *identification* because it neglects the notion of a complete separateness
between "us" and "them"; *partial* because it emphasizes that each person
carries with himself or herself a unique bag of experience that is an individual property and cannot be reduced to patterns shared in common with others. Collective memory and its construction by means of official national
historiography perpetuate the self/Other dichotomization within the Baltic
states as well as between them and the Russian Federation, instead of helping them to define "us" and "them" in terms of different degrees of identification with one another. This perpetuation of difference, in turn, makes the
development of prefriendship (not to mention friendship) patterns between
the Baltic states and Russia difficult or, from the point of view of constructing nation-states, even unwelcome: Others cannot be friends and prefriends, but they are needed as a negative reference point.

As stated at the beginning of this chapter, security communities allow a
period of testing during which time a process of demilitarizing bilateral relations, starting with withdrawing troops from common borders, may unfold. Military developments in the Baltic states, however, are moving in the
opposite direction. These developments show a militarization in its most
basic sense as a growth in military potential. In the present context, it does
not really matter that the actual and future military potential of the Baltic
states is and will be fairly modest. Indeed, the Baltic states failed to realize
their ambition to spend, at the end of the 1990s, 2 percent of GDP on defense
(SIPRI 2003, 355). This failure reflects economic difficulties rather than a
change in thought patterns. Subsequent to the reestablishment of independence, these thought patterns have indeed been quite stable. Nonmilitary security conceptions, discussed and followed in the period leading
to independence, were abandoned once independence was achieved. Neutrality was thought of only in terms of armed neutrality and rejected on
financial and historical grounds. Furthermore, the military was and is understood as a constitutive element of the nation-state. The factual emphasis

put on the buildup of armed forces, intra-Baltic military cooperation, and integration into NATO throughout the 1990s indicates that security is still being thought of predominantly in terms of military security, and this view relativizes the broad understanding of security used in recent security documents. Likewise, the majority of scholarly writings on Baltic security, habitually taking the buildup of armed forces for granted, have focused on such technical issues as operational capabilities, military expenditures, adaptation to NATO structures, military cooperation, and so on. In so doing, they naturalize armed formations and obscure their internal function in terms of strengthening the collective self, constructing domestic order, and imagining historical continuity.

In the initial phase of independence, however, the armed forces did not seem to be particularly appropriate vehicles for the construction of collective identity. Armed forces in general were profoundly delegitimized, and national armed forces were simply nonexistent. Given scarce resources, it appeared advisable to focus on social and economic problems rather than on the buildup of armed forces, the military utility of which in case of an armed aggression would be limited anyway for the time being. For most Baltic decision makers, however, Soviet armed forces served as a negative reference point, whereas armed forces in general did not. Flirting with demilitarization was indeed only a brief intermezzo. Critical populations thus had to be set right and integrated into the state-military nexus through, among other things, universal conscription, total defense concepts, pre-military service training, the cultivation and invention of non-Soviet military traditions, the celebration of violent (albeit unsuccessful) resistance movements, and official contempt for deviating views. Thus, the internal function assigned to the armed forces in terms of imagining internal cohesion, strengthening the collective self, and supporting the idea of state continuity was as important as their operational capabilities.

In the 1990s, equally important to the construction of collective subjectivity were the voluntary territorial defense formations and voluntary paramilitary organizations. They were officially supported and represented as a connecting link between the state and society as well as between the current and the interwar Baltic states. In many cases, however, they were part of the problem rather than part of the solution. The voluntary forces' self-understanding as the real guardians of independence and as representatives of the "true" national values often inhibited civilian and even military control of them. Likewise, rivalries between the national armed forces, the voluntary territorial forces, and the paramilitary forces were quite common. In particular, the territorial forces' self-understanding as guardians of independence collided with the self-esteem of the regular armed forces, which reciprocated by displaying contempt for what they considered hobby soldiers. These rivalries were facilitated by the internal functions officially assigned to the armed formations, which for some time remained unregulated by laws. Strongly politicized armed forces were temporarily a source of insecurity rather than security. Thus, with respect to security, parts of the armed formations were of debatable merit.

The relations between the Baltic states and NATO were hardly as straightforward as most official narratives and retrospective rationalizations try to make one believe. Whatever else membership in the alliance means, it carries with it substantial obligations and narrows a member's freedom of action. Directly after having gained this freedom of action through independence, the Baltic states did not consider this limitation a particularly attractive prospect. Furthermore, the political wisdom of joining the alliance was being called into question. Some officials recommended establishing equally good relations with the "West" and the "East" rather than intensifying the relations with the "West" at the expense of the relations with the "East." Accordingly, approaching NATO was not the only security policy option discussed immediately after independence. Only in late 1993 did NATO membership begin to monopolize the security agenda in the Baltic states, facilitated by the completion of the withdrawal of Russian troops from Baltic territory. The increasing intransigence of some Russian leaders with respect to selected aspects of Russian-Baltic relations and Vladimir Zhirinovsky's electoral success in Russia at that time made the Baltic governments seek refuge with NATO, but made NATO become even more reluctant to consider Baltic membership.

Concealed in technical-operational language emphasizing the Baltic states' insufficient interoperability and compatibility with NATO forces, it became increasingly clear that NATO, although rhetorically adhering to an "open-door" policy toward the Baltic states, did not have the slightest wish to become entangled in Baltic-Russian controversies. Thus, Baltic NATO membership required good relations with Russia. In particular, the recurrent representation of Russia as a threat to Baltic security had to be adapted to the modes prevalent in NATO's representation of Russia as a partner. The

Baltic states had to replace recommending themselves as a bulwark against Russia with patterns of argumentation more easily digestible by NATO. Baltic decision makers thus started emphasizing what the Baltic states had in common with the Western states rather than what distinguished them from Russia. What they found to have in common with the Western states were norms and values. Thus, NATO's character as a community of values rather than as one of defense, as a political rather than a military organization, and as something "new" was being emphasized. Yet only at the NATO Prague summit of November 2002 were the Baltic states finally invited to join NATO in 2004. With Baltic membership in NATO, the period of testing as to security community building in the Baltic Sea region has no doubt entered a new stage, but its influence on security community building seems to be modest as yet.

More important to security community building than NATO enlargement in itself seem to be the initiatives launched throughout the 1990s, such as the PfP, in order to, among other things, cushion the negative effects of NATO enlargement. The enlargement of a military alliance is a rather neutral process that can affect security community building positively, negatively, or negligibly. It may be said that of all the actors analyzed in this book, the Clinton administration seems to have taken the challenges emanating from NATO enlargement most seriously. By launching the NEI, the U.S. administration not only introduced new patterns (at least new to it) of thinking about security in Europe's north. These patterns carried with them a huge potential for both building a security community and thinking about it in a nondominative way: a community living with difference rather than reducing it; a community conceiving of conflict as a path to social change rather than as something that has to be prevented; and a community recognizing different values as equally worthy.

Bypassing governments and focusing on nonstate and nongovernmental actors are most relevant in the present context. To recapitulate, this context is one in which, in principle, "security" is opened up as an issue for nonstate actors. Ordinary citizens can function as a desecuritizing corrective against the security establishment in that they—if not the governments— start expecting and believing in peaceful change. They can do so as a result of individually experienced cooperation across borders in policy areas that are relevant to them, such as environmental protection or health care, resulting in a sense of community and ultimately security. Delegating security only to the "experts" is not compatible with the ideas expressed here. Security can become relevant to everybody not only in the sense that it ultimately concerns everybody, but also in the sense that everybody can contribute to a sense of security. Such a security community requiring the participation of the people would be a rather ambitious construction and will, of course, take time, but integration is a matter of fact, not of time.

APPENDIX

REFERENCES

INDEX

Essential Documents

DOCUMENT 1

Basics of National Security, Republic of Lithuania

Part 2, Chapter 9, First Section

(19 December 1996, No. VIII-49, amended as of 4 June 1998)[1]

Part 2: The Risks and Threats to Lithuania's Security and the Security Ensuring System

Chapter 9: The Risks and Threats to Lithuania's Security

In ensuring Lithuania's national security those dangers and risks shall be considered that definitely may emerge under unfavorable circumstances to Lithuania.

First Section: Potential External Risks and Danger

The specific geopolitical environment, hardly predictable due to existing militarized territories and states of unstable democracy, is an important factor of Lithuania's national security.

The External Risks, Challenges, and Potential Challenges, and Potential Dangers Conditioned by the Geopolitical Environment

Political:

 • political pressure and dictates, attempts to establish zones of special interest and ensure special rights, preventing Lithuania from obtaining international security guarantees;

1. Official translation, made available to the author by the Lithuanian embassy in Berlin in November 1999. This translation of the Basics of National Security of Lithuania can be found as an annex to Ministry of National Defense of the Republic of Lithuania, *Annual Exchange of Information on Defense—1997,* valid as of 1 March 1997.

• threats by foreign states to use force under the pretext of defending their interests; and

• attempts to impose upon Lithuania dangerous and discriminatory international agreements.

Military:

• military capability in close proximity to Lithuanian borders;

• military transit through Lithuania;

• formation of illegal armed gangs or their invasion into the Lithuanian territory; and

• overt aggression.

Specific:

• spying and subversive activities of foreign secret services, including the establishment and activities of undercover organizations; organizing diversions and attempts on life; dissemination of subversive information and other activity undermining the civil society and the State by spreading propaganda and disinformation;

• direct or indirect interference into Lithuania's domestic affairs, attempts to influence government institutions, the Lithuanian domestic policy, and social processes;

• illegal immigration and transit migration; influx of refugees; and incitement of ethnic groups to disloyal behavior toward the State of Lithuania; attempts by other states to impose on Lithuania the principles of dual citizenship.

Economic:

• economic pressure, blockade or other hostile economic actions; dependence of an entire branch of [the] economy on a single country or a group of countries;

• investment of capital with political goals: the take over, through ownership or management control, of energy supply system and enterprises, financial and credit companies, key communication facilities (railway, highway, pipeline, seaports, airports) that are of strategic importance to national security;

• dependence of the energy supply system on the resources of one or a group of foreign countries, the vulnerability of the functioning of the energy supply system;

• high foreign indebtedness destabilizing the State's financial system; indebtedness to countries striving for political influence; and

• destabilizing interventions into Lithuania's financial-banking system, and the influence undermining it.

International Criminal Organizations:

• activities of organized criminal structures, illegal business, and smuggling;

• infiltration of terrorists from abroad and their activities in Lithuania;

• smuggling or illegal transit of or trade in weapons, narcotics, radioactive or other highly dangerous materials; and production and circulation of counterfeit money or laundering of money obtained through illegal means.

DOCUMENT 2
Basics of National Security, Republic of Lithuania
Part 2, Chapter 9, Second Section
(19 December 1996, No. VIII-49, amended as of 4 June 1998)[2]

Second Section: Potential Internal Risks and Domestic Crises

Political:

• political instability or crisis of State power posing a threat to the constitutional order;

• conspiracy against the constitutional democratic order and violation of the Constitution;

• breach of civil rights and freedoms or their restriction in contravention to the Constitution; and

• disregard for long-term national goals in the State policy.

2. Official translation, made available to the author by the Lithuanian embassy in Berlin in November 1999. This translation of the Basics of National Security of Lithuania can be found as an annex to Ministry of National Defense of the Republic of Lithuania, *Annual Exchange of Information on Defense—1997*, valid as of 1 March 1997.

Economic:

    • rise in unemployment, decline in production volume, and decrease of gross national product beyond a critical level;

    • structural and technological backwardness of the economy; and criminalization of [the] economy and uncontrolled economic crime rate;

    • crises inside the banking-financial system and financial panic; destabilization of national currency, depletion of currency reserves of the Bank of Lithuania below the critical margin;

    • internal debt exceeding the State's financial capability.

Social:

    • general impoverishment of the population, unemployment;

    • excessive differentiation in wealth approaching a critical level and threatening to provoke a social conflict, abnormal differences between [*sic*];

    • personal insecurity; and

    • worsening of the health condition of the members of the society; alcoholism, toxic substance abuse, and drug addiction.

National:

    • factors weakening the Nation's immunity and sense of identity; negligence toward national values;

    • national demographic decline; decrement in intellectual potential; large-scale emigration; and

    • spread of antihumanistic, pro-violence pseudo culture; [and]

    • instigation of ethnic hatred.

Criminal:

    • high crime rate;

    • corruption, especially within law enforcement and judiciary institutions, arbitrariness of the State and local government officials; and

    • activities of terrorist, coercive, and other criminal organizations; [and]

    • forging of money, documents, and securities.

Others:

    • accidents, industrial emergencies, fires, and natural disasters; environmental pollution, especially with carcinogens and mutagens; the Ignalina Nuclear Power Plant;

• irrational exploitation of natural resources and environment, its wasteful utilization; and

• especially dangerous epidemics and epizootic occurrences.

DOCUMENT 3
Security Concept of the Republic of Latvia
Approved by the Cabinet of Ministers on 6 May 1997 (Excerpt)[3]

I. General Principles

Latvian national security is protection and preservation of its state sovereignty, territorial integrity, language and national identity, democratic regime as provided in the Satversme (Constitution), market economy and human rights, and protection and guarantees for public and individual interests. In establishing its security, Latvia does not threaten any other state.

The security policy of Latvia is an integral part of state policy; it must guarantee Latvia's sovereignty by all constitutionally and legally available means.

Latvia's security situation to a large degree is determined by its geopolitical situation, economic development, relations with other countries, cultural potential, military and civil defense capabilities, and ecological situation.

National security can be achieved by ensuring political, social, and economic stability, developing effective military structures, establishing crisis management and civil defense systems, developing legislative and administrative structures based on the Constitution, observing international treaties and laws, and integrating into European and Transatlantic political, economic, security, and defense structures. . . .

III. Threats

Threats to Latvia that can be considered real are: activities aimed against the national independence of Latvia and its constitutional system; the political

---

3. From the Ministry of Foreign Affairs of the Republic of Latvia, unofficial translation. Available at: http://www.mfa.gov.lv/eng/policy/security/securconc.htm.

or economic subjugation or other types of dependence to or on other countries; the hindrance of Latvia's integration into European and Transatlantic structures, the unification of different social and ethnic groups into one nation, or economic and social development in Latvia, as well as delaying the improvement of its defense capabilities.

The threat can emerge in political, military, economic, social, ecological spheres, individually or in combination. External pressure, the unfavorable development of international political processes, can destabilize the internal political situation, aggravate conflicts between political forces and social groups within the country, and lead to a crisis. External and internal threat factors are closely related as they interact and combine.

IV. Averting and Reducing Threats

Internal and external threats can only be averted, reduced, and eliminated by coordinated and simultaneous action in all spheres of threat prevention.

Failure to promptly avert or reduce a long-term threat can result in a crisis. A crisis is a very real and initially short-term threat to the state's independence, democratic system, and to the population's health, welfare, and safety. A crisis can lead to the proclamation of a state of emergency. . . .

IV.1. Averting Political Threats

The primary supplementary and complementary mutual goals of Latvia's external security policy are: integration into European and transatlantic political, economic, security, and defense structures, firstly the European Union (EU) and NATO; maintenance of good relations with all neighboring countries, to include Russia and Belarus; cooperation with the most reliable security and defense organization in Europe, NATO; and the graduated realization of the possibility for joint activities and the eventual assumption of the duties and responsibilities of a full member state.

Effective crisis aversion and consultation mechanisms are being developed in conjunction with NATO, the EU, the WEU, the OSCE, the UN, and their member states.

A solid legislative basis for the prevention of any type of conflict and a control mechanism for inflexible adherence to international legal obligations is being developed bilaterally and in conjunction with international organizations. . . .

IV.3. Economic Threat Aversion

The government provides the necessary national economic infrastructure: clear and comprehensive laws, a stable and controlled banking system, a developed system of credit underwriting, and an effective mechanism for attracting investments; while constantly monitoring the balance of federal payments for the purpose of averting the possibility for control of national financial resources by another country.

The government ensures energy conservation measures are in place and that an energy reserve for crisis situations is available, while at the same time diversifying energy import sources.

To ensure economic stability and when resources lend themselves to cooperation, the government must promote enterprises that encourage long-term joint ventures with large foreign companies. The Ministry of the Economy must develop a program for economic actions in the event of possible foreign economic sanctions or blockades.

The government implements measures to prevent smuggling, corruption, money laundering, and other pecuniary crimes.

IV.4. Averting Criminal Threats

Drawing on other country's experiences, Latvia is developing legislation for the war on crime as well as strengthening and developing law enforcement and state security institutions. Preventive measures to uncover and prevent crimes and offences that are dangerous to society are at the improvement stage, along with the application of a punishment system.

Taking note of the greater danger posed to society by organized crime and corruption, special attention is being given to combating this type of crime by developing cooperating with other countries and actively making use of the potential offered by Interpol. . . .

IV.11. Averting Military Threats

Taking into consideration the existing military potential of Latvia, its defense must be based on deterrence and a political resolution of military problems. The National Armed Forces must be small, mobile, well armed, well trained, and interoperable with NATO military structures, as well as under civilian control.

The Ministry of the Interior's Border Guard controls and protects the national border. The navy and air force, in cooperation with applicable civilian structures, controls national airspace and economic zone waters. The ground forces ensure the protection of territory within the country, perform rapid reaction tasks, prepare military reserves, and participate in international operations.

The Ministry of Defense, in the event of an invasion, in accordance with the state defense plan provides for its implementation. The international community is informed of any acts of aggression by means of the global information system. The National Armed Forces, including mobilized reserves, provides military resistance throughout the nation's territory. Activities of the legitimate government and military resistance have to be maintained until international political and economic pressure halts the aggression. The government is provided with the means for communication with its citizens, foreign countries, international security and defense organizations, and the international mass media. Maximum cooperation with NATO, its member states, as well as with other friendly nations is supported in order to develop closer ties and facilitate the fastest possible integration of Latvia into NATO.

DOCUMENT 4
National Security Concept of the Republic of Estonia
Part 1, Estonia's Security Risks (Approved 6 March 2001), Section 4, Estonia's Security Risks[4]

The major risk to Estonia's security is potential instability and developments in the international arena that are politically uncontrollable, as well as international crises. As long as the new post-Cold War Euro-Atlantic security framework has not completed its evolution, the potential for tension exists for Estonia because of delays in an international resolution. Uncertainty in the international arena could well leave Estonia in a security vacuum.

Military Risks
Estonia does not see a direct military threat to its security neither now nor in the foreseeable future. The end of the Cold War has reduced the potential for

4. From the Ministry of Foreign Affairs of the Republic of Estonia. Available at: http://www.vm.ee/eng/policy/Security/ptk1.htm.

military altercation that would encompass all of Europe and has lessened the threat of conflict in the Baltic Sea region, too. A continued lessening of military threat depends on developments in the international arena, including the ongoing process of Euro-Atlantic integration, as well as on development of Estonia's defense capability. In the Baltic Sea region, economic and political developments within the region's states are important, as is an increasing closeness of ties among those states. The only nuclear state of the region, the Russian Federation, has approved a new security policy concept and military doctrine. At the same time, Russia has reduced the size of its forces in the Northwest—that is, in Estonia's immediate vicinity. All other Estonia's neighbor states are also presently reforming their armed forces according to the changed security environment. Estonia will monitor ongoing developments and will maintain a readiness for possible future changes so as to be able to react should the need arise.

The Risk of Outside Political Pressure

Just as there is no direct military threat, there is also no direct danger at present that Estonia would yield to outside political pressure to alter its domestic or foreign policy course. The most effective guarantor that Estonia will be able to avoid—and, should the need arise, to resist—such pressure is Estonia's successful political and economic development, continued integration into the international economic and security system, and the consolidation of society around democratic values.

New Security Risks

Against a backdrop of a reduced military threat, rapid changes in the international arena, in [the] economy, and in technology have brought a number of so-called new, nonmilitary risks to the fore. On a global scale, these include ecological risks, the potential for ethnic conflict, international organized crime, the proliferation of weapons of mass destruction, and the potential volatility of social and economic problems. All of these so-called new risks have the potential to influence Estonia either directly or indirectly. Particularly in the Baltic Sea region, Soviet-era nuclear energy plants heighten the danger of ecological catastrophe, as do other large industries such as chemical-producing plants. In addition to ecological danger, economic and humanitarian catastrophes may result in floods of refugees and widespread in-migration, both of which have the potential to destabilize the states of the region.

A number of factors with a potential to influence Estonia's domestic security stem from the great economic and social transformation that the region is undergoing. The speedy pace of change brings with it the danger of a widening socioeconomic gap. This, in turn, increases social problems such as crime and substance abuse. The explosive growth of post-Cold War phenomena such as international organized crime and terrorism, as well as smuggling of narcotics and arms, among other things, can also influence Estonian society. The increasing use of electronic information systems in Estonia and their interconnectedness with global information systems increases the risk of computer crime as well as the vulnerability of the national information system. These new security risks demand a coordinated response by national institutions and broad international cooperation.

Economic Risks

New economic risks are inherent in increasing globalization. Estonia's economy is strongly integrated into the global economy. For that reason, Estonia is vulnerable to possible global recession or fluctuations in markets important to Estonia. One risk factor is Estonia's dependence on gas and petroleum imports and, in the case of gas, on one single producer. The interconnectedness of Estonia's electricity system with that of Russia is also a risk factor. Estonia must build up a strategic reserve of petroleum products and widen the market to other gas suppliers. With regard to developing East-West transit trade, Estonia must factor in fairly high risks, at least until Russia's economy stabilizes. Estonia's economic ties, including foreign investment in Estonia, must be as diversified as possible, whereby the task of the state is to favor the orientation of exports to stable markets of developed countries. The state must be able to guarantee confidence in the Estonian kroon. Preparations for accession to the EU demand that production continue to be brought into compliance with European norms—this raises the overall competitiveness of the Estonian economy and lessens risks stemming from unstable markets.

DOCUMENT 5
The United States Policy of Non-recognition
Provided by the U.S. Department of State, September 1989[5]

The United States does not recognize the forcible incorporation of Estonia, Latvia, and Lithuania into the USSR which occurred in 1940. We maintain full diplomatic relations with representatives of the last free governments from those states. Those states are represented by three Charges d'Affaires resident in the United States. The non-recognition policy is a broadly supported, bipartisan U.S. stand of almost fifty years existence. The Office of Eastern European and Yugoslav Affairs in the Bureau of European and Canadian Affairs (EUR/EEY) handles the liaison with the Baltic diplomats.

Under the U.S. non-recognition policy, senior [U.S. government] officials do not travel to the Baltic States and do not meet anywhere with senior Baltic officials representing authority exercised by the USSR over the Baltic States. U.S. diplomatic personnel from our missions in Moscow and Leningrad do travel to the Baltic States to carry out consular and reporting functions. From time to time they meet with local officials when required.

The development of the Popular Fronts, Sajudis, and the new democratically oriented political parties has provided an opportunity for us to learn more about events in the Baltics from independent sources. We are encouraging increased contacts with such groups and individuals. Baltic American groups have been very active in keeping the Department [of State] informed of the new political developments. These concerned Americans have also arranged for EUR/EEY officers to meet with visiting Baltic political and human rights activists. We expect to continue close contact with the concerned American groups and to meet with Baltic visitors.

The United States firmly supports the efforts of the Baltic peoples to regain control over their own political and economic destiny. This includes their steps to exercise political self-determination by peaceful means. The United States intends that this longstanding special interest in the Baltic States, and our participation in the Helsinki process, will promote political and social freedoms, and legal safeguards of such rights, for the peoples of

5. U.S. Commission on Security and Cooperation in Europe, *Renewal and Challenge: The Baltic States 1988–1989* (Washington, D.C.: U.S. Government Printing Office, 1990), appendix A, p. 95.

the region. The Soviet Union is fully aware of our policy toward these three countries.

DOCUMENT 6
Charter of Partnership among the United States of America and the Republics of Estonia, Latvia, and Lithuania
Signed on 16 January 1998 (Excerpt)[6]

Security Cooperation

The partners will consult together, as well as with other countries, in the event that a partner perceives that its territorial integrity, independence, or security is threatened or at risk. The Partners will use bilateral and multilateral mechanisms for such consultations.

The United States welcomes and appreciates the contributions that Estonia, Latvia, and Lithuania have already made to European security through the peaceful restoration of independence and their active participation in the Partnership for Peace. The United States also welcomes their contributions to IFOR, SFOR, and other international peacekeeping missions.

Building on the existing cooperation among their respective ministries of defense and armed forces, the United States of America supports the efforts of Estonia, Latvia, and Lithuania to provide for their legitimate defense needs, including development of appropriate and interoperable military forces.

The Partners welcome the establishment of the Baltic Security Assistance Group (BALTSEA) as an effective body for international coordination of security assistance to Estonia's, Latvia's, and Lithuania's defense forces.

The Partners will cooperate further in the development and expansion of defense initiatives such as the Baltic Peacekeeping Battalion (BaltBat), the Baltic Squadron (Baltron), and the Baltic airspace management regime (BALTNET), which provide a tangible demonstration of practical cooperation enhancing the common security of Estonia, Latvia, and Lithuania, and the transatlantic community.

---

6. From *U.S. Information & Texts*, no. 003 (21 Jan. 1998): 14.

The Partners intend to continue mutually beneficial military cooperation and will maintain regular consultations, using the established Bilateral
Working Group on Defense and Military Relations.

DOCUMENT 7
Fact Sheet U.S.-Baltic Relations
Issued by the White House following the signing of the U.S.-Baltic Charter
of Partnership, Washington, 16 January 1998[7]

The United States has greatly expanded its security and military assistance
with the Baltic states. By the end of fiscal year 1998, they will have received
over $29 million under the President's 'Warsaw Initiative' security assis-
tance program, which helps members of the Partnership for Peace enhance
their capabilities and ability to work with NATO. In addition the Baltic governments received over $14 million in equipment and services to develop
the Baltic Peacekeeping Battalion, elements of which are serving with
NATO and SFOR in Bosnia today. A U.S.-sponsored Regional Airspace Initiative, various military education and training programs, and close cooperation with counterparts in the Michigan, Maryland, and Pennsylvania
National Guards have also contributed to the development of their armed
forces.

The United States has provided approximately $8.5 million for the dem-
olition of the former Russian large phased-array radar at Skrunda, Latvia,
and $2 million for cleanup of the former Russian nuclear reactor facility at
Paldiski, Estonia.

7. From *U.S. Information & Texts*, no. 003 (21 Jan. 1998): 16.

References

Aalto, Pami. 2000. "Beyond Restoration: The Construction of Post-Soviet Geopolitics in Estonia." *Cooperation and Conflict* 35, no. 1: 65–88.

Adamkus, Valdas. 1999. *Europe as Unfinished Business: The Role of Lithuania in the 21st Century's Continent.* Bonn: Center for European Integration Studies.

Adler, Emanuel, and Michael Barnett. 1998a. "A Framework for the Study of Security Communities." In *Security Communities,* edited by Emanuel Adler and Michael Barnett, 29–65. Cambridge: Cambridge Univ. Press.

———. 1998b. "Security Communities in Theoretical Perspective." In *Security Communities,* edited by Emanuel Adler and Michael Barnett, 3–28. Cambridge: Cambridge Univ. Press.

Ágh, Attila. 1991. "After the Revolution: A Return to Europe." In *Towards a Future European Peace Order?* edited by Karl E. Birnbaum, Josef B. Binter, and Stephen K. Badzik, 83–97. London: Macmillan.

Albrecht, Ulrich. 1986. *Internationale Politik: Einführung in das System internationaler Herrschaft.* Munich: Oldenbourg.

———. 1994. "Troop Reductions and a New Baltic Security Regime." In *The Future of the Baltic Region: An Analysis of Regional Development,* edited by Lars Rydén, 27–28. Uppsala: Baltic Univ.

———. 1995a. "Die Beziehungen Deutschlands zum Osten." In *Nordeuropa und die deutsche Herausforderung,* edited by Burkhard Auffermann and Pekka Visuri, 69–77. Baden-Baden: Nomos.

———. 1995b. "The Peaceful Unification of Germany." In *Peaceful Changes in World Politics,* edited by Heikki Patomäki, 168–207. Tampere, Finland: Tampere Peace Research Institute.

Allison, Roy. 1993. *Military Forces in the Soviet Successor States.* London: International Institute for Strategic Studies and Brassey's.

Archer, Clive. 1996. "The Nordic Area as a 'Zone of Peace.' " *Journal of Peace Research* 33, no. 4: 451–67.

Arms Control Today. 1992. Vol. 22, no. 1: 44.

Arnswald, Sven. 2000. *EU Enlargement and the Baltic States: The Incremental Making of*

New Members. Helsinki: Finnish Institute of International Affairs and Institut für Europäische Politik.

Asmus, Ronald D. 1998. "Towards Better Security: A *City Paper* Interview with U.S. Deputy Secretary of State Ron Asmus." *City Paper: The Baltic States*, no. 34: 22–24.

———. 1999. "Address at the 4th Annual Stockholm Conference on Baltic Sea Region Security and Cooperation, November 4, 1999." Available at: http://www .usemb.se/bsconf/1999/text/11asmus.html.

Asmus, Ronald D., and Robert C. Nurick. 1996a. *NATO Enlargement and the Baltic States*. Santa Monica, Calif.: RAND.

———. 1996b. "NATO Enlargement and the Baltic States." *Survival* 38, no. 2: 121–42.

Baev, Pavel K. 1996. *The Russian Army in a Time of Troubles*. London: Sage.

———. 1997. "Russia's Departure from Empire: Self-assertiveness and a New Retreat." In *Geopolitics in Post-Wall Europe: Security, Territory, and Identity*, edited by Ola Tunander, Pavel Baev, and Victoria Ingrid Einagel, 174–95. London: Sage.

Bajarunas, Eitvydas. 1995. "Lithuania's Security Dilemma." In *The Baltic States: Security and Defense after Independence*, edited by Peter van Ham, 9–36. Paris: Institute for Security Studies of Western European Union.

Baldwin, David A. 1995. "Security Studies and the End of the Cold War." *World Politics* 48, no. 1: 117–41.

———. 1997. "The Concept of Security." *Review of International Studies* 23, no. 1: 5–26.

Baltic Assembly. 1995. *Session Documents 1991–1994*. Riga, Latvia: Baltic Assembly.

Baltic Times. 1996. Vol. 1, no. 4, 11–17 Apr., 5.

———. 1997. Vol. 2, no. 83, 30 Oct.–5 Nov., 8.

———. 1999. Vol. 4, no. 157, 6–12 May, 2.

Baranovsky, Vladimir. 1995. "Russia and Its Neighborhood: Conflict Developments and Settlement Efforts." In Stockholm International Peace Research Institute (SIPRI), *SIPRI Yearbook 1995: Armaments, Disarmament, and International Security*, 231–64. Oxford: Oxford Univ. Press.

———. 1998. "Russia: Conflicts and Peaceful Settlement of Disputes." In Stockholm International Peace Research Institute (SIPRI), *SIPRI Yearbook 1998: Armaments, Disarmament, and International Security*, 111–39. Oxford: Oxford Univ. Press.

Barnes, Trevor, and Derek Gregory. 1997. "Introduction." In *Reading Human Geography: The Politics and Poetics of Inquiry*, edited by Trevor Barnes and Derek Gregory, 1–12. London: Arnold.

Barnett, Michael, and Emanuel Adler. 1998. "Studying Security Communities in Theory, Comparison, and History." In *Security Communities*, edited by Emanuel Adler and Michael Barnett, 413–41. Cambridge: Cambridge Univ. Press.

Batt, Judy. n.d. "ESRC The Edge-Concept of the Moment." Available at: http://www.bham.ac.uk/crees/statehood/concept.htm.

Bebriss, Peteris. 1996a. "Cabinet Hopes to Pass New Military Service Law." *Baltic Times* 1, no. 4, 11–17 April, 5.

———. 1996b. "Military Service Law Vetoed." *Baltic Times* 1, no. 5, 18–24 April, 5.

Bengtsson, Rikard. 2000. "Towards a Stable Peace in the Baltic Sea Region?" *Cooperation and Conflict* 35, no. 4: 355–88.

Berger, Peter, and Thomas Luckmann. 1967. *The Social Construction of Reality: A Treatise in the Sociology of Knowledge.* London: Penguin.

Berzins, Indulis. 1999. "Latvia's Membership: Good for Latvia, Good for NATO." *NATO'S Nations and Partners for Peace* (special issue): 56–57.

Bially Mattern, Janice. 2000. "Taking Identity Seriously." *Cooperation and Conflict* 35, no. 3: 299–308.

Bildt, Carl. 1994. "The Baltic Litmus Test." *Foreign Affairs* 73, no. 5: 72–85.

Binnendijk, Hans, and Richard L. Kugler. 1999. "Open NATO's Door Carefully." *Washington Quarterly* 22, no. 2: 125–38.

Birckenbach, Hanne-Margret. 1997. *Preventive Diplomacy Through Fact-Finding: How International Organizations Review the Conflict over Citizenship in Estonia and Latvia.* Hamburg: Lit.

Birkavs, Valdis. 1997a. "Presentation." In *NATO and the Baltic States: Quo Vadis?* edited by Paulis Apinis and Atis Lejins, 12–23. Riga: Latvian Institute of International Affairs.

———. 1997b. "Remarks at Presidential Palace." Available at: http://secretary.state.gov/www/statements/970713a.html.

Bitzinger, Richard A. 1991. "Neutrality for Eastern Europe: Problems and Prospects." *Bulletin of Peace Proposals* 22, no. 3: 281–89.

Black, J. L. 2000. *Russia Faces NATO Expansion: Bearing Gifts or Bearing Arms?* Lanham, Md.: Rowman and Littlefield.

Bohlen, Avis. 2003. "The Rise and Fall of Arms Control." *Survival* 45, no. 3: 7–34.

Bollow, Undine. 1993. *Die baltische Frage in der internationalen Politik nach 1945.* Berlin: Institut für Internationale Politik und Regionalstudien des Fachbereichs Politische Wissenschaft der Freien Universität Berlin.

Bolving, Klaus. 2001. *Baltic CFE Membership.* Copenhagen: Danish Institute of International Affairs.

Bonn International Center for Conversion. 1996. *Conversion Survey 1996: Global Disarmament, Demilitarization, and Demobilization.* Oxford: Oxford Univ. Press.

Booth, Ken. 1991a. "Introduction: The Interregnum: World Politics in Transition." In *New Thinking about Strategy and International Security,* edited by Ken Booth, 1–28. London: Harper Collins Academic.

———. 1991b. "War, Security, and Strategy: Towards a Doctrine for Stable Peace." In

New Thinking about Strategy and International Security, edited by Ken Booth, 335–76. London: Harper Collins Academic.

———. 1997. "Security and Self: Reflections of a Fallen Realist." In *Critical Security Studies: Concepts and Cases,* edited by Keith Krause and Michael C. Williams, 83–119. London: University College London Press.

Brill Olcott, Martha. 1990. "The Lithuanian Crisis." *Foreign Affairs* 69, no. 3: 30–46.

Brock, Lothar. 1991. "Peace Through Parks: The Environment on the Peace Research Agenda." *Journal of Peace Research* 28, no. 4: 407–23.

Broks, Janis, Uldis Ozolins, Gunars Ozolzile, Aivars Tabuns, and Talis Tisenkopfs. 1996–97. "The Stability of Democracy in Latvia: Pre-requisites and Prospects." *Humanities and Social Sciences Latvia,* issue 4(13)–1(14): 103–34.

Browning, Christopher S. 2001. "A Multi-dimensional Approach to Regional Cooperation: The United States and the Northern European Initiative." *European Security* 10, no. 4: 84–108.

Browning, Christopher S., and Pertti Joenniemi. 2004. "Regionality beyond Security? The Baltic Sea Region after Enlargement." *Cooperation and Conflict* 39, no. 3: 233–53.

Brubaker, Rogers. 1996. *Nationalism Reframed: Nationhood and the National Question in the New Europe.* Cambridge: Cambridge Univ. Press.

Brzezinski, Zbigniew. 1995. "U.S. Foreign Policy and the Baltic States." In *The Baltic States on Their Way to the European Union: Security Aspects,* edited by Atis Lejins and Paulis Apinis, 9–34. Riga: Latvian Institute of International Affairs.

Budryte, Migle. 2000. "Lithuanian Foreign Policy and Dual Enlargement." In *Democratic Security Building: Cases from the Baltic and Black Sea Regions,* edited by Unto Vesa, 125–50. Tampere, Finland: Tampere Peace Research Institute.

Bull, Hedley. 1995. "The Theory of International Politics, 1919–1969." In *International Theory: Critical Investigations,* edited by James Der Derian, 181–211. Houndmills, England: Macmillan.

Bungs, Dzintra. 1998. *The Baltic States: Problems and Prospects of Membership in the European Union.* Baden-Baden: Nomos.

Burant, Stephen R., and Voytek Zubek. 1993. "Eastern Europe's Old Memories and New Realities: Resurrecting the Polish-Lithuanian Union." *East European Politics and Societies* 7, no. 2: 370–93.

Burke, Anthony. 2002. "Aporias of Security." *Alternatives* 27, no. 1: 1–27.

Butkevicius, Audrius. 1993a. "The Baltic Region in the New Europe." In *Brassey's Defense Yearbook 1993,* edited by Center for Defense Studies, King's College London, 169–78. London: Brassey's.

———. 1993b. "The Baltic Region in Post–Cold War Europe." *NATO Review* 41, no. 1: 7–11.

Butrimas, Vytautas. 1999. "Presentation." In *Rebuilding the Armed Forces for the XXIst Century*, edited by Jeffrey Simon, Nicolae Uscoi, and Constantin Mostoflei, 44–49. Bucharest: Institute for Political Studies and Military History.

Buzan, Barry. 1991. *People, States, and Fear: An Agenda for International Security Studies in the Post–Cold War Era*. 2d ed. Hemel Hempstead, England: Harvester Wheatsheaf.

———. 2000a. "The Logic of Regional Security in the Post–Cold War World." In *The New Regionalism and the Future of Security and Development*, edited by Björn Hettne, András Inotai, and Osvaldo Sunkel, 4:1–25. London: Macmillan.

———. 2000b. Review of *Security Communities* by Emanuel Adler and Michael Barnett. *International Affairs* 76, no. 1: 154.

Buzan, Barry, and Ole Wæver. 2003. *Regions and Powers: The Structure of International Security*. Cambridge: Cambridge Univ. Press.

Buzan, Barry, Ole Wæver, and Jaap de Wilde. 1998. *Security: A New Framework for Analysis*. Boulder, Colo.: Lynne Rienner.

Campbell, David. 1998. *Writing Security: United States Foreign Policy and the Politics of Identity*. Rev. ed. Manchester: Manchester Univ. Press.

Carrafiello, Lewis J., and Nico Vertongen. 1998. *Pivotal States, Pivotal Region: Security in Estonia, Latvia, Lithuania, and the Baltic Sea Region*. Leuven, Belgium: Centrum voor Vredesonderzoek and Strategische Studies.

Cengel, Katya. 1999. "Lost Souls from a Lost Empire." *Baltic Times* 4, no. 168, 29 July–4 Aug., 1, 8.

Chaise, Christian. 1998. "Symbolic Charter Signed." *Baltic Times* 3, no. 93, 22–28 Jan:, 8.

Chaturvedi, Sanjay. 1996. "The Arctic Today: New Thinking, New Visions, Old Power Structures." In *Dreaming of the Barents Region: Interpreting Cooperation in the Euro-Arctic Rim*, edited by Jyrki Käkönen, 23–54. Tampere, Finland: Tampere Peace Research Institute.

Checkel, Jeffrey T. 1997. *Ideas and International Political Change: Soviet/Russian Behavior and the End of the Cold War*. New Haven, Conn.: Yale Univ. Press.

Christophe, Barbara. 1997. *Staat versus Identität. Zur Konstruktion von "Nation" und "nationalem Interesse" in den litauischen Transformationsdiskursen von 1987 bis 1995*. Cologne: Verlag Wissenschaft und Politik.

City Paper: The Baltic States. 2001. No. 53, July–August, 7.

Clemens, Walter C., Jr. 1991. *Baltic Independence and Russian Empire*. Basingstoke, England: Macmillan.

———. 2001. *The Baltic Transformed: Complexity Theory and European Security*. Lanham, Md.: Rowman and Littlefield.

Clemmesen, Michael. 1998. "Foreign Military Assistance." In *Bordering Russia: The-*

ory and Prospects for Europe's Baltic Rim, edited by Hans Mouritzen, 227–57. Aldershot, England: Ashgate.

Clinton, William J. 2000. "Statement by the President: Cross-Border Cooperation and Environmental Safety in Northern Europe Act of 2000." Available at: http://www.state.gov/www/regions/eur/nei/stmnt_000803_nei_bill.html.

Conley, Heather. 2003. "Building on Success: The Enhanced Partnership in Northern Europe." Remarks to the School for Advanced International Studies, Washington, D.C., 15 Oct. Available at: http://www.state.gov/p/eur/rls/rm/2003/25286pf.htm.

Connerton, Paul. 1989. *How Societies Remember.* Cambridge: Cambridge Univ. Press.

Coser, Lewis A. 1956. *The Functions of Social Conflict: An Examination of the Concept of Social Conflict and Its Use in Empirical Sociological Research.* New York: Free Press.

Couldry, Nick. 2000. *Inside Culture: Re-imagining the Method of Cultural Studies.* London: Sage.

Cunningham, Keith B. 1997. *Base Closure and Redevelopment in Central and Eastern Europe.* Bonn: Bonn International Center for Conversion.

Czempiel, Ernst-Otto. 1996. "Kants Theorem: Oder: Warum sind Demokratien (noch immer) nicht friedlich?" *Zeitschrift für Internationale Beziehungen* 3, no. 1: 79–101.

Daalder, Ivo H., and James M. Goldgeier. 2001. "Putting Europe First." *Survival* 43, no. 1: 71–91.

Dalbins, Juris. 1996. "Die Zusammenarbeit der baltischen Staaten als Schlüssel zu mehr Sicherheit." *NATO-Brief* 44, no. 1: 7–10.

Dalby, Simon. 1997. "Contesting an Essential Concept: Reading the Dilemmas in Contemporary Security Discourse." In *Critical Security Studies: Concepts and Cases,* edited by Keith Krause and Michael C. Williams, 3–31. London: University College London Press.

Daniels, Stephen. 1988. "The Political Iconography of Woodland in Later Georgian England." In *The Iconography of Landscape: Essays on the Symbolic Representation, Design, and Use of Past Environments,* edited by Denis Cosgrove and Stephen Daniels, 43–82. Cambridge: Cambridge Univ. Press.

Danish Ministry of Defense. 1995. *The Baltbat-Project: Documentation from the Danish Ministry of Defense.* Copenhagen: Danish Ministry of Defense.

Der Derian, James. 1995. "The Values of Security: Hobbes, Marx, Nietzsche, and Baudrillard." In *On Security,* edited by Ronnie D. Lipschutz, 24–45. New York: Columbia Univ. Press.

Deutsch, Karl W. 1954. *Political Community at the International Level: Problems of Definition and Measurement.* Garden City, N.Y.: Doubleday.

Deutsch, Karl W., Sidney A. Burrell, Robert A. Kann, Maurice Lee Jr., Martin Lichterman, Raymond E. Lindgren, Francis L. Loewenheim, and Richard W. Van Wagenen. 1957. *Political Community and the North Atlantic Area: International Organization in the Light of Historical Experience.* Princeton, N.J.: Princeton Univ. Press.

Deutsch, Morton. 1987. "A Theoretical Perspective on Conflict and Conflict Resolution." In *Conflict Management and Problem Solving: Interpersonal to International Applications,* edited by Dennis J. D. Sandole and Ingrid Sandole-Staroste, 38–49. London: Francis Pinter.

Donnelly, Christopher. 1992. "Evolutionary Problems in the Former Soviet Armed Forces." *Survival* 34, no. 3: 28–42.

Dreifelds, Juris. 1996. *Latvia in Transition.* Cambridge: Cambridge Univ. Press.

Duke, Simon. 1996. "The United States." In *The Baltic Sea Region: National and International Security Perspectives,* edited by Axel Krohn, 183–204. Baden-Baden: Nomos.

Dunlop, John. 1993. "Russia: Confronting a Loss of Empire." In *Nations and Politics in the Soviet Successor States,* edited by Ian Bremmer and Ray Taras, 43–72. Cambridge: Cambridge Univ. Press.

Eagleton, Terry. 2000. *The Idea of Culture.* London: Blackwell.

The Economist. 1996. 4 May, 26.

Edelman, Eric S. 1999. "Plenary Speech." In *Envisioning the Northern Dimension: Toward an Arctic of Regions,* edited by Janne Hukkinen, 47–52. Rovaniemi, Finland: Arctic Center Univ. of Lapland.

Ekedahl, Carolyn McGiffert, and Melvin A. Goodman. 2001. *The Wars of Eduard Shevardnadze.* 2d ed., rev. and updated. London: Brassey's.

Ellis, Jason. 1997. "Nunn-Lugar's Mid-life Crisis." *Survival* 39, no. 1: 84–110.

Engberg, Katarina, Björn Hagelin, Lena Jonson, Michael Karlsson, Claes Levinsson, Erik Melander, Kjell-Åke Nordqvist, and Peter Wallensteen. 2002. "Peace and Security." In *The Baltic Sea Region: Cultures, Politics, Societies,* edited by Witold Maciejewski, 432–84. Uppsala, Sweden: Baltic Univ. Press.

Estonian Review. 1992a. 2, 11–16 July.

Estonian Review. 1992b. Vol. 2, no. 68, 7–13 Sep.

———. 1992c. Vol. 2, no. 80, 30 Nov.–6 Dec.

———. 1993a. Vol. 3, no. 89, 8–14 Feb.

———. 1993b. Vol. 3, no. 102, 10–16 May.

———. 1993c. Vol. 3, no. 110, 5–11 July.

———. 1993d. Vol. 3, no. 129, 15–21 Nov.

———. 1998. Vol. 8, no. 3, 11–17 Jan.

Estonia Today. 1998. 27 Aug.

Evangelista, Matthew. 1996. "Historical Legacies and the Politics of Intervention in the Former Soviet Union." In *The International Dimension of Internal Conflict,* edited by Michael E. Brown, 107–41. Cambridge, Mass.: MIT Press.

Eyre, Dana P., and Mark C. Suchman. 1996. "Status, Norms, and the Proliferation of Conventional Weapons: An Institutional Theory Approach." In *The Culture of National Security: Norms and Identity in World Politics,* edited by Peter J. Katzenstein, 79–113. New York: Columbia Univ. Press.

Ferm, Ragnhild. 1997. "Chronology 1996." In Stockholm International Peace Research Institute (SIPRI), *SIPRI Yearbook 1997: Armaments, Disarmament, and International Security,* 547–53. Oxford: Oxford Univ. Press.

Fierke, Karin M. 1997. "Changing Worlds of Security." In *Critical Security Studies: Concepts and Cases,* edited by Keith Krause and Michael C. Williams, 223–52. London: University College London Press.

Frankfurter Allgemeine Zeitung. 1997. No. 263, 12 Nov., 9.

Garthoff, Raymond L. 1994. *The Great Transition: American-Soviet Relations and the End of the Cold War.* Washington, D.C.: Brookings Institution.

Gaskaite-Zemaitiene, Nijole. 1999. "War in Lithuania from 1944 to 1953." In *The Anti-Soviet Resistance in the Baltic States,* edited by Arvydas Anusauskas, 23–45. Vilnius, Lithuania: Du Ka.

Geller, Daniel S., and J. David Singer. 1998. *Nations at War: A Scientific Study of International Conflict.* Cambridge: Cambridge Univ. Press.

George, Jim, and David Campbell. 1990. "Patterns of Dissent and the Celebration of Difference: Critical Social Theory and International Relations." *International Studies Quarterly* 34, no. 3: 269–93.

Geyer, Michael. 1984. *Deutsche Rüstungspolitik, 1860–1980.* Frankfurt: Suhrkamp.

Gießmann, Hans-Joachim. 1997–98. "Estland." In *Handbuch Sicherheit 1997: Militär und Sicherheit in Mitteleuropa im Spiegel der NATO-Erweiterung. Daten—Fakten—Analysen,* edited by Hans-Joachim Gießmann, 149–71. Baden-Baden: Nomos.

Gill, Graeme. 1994. *The Collapse of a Single-Party System: The Disintegration of the Communist Party of the Soviet Union.* Cambridge: Cambridge Univ. Press.

Girnius, Kestutis. 1991. "The Party and Popular Movements in the Baltic." In *Toward Independence: The Baltic Popular Movements,* edited by Jan Arveds Trapans, 57–69. Boulder, Colo.: Westview Press.

Girnius, Saulius. 1993. "Problems in the Lithuanian Military." *Radio Free Europe/Radio Liberty Research Report* 2, no. 42: 44–47.

———. 1995. "Compromise at the Crossroads." *Transition* 1, no. 4: 44–46, 63.

Goble, Paul. 2000a. "Another Precedent from Kosovo?" *Baltic Times* 5, no. 233, 16–22 Nov., 19.

———. 2000b. "A Red-Brown Coalition in Lithuania." *Radio Free Europe/Radio Liberty Baltic States Report* 1, no. 14. Available at: http://www.rferl.org/balticreport/2000/04/14-240400.html.

Gong, Gerrit W. 2001. "The Beginning of History: Remembering and Forgetting as Strategic Issues." *Washington Quarterly* 24, no. 2: 45–57.

Gorohhov, Sergei. 1997. "Integration in Practice: The Case of Narva." In *The Integration of Non-Estonians into Estonian Society: History, Problems, and Trends*, edited by Aksel Kirch, 122–41. Tallinn: Estonian Academy Publishers.

Götz-Coenenburg, Roland. 1990. *Die Wirtschaft des Baltikums*. Cologne: Bundesinstitut für Ostwissenschaftliche und internationale Studien.

Graube, Raimonds. 1999. "The Latvian Armed Forces Today." *NATO's Nations and Partners for Peace* (special issue): 63–68.

Gvosdev, Nikolas K. 1995. "The Formulation of an American Response to Lithuanian Independence, 1990." *East European Quarterly* 29, no. 1: 17–41.

Haab, Mare. 1994. "Supporting Common Security or Favoring Instability? The Military Aspect of Contemporary Security Policy in Estonia." In *Common Security in Northern Europe after the Cold War: The Baltic Sea Region and the Barents Sea Region*, edited by Göran Baecklund, 148–55. Stockholm: Olof Palme International Center.

———. 1995. "Estonia and Europe: Security and Defense." In *The Baltic States: Security and Defense after Independence*, edited by Peter van Ham, 37–60. Paris: Institute for Security Studies of Western European Union.

———. 1998. "Estonia." In *Bordering Russia: Theory and Prospects for Europe's Baltic Rim*, edited by Hans Mouritzen, 109–29. Aldershot, England: Ashgate.

Haas, Ernst. 1958. "Persistent Themes in Atlantic and European Unity." *World Politics* 10, no. 4: 614–28.

Haftendorn, Helga. 1991. "The Security Puzzle: Theory-Building and Discipline-Building in International Security." *International Studies Quarterly* 35, no. 1: 3–17.

Halliday, Fred. 2000. "Culture and International Relations: A New Reductionism?" In *Confronting the Political in International Relations*, edited by Michi Ebata and Beverly Neufeld, 47–71. London: Macmillan.

Hartmann, Rüdiger, Wolfgang Heydrich, and Nikolaus Meyer-Landrut. 1994. *Der Vertrag über konventionelle Streitkräfte in Europe: Vertragswerk, Verhandlungsgeschichte, Kommentar, Dokumentation*. Baden-Baden: Nomos.

Harvey, David. 1997. "Between Space and Time: Reflections on the Geographical Imagination." In *Reading Human Geography: The Poetics and Politics of Inquiry*, edited by Trevor Barnes and Derek Gregory, 256–79. London: Arnold.

Heinemann-Grüder, Andreas. 2002. *Small States—Big Worries: Choice and Purpose in the Security Policies of the Baltic States*. Bonn: Bonn International Center for Conversion.

Heininen, Lassi. 1999. *Euroopan pohjoinen 1990-luvulla: Moniulotteisten ja ristiriitaisten intressien alue.* Rovaniemi: Univ. of Lapland.

Helme, Rein. 1997. "Some Military Aspects of Estonian Security Policy." In *St. Petersburg, the Baltic Sea, and European Security: Ideas and Perspectives in the New Situation,* edited by Ritva Grönick, Katarina Sehm, Marjut Pukarinen, and Mark Waller, 106–15. Helsinki: Finnish Committee for European Security.

Herd, Graeme P. 1997a. "Baltic Security: A Crisis Averted?" In *Brassey's Defense Yearbook 1997,* edited by Center for Defense Studies, King's College London, 93–118. London: Brassey's.

———. 1997b. "Baltic Security Politics." *Security Dialogue* 28, no. 2: 251–53.

———. 1999. "Russia's Baltic Policy after the Meltdown." *Security Dialogue* 30, no. 2: 197–212.

Herman, Robert G. 1996. "Identity, Norms, and National Security: The Soviet Foreign Policy Revolution and the End of the Cold War." In *The Culture of National Security: Norms and Identity in World Politics,* edited by Peter J. Katzenstein, 271–316. New York: Columbia Univ. Press.

Herz, John. 1950. "Idealist Internationalism and the Security Dilemma." *World Politics* 2: 157–80.

———. 1951. *Political Realism and Political Idealism: A Study in Theories and Realities.* Chicago: Univ. of Chicago Press.

Hillingsø, Kjeld, and Robert Dalsjö. 1999. "Can the Baltics Be Defended?" *NATO's Nations and Partners for Peace* (special issue): 26–28.

Hirsch, Steve, ed. 1989. *New Soviet Voices on Foreign and Economic Policy.* Washington, D.C.: Bureau of National Affairs.

Hoffmann, Stanley. 1959. "Vers l'etude systématique des mouvements d'intégration internationale." *Revue Française de Science Politique* 9, no. 2: 474–85.

Holloway, David. 1989–90. "State, Society, and the Military under Gorbachev." *International Security* 14, no. 3: 5–24.

Holsti, Kalevi J. 1985. *The Dividing Discipline: Hegemony and Diversity in International Theory.* Boston: Allen and Unwin.

———. 1991a. *Change in the International System: Essays on the Theory and Practice of International Relations.* Aldershot, England: Edward Elgar.

———. 1991b. *Peace and War: Armed Conflict and International Order 1648–1989.* Cambridge: Cambridge Univ. Press.

———. 1995. *International Politics: A Framework for Analysis.* 7th ed. Englewood Cliffs, N.J.: Prentice-Hall.

———. 1996. *The State, War, and the State of War.* Cambridge: Cambridge Univ. Press.

———. 1999. "The Coming Chaos? Armed Conflict in the World's Periphery." In *In-*

ternational Order and the Future of World Politics, edited by T. V. Paul and John A. Hall, 282–310. Cambridge: Cambridge Univ. Press.

Huang Mel. 2001. "Estonia's Year in Defense: 2000." In *Baltic Security in 2000,* by Graeme P. Herd and Mel Huang, 29–44. Sandhurst, England: Conflict Studies Research Center.

Hubel, Helmut, and Stefan Gaenzle. 2001. *The Council of the Baltic Sea States (CBSS) as a Sub-regional Organisation for 'Soft Security Risk Management' in the North-East of Europe.* Report to the Presidency of the CBSS, 18 May. Jena, Germany: Friedrich Schiller Universität.

Huber, Mária. 2002. *Moskau, 11. März 1985. Die Auflösung des sowjetischen Imperiums.* Munich: dtv.

Human Rights Watch. 2000. *Landmine Monitor Report 2000: Toward a Mine-Free World.* New York: Human Rights Watch.

Huntington, Samuel. 1997. *The Clash of Civilizations and the Remaking of World Order.* New York: Simon and Schuster.

Hurd, Ian. 1999. "Legitimacy and Authority in International Politics." *International Organization* 53, no. 2: 379–408.

Ilves, Toomas Hendrik. 1991. "Reaction: The Intermovement in Estonia." In *Toward Independence: The Baltic Popular Movements,* edited by Jan Arveds Trapans, 71–83. Boulder, Colo.: Westview Press.

———. 1998. "Address by the Minister of Foreign Affairs, Toomas Hendrik Ilves, in the name of the Government to the Riigikogu, 12 Feb." Available at: http://www.vm.ee/eng/pressreleases/speeches/1998/0112parl.html.

———. 2000. "On the Developments in the Main Directions of Estonia's Foreign Policy: Address by Toomas Hendrik Ilves, Minister of Foreign Affairs, on Behalf of the Government of Estonia to the Riigikogu, 12 Oct." Available at: http://www.vm.ee/eng/pressreleases/speeches/2000/Riigikogu_okt2000 .htm.

———. 2001. "Speech at the Commemoration of Estonian Foreign Ministry Colleagues Deported and Killed by the Soviets, 15 Jan." Available at: http://www.vm.ee/eng/pressreleases/speeches/1201malestustahvel.htm.

Independent Task Force U.S. Policy toward Northeastern Europe. 1999. *U.S. Policy Toward Northeastern Europe.* Sponsored by the Council on Foreign Relations, chaired by Zbigniew Brzezinski, and directed by F. Stephen Larrabee. Available at: http://www.cfr.org/public/pubs/baltics.html.

International Institute for Strategic Studies (IISS). 1992. *The Military Balance 1992–1993.* London: Brassey's.

———. 1993. *The Military Balance 1993–1994.* London: Brassey's.

———. 1995. *The Military Balance 1995–1996.* Oxford: Oxford Univ. Press.

———. 1999. *The Military Balance 1999–2000.* Oxford: Oxford Univ. Press.

———. 2000. *The Military Balance 2000–2001.* Oxford: Oxford Univ. Press.

———. 2003. *The Military Balance 2003–2004.* Oxford: Oxford Univ. Press.

———. 2004. *The Military Balance 2004–2005.* Oxford: Oxford Univ. Press.

Jackson, Peter. 1989. *Maps of Meaning: An Introduction to Cultural Geography.* London: Routledge.

Jæger, Øjvind. 1997. *Securitizing Russia: Discursive Practices of the Baltic States.* Copenhagen: Copenhagen Peace Research Institute.

Jakobson, Max. 1996. "Finland: A Nation That Dwells Alone." *Washington Quarterly* 19, no. 4: 37–57.

Jakowlew, Alexander. 2003. *Die Abgründe meines Jahrhunderts: Eine Autobiographie.* Translated by Friedrich Hitzer. Leipzig: Faber and Faber.

Jalonen, Olli-Pekka. 1988. "The Strategic Significance of the Arctic." In *The Arctic Challenge: Nordic and Canadian Approaches to Security and Cooperation in an Emerging International Region,* edited by Kari Möttölä, 157–81. Boulder, Colo.: Westview Press.

Järve, Priit. 1996. "Security Choices of a Re-independent Small State: An Estonian Case." In *Small States and the Security Challenge in the New Europe,* edited by Werner Bauwens, Armand Clesse, and Olav F. Knudsen, 222–35. London: Brassey's.

Jervis, Robert. 1976. *Perception and Misperception in International Politics.* Princeton, N.J.: Princeton Univ. Press.

———. 1978. "Cooperation under the Security Dilemma." *World Politics* 30, no. 2: 167–214.

———. 1982. "Security Regimes." *International Organization* 36, no. 2: 357–78.

Joenniemi, Pertti. 1998. "Norden, Europe, and Post-security." In *Northern Dimensions 1998,* edited by Tuomas Forsberg, 62–72. Helsinki: Finnish Institute of International Affairs.

Johnson, Steven C. 1999. "NATO Bombs, Russian Rhetorics, and the Baltics." *Baltic Times* 4, no. 155, 22–28 April, 1, 10.

Joint Baltic-American National Committee (JBANC). 2001. "JBANC Statement of Policy 2001–2002." Available at: http://www.jbanc.org/policy2001.html.

Jones, Christopher. 1998. "Estonian Participation in International Military Co-operation." Paper presented at the Second Baltic-Nordic Conference, "Regionalism and Conflict Resolution," Vilnius, Lithuania, 24–27 Sept.

Jonson, Lena. 1991. "The Role of Russia in Nordic Regional Cooperation." *Cooperation and Conflict* 26, no. 3: 129–44.

———. 1992. "Russia in the Nordic Region in a Period of Change (1990–1992)." In *The Baltic Sea Region—A Region in the Making,* edited by Mare Kukk, Sverre Jervell, and Pertti Joenniemi, 79–106. Oslo: Europa-programmet.

Jundzis, Talavs. 1995. *Latvijas drosiba un aizsardziba.* Riga, Latvia: Junda.

———. n.d. "Latvia's Security and Relations with Russia." Unpublished manuscript.

Jurkynas, Mindaugas. 1997. "Some Aspects of National Minority Integration: The Case of Lithuania." In *Ethnicity and Politics in Estonia, Latvia, and Lithuania,* edited by Anton Steen, 116–34. Oslo: Department of Political Science, Univ. of Oslo.

Jutila, Matti Antero. 2002. "A Bump on the Road to Europe: An Iconic Model of the International Organizations' Pressure on Latvia's Minority Policy." M.A. thesis, Univ. of Lapland.

Kahar, Andres. 1996a. "Student Rally Protests Draft." *Baltic Times* 1, no. 7, 2–8 May, 1.

———. 1996b. "Ulmanis Vetoed Undergrad Conscription Law." *Baltic Times* 1, no. 35, 14–20 Nov., 5.

———. 1997a. "CBSS Commissioner Targets Latvian Military." *Baltic Times* 2, no. 59, 15–21 May, 8.

———. 1997b. "Skrunda 'Monster' Still Haunts Latvia." *Baltic Times* 2, no. 60, 22–28 May, 1, 8.

Kant, Immanuel. 1984. *Zum ewigen Frieden: Ein philosophischer Entwurf.* 1781. Reprint. Stuttgart: Reclam.

Kaplan, Cynthia. 1993. "Estonia: A Plural Society on the Road to Independence." In *Nations and Politics in the Soviet Successor States,* edited by Ian Bremmer and Ray Taras, 206–21. Cambridge: Cambridge Univ. Press.

Karklins, Rasma. 1994. *Ethnopolitics and Transition to Democracy: The Collapse of the USSR and Latvia.* Washington, D.C.: Woodrow Wilson Center Press.

Katzenstein, Peter J. 1996. "Introduction: Alternative Perspectives on National Security." In *The Culture of National Security: Norms and Identity in World Politics,* edited by Peter J. Katzenstein, 1–32. New York: Columbia Univ. Press.

Kerner, Manfred. 1994. "Baltische Sicherheit: Funken im Pulverfaß." *Nordeuropaforum* 4, no. 1: 31–34.

———. 1995. "Sicherheitsprobleme mit den Sicherheitskräften. Teil II: Estland und Litauen." *Antimilitarismus Information* 25, no. 2: 23–25.

Kerner, Manfred, Robert Ernecker, Marcis Gobins, and Frank Möller. 1998. "Außen- und Sicherheitspolitik." In *Handbuch Baltikum heute,* edited by Heike Graf and Manfred Kerner, 113–47. Berlin: Berlin Verlag Arno Spitz.

Kerner, Manfred, Sigmar Stopinski, and Felix Weiland. 1993. *Lettland und Litauen zu Beginn der neuen Unabhängigkeit: Gespräche im Herbst 1991 mit baltischen Politikern in Riga und Vilnius.* Berlin: Institut für Internationale Politik und Regionalstudien des Fachbereichs Politische Wissenschaft der Freien Universität Berlin.

Kionka, Riina, and Raivo Vetik. 1996. "Estonia and the Estonians." In *The Nationalities Question in the Post-Soviet States*, 2d ed., edited by Graham Smith, 129–46. London: Longman.

Kirby, David. 1995. *The Baltic World 1772–1993: Europe's Northern Periphery in an Age of Change*. London: Longman.

Kis, Janos. 1999–2000. "Das Erbe der demokratischen Opposition." *Transit*, no. 18: 17–39.

Klaar, Toivo. 1997. "Estonia's Security Policy Priorities." In *Baltic Security: Looking Towards the 21st Century*, edited by Gunnar Artéus and Atis Lejins, 18–32. Riga: Latvian Institute of Foreign Affairs and Försvarshögskolan.

Klinge, Matti. 1997. *The Baltic World*. Keuruu, Finland: Otava.

Knudsen, Olav F. 1993. "The Foreign Policies of the Baltic States: Interwar Years and Restoration." *Cooperation and Conflict* 28, no. 1: 47–72.

———. 1998. *Cooperative Security in the Baltic Sea Region*. Paris: Institute for Security Studies of Western European Union.

Knudsen, Olav F., and Iver B. Neumann. 1995. *Subregional Security Cooperation in the Baltic Sea Area: An Exploratory Study*. Oslo: Norwegian Institute of International Affairs.

Kolodziej, Edward A. 1992. "Renaissance in Security Studies? Caveat Lector!" *International Studies Quarterly* 36, no. 4: 421–38.

Korbonski, Andrzej. 1995. "United States Policy toward Eastern Europe and the Baltics." In *Baltic Europe in the Perspective of Global Change*, edited by Antoni Kuklinski, 130–36. Warsaw: Oficyna Naukowa.

Kowert, Paul, and Jeffrey Legro. 1996. "Norms, Identity, and Their Limits: A Theoretical Reprise." In *The Culture of National Security: Norms and Identity in World Politics*, edited by Peter J. Katzenstein, 451–97. New York: Columbia Univ. Press.

Kozaryn, Linda D. 1996. "Perry Says Baltic Nations Not Yet Ready for NATO-Membership." *American Forces Information Service*, news article, 3 Oct.

Kramer, Mark. 2002. "NATO, the Baltic States, and Russia: A Framework for Sustainable Enlargement." *International Affairs* 78, no. 4: 731–56.

Krause, Keith, and Michael C. Williams. 1996. "Broadening the Agenda of Security Studies: Politics and Methods." In *Mershon International Studies Review* 40, supplement 2: 229–54.

Krickus, Richard. 1993. "Lithuania: Nationalism in the Modern Era." In *Nations and Politics in the Soviet Successor States*, edited by Ian Bremmer and Ray Taras, 157–81. Cambridge: Cambridge Univ. Press.

———. 1999. "U.S. Support Might Not Mean NATO Membership." *Baltic Times* 4, no. 161, 3–9 June, 19.

Krippendorff, Ekkehard. 1985. *Staat und Krieg: Die historische Logik politischer Unvernunft*. Frankfurt: Suhrkamp.

Kristovskis, Girts Valdis. 2000. *Report of the Minister of Defense to the Parliament (Saeima) on State Defense Policy and Armed Forces Development for the Year 2000.* Available at: http://www.mod.lv/English/sec_inform/00.html.

Kronkaitis, Jonas. 1999. "Focusing on Communications." *Military Technology*, no. 10: 116–18.

Krupavicius, Algis. 1998. "The Development of Lithuania's Parties and Their International Contacts." In *The Baltic States at Historical Crossroads: Political, Economic, and Legal Problems in the Context of International Cooperation on the Doorstep of the 21th Century*, edited by Talavs Jundzis, 163–86. Riga: Academy of Sciences of Latvia.

Kuodyte, Dalia. 1999. "The Contacts Between the Lithuanian Resistance and the West." In *The Anti-Soviet Resistance in the Baltic States*, edited by Arvydas Anusauskas, 71–83. Vilnius, Lithuania: Du Ka.

Kupchan, Charles. 1996. "Reviving the West." In *Visions of European Security—Focal Point Sweden and Northern Europe*, edited by Anders Orrenius and Lars Truedson, 264–77. Stockholm: Olof Palme International Center.

Kuus, Merje. 2002. "Toward Cooperative Security? International Integration and the Construction of Security in Estonia." *Millennium* 31, no. 2: 297–317.

———. 2003a. "Borders of Security in Estonia." In *Routing Borders Between Territories, Discourses, and Practices*, edited by Eiki Berg and Henk van Houtum, 35–49. Aldershot, England: Ashgate.

———. 2003b. "From Threats to Risks: The Reconfiguration of Security Debates in the Context of Regional Co-operation." In *The Estonian Foreign Policy Yearbook 2003*, edited by Andres Kasekamp, 9–20. Tallinn: Estonian Foreign Policy Institute.

Laar, Mart. 1999. "The Armed Resistance Movement in Estonia from 1944 to 1956." In *The Anti-Soviet Resistance in the Baltic States*, edited by Arvydas Anusauskas, 209–41. Vilnius, Lithuania: Du Ka.

Lachowski, Zdzislaw. 1994. "Conventional Arms Control and Security Co-operation in Europe." In Stockholm International Peace Research Institute (SIPRI), *SIPRI Yearbook 1994*, 565–94. Oxford: Oxford Univ. Press.

———. 1995. "Conventional Arms Control and Security Dialogue in Europe." In Stockholm International Peace Research Institute (SIPRI), *SIPRI Yearbook 1995: Armaments, Disarmament, and International Security*, 761–90. Oxford: Oxford Univ. Press.

———. 2000. "Building Military Stability in the Baltic Sea Region." In *The NEBI Yearbook 2000: North European and Baltic Sea Integration,* edited by Lars Hedegaard and Bjarne Lindström, 259–73. Berlin: Springer.

———. 2002. *The Adapted CFE Treaty and the Admission of the Baltic States to NATO.* Stockholm: Stockholm International Peace Research Institute.

Lachowski, Zdzislaw, and Martin Sjögren. 2004. "Conventional Arms Control." In Stockholm International Peace Research Institute (SIPRI), *SIPRI Yearbook 2004: Armaments, Disarmament, and International Security,* 713–36. Oxford: Oxford Univ. Press.

Lagerspetz, Mikko. 1999. "The Cross of Virgin Mary's Land: A Study in the Construction of Estonia's 'Return to Europe.' " *Idäntutkimus* 6, no. 3–4: 17–28.

Landsbergis, Vytautas. 1997. "Die Zukunft Europas?" Speech given at Europäische Akademie Berlin, 24 March.

Lange, Peer H. 1995. "Estonia's Security: Consolidating Forces and Growing Uncertainties." *World Affairs* 157, no. 3: 131–36.

Latvian Human Rights Committee. 1999. *National Minorities in Latvia and Human Rights.* Riga: Latvian Human Rights Committee.

Laue, James. 1987. "The Emergence and Institutionalization of Third Party Roles in Conflict." In *Conflict Management and Problem Solving: Interpersonal to International Applications,* edited by Dennis J. D. Sandole and Ingrid Sandole-Staroste, 17–29. London: Francis Pinter.

Lauristin, Marju. 1997. "Contexts of Transition." In *Return to the Western World: Cultural and Political Perspectives on the Estonian Post-Communist Transition,* edited by Marju Lauristin and Peeter Vihalemm, with Karl Erik Rosengren and Lennart Weibull, 25–40. Tartu, Estonia: Tartu Univ. Press.

Lauristin, Marju, and Peeter Vihalemm. 1997. "Recent Historical Developments in Estonia: Three Stages of Transition (1987–1997)." In *Return to the Western World: Cultural and Political Perspectives on the Estonian Post-Communist Transition,* edited by Marju Lauristin and Peeter Vihalemm, with Karl Erik Rosengren and Lennart Weibull, 73–126. Tartu, Estonia: Tartu Univ. Press.

Leffler, Melvyn P. 1994. *The Specter of Communism: The United States and the Origins of the Cold War, 1917–1953.* New York: Hill and Wang.

Lehti, Marko. 1999. "Sovereignty Redefined: Baltic Cooperation and the Limits of National Self-determination." *Cooperation and Conflict* 34, no. 4: 413–43.

Lehtinen, Ari Aukusti. 2003. "Mnemonic North: Multilayered Geographies of the Barents Region." In *Encountering the North: Cultural Geography, International Relations, and Northern Landscapes,* edited by Frank Möller and Samu Pehkonen, 31–56. Aldershot, England: Ashgate.

Lejins, Atis. 1996. "Latvia." In *The Baltic Sea Region: National and International Security Perspectives,* edited by Axel Krohn, 40–60. Baden-Baden: Nomos.

———. 1998. "The 'Twin Enlargements' and Baltic Security." *Humanities and Social Sciences Latvia,* issue 2(19)–3(20): 8–40.

Levi, Primo. 1989. *The Drowned and the Saved.* New York: Vintage International.

Lewada, Juri. 1993. *Die Sowjetmenschen 1989–1991: Soziogramm eines Verfalls.* Munich: dtv.

Lieven, Anatol. 1994. *The Baltic Revolution: Estonia, Latvia, Lithuania, and the Path to Independence.* New Haven, Conn.: Yale Univ. Press.

———. 1996. "Baltic Iceberg Dead Ahead: NATO Beware." *The World Today* 52, no. 7: 175–79.

Light, Margot. 1988. *The Soviet Theory of International Relations.* New York: St. Martin's Press.

Linden, R. Ruth. 1993. *Making Stories, Making Selves: Feminist Reflections on the Holocaust.* Columbus: Ohio State Univ. Press.

Linkevicius, Linas. 1995. "Priorities of Lithuanian Security and Defense Policy." In *Vilnius—Kaliningrad: Ideas on Cooperative Security in the Baltic Sea Region,* edited by Ritva Grönick, Mia Grönqvist, and Nina Granlund, 100–108. Helsinki: Nordic Forum for Security Policy and Finnish Committee for European Security.

Lithuanian-American Community. 2001. "Classification of the U.S. Senators with Reference to Their Opinion on NATO Enlargement." Available at: http://www.lithuanian-american.org.

Liulevicius, Vejas Gabriel. 1995. "As Go the Baltics, So Goes Europe." *Orbis* 39, no. 3: 387–402.

Loth, Wilfried. 1987. *Die Teilung der Welt: Geschichte des Kalten Krieges 1941–1955.* Munich: dtv.

Lukic, Reneo, and Allen Lynch. 1996. *Europe from the Balkans to the Urals: The Disintegration of Yugoslavia and the Soviet Union.* Oxford: Oxford Univ. Press.

Lukin, Vladimir. 1995. "Krankhaft antirussisch." Interview. *Der Spiegel,* no. 39: 186–89.

Luongo, Kenneth N. 2001. "The Uncertain Future of U.S.-Russian Cooperative Nuclear Security." *Arms Control Today* 31, no. 1: 3–10.

Lydeka, Arminas. 1995. "The Guarantee of Lithuanian Security—Objective Estimation of the Geopolitical Situation." In *Vilnius—Kaliningrad: Ideas on Cooperative Security in the Baltic Sea Region,* edited by Ritva Grönick, Mia Grönqvist, and Nina Granlund, 109–16. Helsinki: Nordic Forum for Security Policy and Finnish Committee for European Security.

MacLaury, Bruce. 1994. "Foreword." In *The Great Transition: American-Soviet Relations and the End of the Cold War,* by Raymond Garthoff, ix–x. Washington, D.C.: Brookings Institution Press.

Main, Steven J. 1998. "Instability in the Baltic Region." In *Central and Eastern Europe: Problems and Prospects,* edited by Charles Dick and Anne Aldis, 180–98. Bristol, England: Strategic and Combat Studies Institute.

Maley, William. 1995. "Does Russia Speak for Baltic Russians?" *The World Today* 51, no. 1: 4–6.

Maniokas, Klaudijus. 1998. "Lithuania and EU Common Foreign and Security Policy." *Lithuania in the World* 6, no. 1: 28–31.

Mantenieks, Maris. 1990. "The Baltic Dilemma." *Foreign Affairs* 69, no. 3: 167–69.

Marsh, Christopher. 1998. "Realigning Lithuanian Foreign Relations." *Journal of Baltic Studies* 29, no. 2: 149–64.

Marshall, Monty G. 1999. *Third World War: System, Process, and Conflict Dynamics.* Lanham, Md.: Rowman and Littlefield.

Mason, Henry L. 1959. "The Process of Integration: A Critique of Four Recent Books on European Unification." *Journal of Conflict Resolution* 3, no. 2: 173–81.

Mazower, Mark. 1999. *Dark Continent: Europe's Twentieth Century.* London: Penguin.

McCausland, Jeffrey D. 1996. *Conventional Arms Control and European Security: Conventional Arms-Control Agreements and Their Role in the Emerging European Security Architecture.* London: International Institute for Strategic Studies and Brassey's.

MccGwyer, Michael. 1991. *Perestroika and Soviet National Security.* Washington, D.C.: Brookings Institution Press.

———. 1997. "Russia, the Expansion of NATO, and Security in Europe." In *Brassey's Defense Yearbook 1997,* edited by Center for Defense Studies, King's College London, 138–60. London: Brassey's.

McSweeney, Bill. 1999. *Security, Identity, and Interests: A Sociology of International Relations.* Cambridge: Cambridge Univ. Press.

Medearis, Sandra. 1998. "Russians Throw Skrunda Switch." *Baltic Times* 3, no. 124, 3–9 Sep., 1, 7.

Medvedev, Sergei. 2000. *Russia's Futures: Implications for the EU, the North, and the Baltic Region.* Helsinki: Finnish Institute of International Affairs and Institut für Europäische Politik.

Meri, Lennart. 1991. "Estonia's Role in the New Europe." *International Affairs* 67, no. 1: 107–10.

———. 2000a. "Estonia's Security and Defense Policy—New Steps Towards NATO

Membership. The President of the Republic of Estonia at the Royal United Services Institute London, 10 March." Available at: http://www.president.ee/eng/e_speeches. html?DOCUMENT_ID=4048.

———. 2000b. "Speech of the President of the Republic on the 82nd Anniversary of the Republic of Estonia in the *Estonia* Hall, 24 Feb." Available at: http://www.president.ee/eng/e_speeches.html?DOCUMENT_ID=4049.

———. 2004. "Mr Western Sea, President Lennart Meri." Interview. *Nordicum—Scandinavian Business Magazine,* no. 3: 6–8.

Meyer, Bertold, Harald Müller, and Hans-Joachim Schmidt. 1996. *NATO 96: Bündnis im Widerspruch.* Frankfurt: Hessische Stiftung Friedens- und Konfliktforschung.

Meyer, Stephen M. 1988. "The Sources and Prospects of Gorbachev's New Political Thinking on Security." *International Security* 13, no. 2: 124–63.

———. 1991–92. "How the Threat (and the Coup) Collapsed: The Politicization of the Soviet Military." *International Security* 16, no. 3: 5–38.

Mihalisko, Kathleen. 1991. "The Popular Movement in Belorussia and Baltic Influences." In *Toward Independence: The Baltic Popular Movements,* edited by Jan Arveds Trapans, 123–32. Boulder, Colo.: Westview Press.

Miniotaite, Grazina. 1997. "Lithuania: From Nonviolent Liberation to Nonviolent Defense?" *Gandhi Marg* 19, no. 2: 199–209.

———. 1998. "Lithuania." In *Bordering Russia: Theory and Prospects for Europe's Baltic Rim,* edited by Hans Mouritzen, 165–93. Aldershot, England: Ashgate.

Ministry of Defense of the Republic of Estonia. 1999. *Eesti NATO lävepakul: 80-aastane Eesti Kaitsevägi.* Tallinn: Ministry of Defense of the Republic of Estonia.

Ministry of Defense of the Republic of Latvia. n.d. "Premilitary Service Training." Available at: http://www.mod.lv/English/sec_darbs/apmaciba.htm.

Ministry of Foreign Affairs of the Republic of Estonia. 1999. *The Baltic Battalion (BaltBat): Regional and International Co-operation in Action.* Fact sheet, updated 25 Oct. Tallinn: Ministry of Foreign Affairs of the Republic of Estonia.

———. 2001a. "National Security Concept of the Republic of Estonia, approved 6 March." Available at: http://www.vm.ee/eng/policy/Security/index.htm.

———. 2001b. "63% of Estonians Support Joining NATO." Press Release no. 9-E, 15 Mar.

Ministry of Foreign Affairs of the Republic of Lithuania. 1998. *Fact Sheet: Preparing for NATO.* Vilnius: Ministry of Foreign Affairs of the Republic of Lithuania, 11 Nov.

———. 1999. *Fact Sheet: Military Cooperation with NATO.* Vilnius: Ministry of Foreign Affairs of the Republic of Lithuania.

Ministry of National Defense of the Republic of Lithuania. 1999. "White Paper '99." Available at: http://www.kam.lt/English.

Misiunas, Romuald, and Rein Taagepera. 1993. *The Baltic States: Years of Dependence 1940–1990.* Expanded and updated ed. Berkeley: Univ. of California Press.

Missiroli, Antonio. 2004. "Central Europe Between the EU and NATO." *Survival* 46, no. 4: 121–36.

Mladineo, Peter J. 2000. "For Baltics, It's Military Finance Time Again." *Baltic Times* 5, no. 200, 23–29 Mar., 6.

Möller, Frank. 2000. "Towards a Post–Security Community in Northeastern Europe—Policy Initiatives of the Clinton Administration toward the Baltic Sea Region." Paper presented at the 18th International Peace Research Association General Conference, Tampere, Finland, 5–9 Aug.

———. 2002a. "Peaceful Change but Not Yet Stable Peace: Military Developments in the Baltic Sea Region, 1990–2000." In *The NEBI Yearbook 2001–2002: North European and Baltic Sea Integration,* edited by Lars Hedegaard and Bjarne Lindström, 245–67. Berlin: Springer.

———. 2002b. "Reconciling International Politics with Local Interests: The United States in Northern Europe." In *The New North of Europe: Policy Memos,* edited by Teresa Pohjola and Johanna Rainio, 77–81. Helsinki: Finnish Institute of International Affairs and Institut für Europäische Politik.

———. 2003a. "Capitalizing on Difference: A Security Community or/as a Western Project." *Security Dialogue* 34, no. 3: 315–28.

———. 2003b. "Cognition, Representation, and Security Community Building in the Baltic Sea Region." *Nordeuropaforum* 13, no. 2: 61–85.

———. 2003c. "Gefahrendiskurse und baltische Sicherheitspolitik." *WeltTrends* 12, no. 42: 89–100.

———. 2004. "Security Communities and Communities of Memory—Taking Historical Experience Seriously." *Kosmopolis* 34 (Special Celebratory Issue for Unto Vesa's 60th Birthday): 91–102.

Möller, Frank, and Hendrik Ehrhardt. 2005. "Sicherheit als Sprechakt: Legitimation von Gewalt durch die Artikulation von Sicherheit." In *Diskurse der Gewalt—Gewalt der Diskurse,* edited by Michael Schultze, Jörg Meyer, Britta Kraus, and Dietmar Fricke, 47–58. Frankfurt: Peter Lang.

Möller, Frank, and Samu Pehkonen. 2003. "Discursive Landscapes of the European North." In *Encountering the North: Cultural Geography, International Relations, and Northern Landscapes,* edited by Frank Möller and Samu Pehkonen, 1–30. Aldershot, England: Ashgate.

Möller, Frank, and Arend Wellmann. 2001. "Baltic States." In *Security Handbook 2001: Security and Military in Central and Eastern Europe,* edited by Hans J. Gießmann and Gustav E. Gustenau, 77–117. Baden-Baden: Nomos.

Morgenthau, Hans J. 1984. "Human Rights and Foreign Policy." In *Moral Dimensions of American Foreign Policy,* edited by Kenneth W. Thompson, 341–48. New Brunswick, N.J.: Transaction Books.

Motulaite, Violeta. 1996. "Sources of National Security in Lithuania." In *Visions of European Security—Focal Point Sweden and Northern Europe,* edited by Anders Orrenius and Lars Truedson, 162–70. Stockholm: Olof Palme International Center.

Mouritzen, Hans. 1998a. "Focus and Axioms." In *Bordering Russia: Theory and Prospects for Europe's Baltic Rim,* edited by Hans Mouritzen, 1–13. Aldershot, England: Ashgate.

———. 1998b. "Thule and Theory: Democracy vs. Elitism in Danish Foreign Policy." In *Danish Foreign Policy Yearbook 1998,* edited by Bertel Heurlin and Hans Mouritzen, 79–101. Copenhagen: Danish Institute of International Affairs.

———. 2001. "Security Communities in the Baltic Sea Region: Real and Imagined." *Security Dialogue* 33, no. 3: 297–310.

Muiznieks, Nils. 1993. "Latvia: Origins, Evolution, and Triumph." In *Nations and Politics in the Soviet Successor States,* edited by Ian Bremmer and Ray Taras, 182–205. Cambridge: Cambridge Univ. Press.

———. 2002. "Latvia's Faux Pas." *Foreign Policy* (Jan.–Feb.): 88–89.

Müllerson, Rein. 1994. *International Law, Rights, and Politics: Developments in Eastern Europe and the CIS.* London: Routledge.

Mulloy, Patrick A. 1998. "The U.S. View of the Baltic Regional Market." Presentation at the Third Annual Baltic Sea Region Conference, "The Baltic Sea Region: Building an Inclusive System of Security and Cooperation," Stockholm, 19 Nov. Available at: http://www.usemb.se/bscont/1998/text/mulloy.html.

Nahaylo, Bohdan. 1991. "Baltic Echoes in Ukraine." In *Toward Independence: The Baltic Popular Movements,* edited by Jan Arveds Trapans, 109–22. Boulder, Colo.: Westview Press.

Narusk, Anu. 1997. "Perception of Social Problems and the 'Real Life.'" *Social Studies* 3: 9–23.

Nekrasas, Evaldas. 1996. "Lithuania's Security Concerns and Responses." In *The Baltic States: Search for Security,* edited by Atis Lejins and Daina Bleiere, 58–74. Riga: Latvian Institute of International Affairs.

Neocleous, Mark. 2000. "Against Security." *Radical Philosophy,* no. 100: 7–15.

Neumann, Iver B. 1992. *Regions in International Relations Theory: The Case for a Region-Building Approach.* Oslo: Norwegian Institute of International Affairs.

———. 1997. "The Geopolitics of Delineating 'Russia' and 'Europe': The Creation of the 'Other' in European and Russian Tradition." In *Geopolitics in Post-Wall Europe: Security, Territory, and Identity,* edited by Ola Tunander, Pavel Baev, and Victoria Ingrid Einagel, 147–73. London: Sage.

Newsfile Lithuania. 2000. No. 629, 3–16 July.

———. 2001. No. 648, 19–25 Feb.

Nora, Pierre. 1998. *Zwischen Geschichte und Gedächtnis.* Translated by Wolfgang Kaiser. Frankfurt: Fischer.

———. 2002. "The Reasons for the Current Upsurge in Memory." *Tr@nsit-Virtuelles Forum,* no. 22. Available at: http://www.univie.ac.at/iwm/t-22txt3.htm.

Noreen, Erik, and Roxanna Sjöstedt. 2004. "Estonian Identity Formations and Threat Framing in the Post–Cold War Era." *Journal of Peace Research* 41, no. 6: 733–50.

Norkus, Renatas. 1998. "Preventing Conflict in the Baltic States: A Success Story That Will Hold?" In *Preventing Violent Conflict: Issues from the Baltic and the Caucasus,* edited by Gianni Bonvicini, Ettore Greco, Bernard von Plate, and Reinhardt Rummel, 135–67. Baden-Baden: Nomos.

———. 1999. "The U.S. Role in Lithuania's Foreign and Security Policy." *Lithuanian Foreign Policy Review,* no. 3, electronic version. Available at: http://www .urm.lt/lfpr/usrole.htm.

North Atlantic Treaty Organization (NATO). 1995. *Study on NATO Enlargement.* Brussels: NATO.

Novick, Peter. 2001. *The Holocaust and Collective Memory: The American Experience.* London: Bloomsbury.

Nurick, Robert C. 1997. "Presentation." In *NATO and the Baltic States: Quo Vadis?* edited by Paulis Apinis and Atis Lejins, 51–64. Riga: Konrad Adenauer Stiftung and Latvian Institute of International Affairs.

Nye, Joseph S., Jr. 1988. "Gorbachev's Russia and U.S. Options." In *Gorbachev's Russia and American Foreign Policy,* edited by Seweryn Bialer and Michael Mandelbaum, 385–408. Boulder, Colo.: Westview Press.

Oldberg, Ingmar. 1998. "Kaliningrad: Problems and Prospects." In *Kaliningrad: The European Amber Region,* edited by Pertti Joenniemi and Jan Prawitz, 1–31. Aldershot, England: Ashgate.

———. 2003. "Kaliningrad in der Militär- und Sicherheitspolitik Rußlands." *Osteuropa* 53, no. 2–3: 270–85.

Olson, William, and A. J. R. Groom. 1991. *International Relations Then and Now: Origins and Trends in Interpretation.* London: Harper Collins Academic.

Öövel, Andrus. 1996. "Estonian Defense Policy, NATO, and the European Union." *Security Dialogue* 27, no. 1: 65–68.

Österreichische Militärische Zeitschrift. 1990. Vol. 28, no. 4: 333–37.

———. 1991a. Vol. 29, no. 2: 170–74.

———. 1991b. Vol. 29, no. 3: 266–67.

Osteuropa-Archiv. 1994. June: A311–A322.

Ozolina, Zaneta. 1996. "Latvian Security Policy." In *The Baltic States: Search for Secu-*

rity, edited by Atis Lejins and Daina Bleiere, 29–57. Riga: Latvian Institute of International Affairs.

———. 1998. "Latvia." In *Bordering Russia: Theory and Prospects for Europe's Baltic Rim,* edited by Hans Mouritzen, 131–63. Aldershot, England: Ashgate.

———. 1999. *The Regional Dimension in Latvian Security Policy.* Groningen, Netherlands: Center for European Security Studies.

Ozolins, Andris. 1994. "The Policies of the Baltic Countries vis-à-vis the CSCE, NATO, and WEU." In *The Foreign Policies of the Baltic Countries: Basic Issues,* edited by Pertti Joenniemi and Juris Prikulis, 49–74. Riga, Latvia: Center of Baltic-Nordic History and Political Studies, and Tampere Peace Research Institute.

Paasi, Anssi. 1996. *Territories, Boundaries, and Consciousness: The Changing Geographies of the Finnish-Russian Border.* Chichester, England: John Wiley and Sons.

Palosaari, Teemu, and Frank Möller. 2004. "Security and Marginality: Arctic Europe after the Double Enlargement." *Cooperation and Conflict* 39, no. 3: 255–81.

Park, Andrus. 1991. "From Perestroika to Cold Civil War: Reflections on the Soviet Disintegration Crisis." *Bulletin of Peace Proposals* 22, no. 3: 257–64.

Patomäki, Heikki, and Ole Wæver. 1995. "Introducing Peaceful Changes." In *Peaceful Changes in World Politics,* edited by Heikki Patomäki, 3–27. Tampere, Finland: Tampere Peace Research Institute.

Penikis, Janis. 1996. "Five Years of Independence." *Baltic Times* 1, no. 30, 10–16 Oct., 23.

Petersen, Phillip. 1992. "Security Policy in the Post-Soviet Baltic States." *European Security* 1, no. 1: 13–49.

Plakans, Andrejs. 1998. "The Baltic Region and the Baltic States in Contemporary American Commentary." In *The Baltic States at Historical Crossroads: Political, Economic, and Legal Problems in the Context of International Cooperation on the Doorstep of the 21st Century,* edited by Talavs Jundzis, 641–50. Riga: Academy of Sciences of Latvia.

Poleshuk, Vadim. 2001. *Advice Not Welcomed: Recommendations of the OSCE High Commissioner to Estonia and Latvia and the Response.* Münster, Germany: Lit.

Polkowski, Andreas, on behalf of the Hamburg Chamber of Commerce. 2000. *Mare Balticum: Economic Situation, Trading Relations, and Direct Investment in the Baltic Sea Region.* Hamburg: Hamburg Chamber of Commerce.

Puheloinen, Ari. 1999. *Russia's Geopolitical Interests in the Baltic Area.* Helsinki: National Defense College.

Putin, Vladimir. 2001. "Interview Granted by the Russian President, Vladimir Putin, to the Finnish Newspaper Helsingin Sanomat, 1 Sept." Available at: http://www.mid.ru.

Raid, Aare. 1993. "The Baltic States and Security Strategies." In *New Actors on the In-*

ternational Arena: The Foreign Policies of the Baltic Countries, edited by Pertti Joenniemi and Peeter Vares, 39–50. Tampere, Finland: Tampere Peace Research Institute.

———. 1996. "Security Policy of the Baltic States: The Case of Estonia." In *The Baltic States: Search for Security,* edited by Atis Lejins and Daina Bleiere, 8–28. Riga: Latvian Institute of International Affairs.

Raitviir, Tiina. 1996. *Eesti üleminekuperioodi valimiste (1989–1993): Võrdlev uurimine.* Tallinn, Estonia: Teaduste Akadeemia Kirjastus.

Rankin, Ian. 2001. *Mortal Causes.* London: Orion.

RFE/RL Baltic States Report. 2000a. Vol. 1, no. 5 (21 Feb.).

———. 2000b. Vol. 1, no. 6 (28 Feb.).

———. 2000c. Vol. 1, no. 18 (22 May).

———. 2000d. Vol. 1, no. 22 (27 June).

———. 2000e. Vol. 1, no. 33 (16 Oct.).

———. 2000f. Vol. 1, no. 36 (29 Nov.).

———. 2001a. Vol. 2, no. 4 (19 Feb.).

———. 2001b. Vol. 2, no. 7 (23 Mar.).

Rikmann, Erle. 1999. "Retroactive History and Personal Memory." *Idäntutkimus* 6, no. 3–4: 60–73.

Rislakki, Jukka. 2001. "Nuorten teksteistä koottu kirja venäläisistä kuohuttaa Latviassa." *Helsingin Sanomat,* 14 Aug., C2.

Risse-Kappen, Thomas. 1996. "Collective Identity in a Democratic Community: The Case of NATO." In *The Culture of National Security: Norms and Identity in World Politics,* edited by Peter J. Katzenstein, 357–99. New York: Columbia Univ. Press.

Roberts, Adam. 1994. "Civil Resistance in the East European and Soviet Revolutions of 1989–91." In *The Soviet Union in Eastern Europe, 1945–89,* edited by Odd Arne Westad, Sven Holtsmark, and Iver B. Neumann, 175–206. New York: St. Martin's Press.

Roeder, Philip G. 1991. "Soviet Federalism and Ethnic Mobilization." *World Politics* 43, no. 2: 196–232.

Rogov, Sergei M. 2000. "Presentation." In *Facing the New Millennium: Russian-Norwegian Relations in a Changing World,* edited by Helge Blakkisrud and Christina Brookes, 25–31. Oslo: Norwegian Institute of International Affairs.

Rosati, Jerel A. 1995. "A Cognitive Approach to the Study of Foreign Policy." In *Foreign Policy Analysis: Continuity and Change in Its Second Generation,* edited by Laura Neack, Jeanne A. K. Hey, and Patrick J. Haney, 49–70. Englewood Cliffs, N.J.: Prentice-Hall.

Rose, Richard. 1997. *New Baltic Barometer III: A Survey Study*. Glasgow: Center for the Study of Public Policy, Univ. of Strathclyde.

Rotfeld, Adam Daniel. 1998. "Europe: The Transition to Inclusive Security." In Stockholm International Peace Research Institute (SIPRI), *SIPRI Yearbook 1998: Armaments, Disarmament, and International Security*, 141–67. Oxford: Oxford Univ. Press.

Ruggie, John Gerard. 1998. *Constructing the World Polity: Essays on International Institutionalization*. London: Routledge.

Ruhala, Kalevi. 1988. "Finland's Security Policy: The Arctic Dimension." In *The Arctic Challenge: Nordic and Canadian Approaches to Security and Cooperation in an Emerging International Region*, edited by Kari Möttölä, 117–29. Boulder, Colo.: Westview Press.

Russett, Bruce. 1993. *Grasping the Democratic Peace: Principles for a Post–Cold War World*. Princeton, N.J.: Princeton Univ. Press.

Sagan, Scott D. 1993. *The Limits of Safety: Organizations, Accidents, and Nuclear Weapons*. Princeton, N.J.: Princeton Univ. Press.

Said, Edward. 1994. *Culture and Imperialism*. London: Vintage.

Saudargas, Algirdas. 1999. "Not Only a Consumer of Security." *NATO's Nations and Partners for Peace* (special issue): 77–78.

Sawhill, Steven G. 2000. "Cleaning-up the Arctic's Cold War Legacy: Nuclear Waste and Arctic Military Environmental Cooperation." *Cooperation and Conflict* 35, no. 1: 5–35.

Sawhill, Steven G., and Anne-Kristin Jørgensen. 2001. *Military Nuclear Waste and International Cooperation in Northwest Russia*. Oslo: Fridtjof Nansen Institute.

Schama, Simon. 1995. *Landscape and Memory*. London: Fontana Press.

Schmidt, Hans-Joachim. 1998. *Die Anpassung des KSE-Vertrages: Konventionelle Rüstungskontrolle zwischen Bündnisverteidigung und Kooperativer Sicherheit*. Frankfurt: Hessische Stiftung Friedens- und Konfliktforschung.

———. 2000. *Die Anpassung des KSE-Vertrages und die Gefährdung der globalen Rüstungskontrolle*. Frankfurt: Hessische Stiftung Friedens- und Konfliktforschung.

Schöpflin, George. 1999. "Uses of the Past in Inter-ethnic Relations." *Idäntutkimus* 6, no. 3–4: 7–16.

Schröder, Hans-Henning. 1996. "Rußlands Armee in der Politik: Risikofaktor oder Garant politischer Stabilität?" In *Eine Welt oder Chaos?* edited by Berthold Meyer, 132–52. Frankfurt: Suhrkamp.

Senn, Alfred Erich. 1996. "Lithuania and the Lithuanians." In *The Nationalities Question in the Post-Soviet States*, 2d ed., edited by Graham Smith, 170–83. London: Longman.

Sergounin, Alexander. 1997. "In Search of a New Strategy in the Baltic/Nordic Area." In *Russia and Europe: The Emerging Security Agenda,* edited by Vladimir Baranovsky, 325–49. Oxford: Oxford Univ. Press.

———. 2000. "Russian Post-Communist Foreign Policy Thinking at the Cross-roads: Changing Paradigms." *Journal of International Relations and Development* 3, no. 3: 216–55.

Shapiro, Michael J. 1997. *Violent Cartographies: Mapping the Cultures of War.* Minneapolis: Univ. of Minnesota Press.

———. 2004. *Methods and Nations: Cultural Governance and the Indigenous Subject.* New York: Routledge.

Sharp, Jane M. O. 1990. "Conventional Arms Control in Europe." In Stockholm International Peace Research Institute (SIPRI), *SIPRI Yearbook 1990: World Armaments and Disarmament,* 459–507. Oxford: Oxford Univ. Press.

———. 1991. "Conventional Arms Control in Europe." In Stockholm International Peace Research Institute (SIPRI), *SIPRI Yearbook 1991: World Armaments and Disarmament,* 407–60. Oxford: Oxford Univ. Press.

———. 1992. "Conventional Arms Control in Europe: Developments and Prospects in 1991." In Stockholm International Peace Research Institute (SIPRI), *SIPRI Yearbook 1992: World Armaments and Disarmament,* 459–79. Oxford: Oxford Univ. Press.

———. 1993. "Conventional Arms Control in Europe." In Stockholm International Peace Research Institute (SIPRI), *SIPRI 1993: World Armaments and Disarmament,* 591–631. Oxford: Oxford Univ. Press.

———. 1998. "CFE and the Baltic Rim." In *The NEBI Yearbook 1998: North European and Baltic Sea Integration,* edited by Lars Hedegaard and Bjarne Lindström, 423–35. Berlin: Springer.

Shevardnadze, Eduard. 1989. "Foreign Policy and Perestroika." Speech by Soviet foreign minister Eduard Shevardnadze to the Plenary Session of the USSR Supreme Soviet, Moscow, 23 Oct. Available at: http://projects.sipri.se/SAC/891023.01.html.

Shore, Sean M. 1998. "No Fences Make Good Neighbors: The Development of the U.S.-Canadian Security Community, 1871–1940." In *Security Communities,* edited by Emanuel Adler and Michael Barnett, 333–67. Cambridge: Cambridge Univ. Press.

Simenas, Albertas. 1997. "Formation of the Market Economy in Lithuania." In *Lithuanian Economic Reforms: Practice and Perspectives,* edited by Antanas Buracas, 17–62. Vilnius, Lithuania: Margi Rastai.

Simm, Lesley. 1995. *Developing a National Security Concept.* Riga, Latvia: Baltic Center for Peace and Security Studies.

Simon, Jeffrey. 2000. "Transforming the Armed Forces of Central and East Europe." *Strategic Forum*, no. 172: 1–4.

Simonsen, Sven Gunnar. 1996. *Politics and Personalities: Key Actors in the Russian Opposition*. Oslo: International Peace Research Institute Oslo.

Sindrich, Jaclyn M. 2000. "Students Face Barracks before Books." *Baltic Times* 5, no. 194, 10–16 Feb., 1, 8.

SINUS Moskau, Sozialforschung und Marktforschung. 1994. *Militäreliten in Rußland 1994*. Munich: SINUS.

Skak, Mette. 1996. *From Empire to Anarchy: Postcommunist Foreign Policy and International Relations*. London: Hurst.

Skujins, Maris. 1996. "Die strategische Lage der baltischen Staaten." *Österreichische Militärische Zeitschrift* 34, no. 6: 643–50.

Skultans, Vieda. 1998. *The Testimony of Lives: Narrative and Memory in Post-Soviet Latvia*. London: Routledge.

Smith, David. 2001. *Cultural Autonomy in Estonia: A Relevant Paradigm for the Post-Soviet Era?* Bradford: Univ. of Bradford, Department of European Studies and Baltic Research Unit.

———. 2002. "Narva Region within the Estonian Republic: From Autonomism to Accommodation?" *Regional and Federal Studies* 12, no. 2: 89–110.

Smith, Graham. 1996. "Latvia and the Latvians." In *The Nationalities Question in the Post-Soviet States*, 2d ed., edited by Graham Smith, 147–69. London: Longman.

Smolar, Aleksander. 1999–2000. "Vergangenheitspolitik seit 1989: Eine vergleichende Zwischenbilanz." *Transit*, no. 18: 81–101.

Snyder, Jack. 1991. *Myths of Empire: Domestic Politics and International Ambitions*. Ithaca, N.Y.: Cornell Univ. Press.

Snyder, Richard C. 1954. "Editor's Foreword." In *Political Community at the International Level: Problems of Definition and Measurement*, by Karl W. Deutsch, v–vii. Garden City, N.Y.: Doubleday.

Snyder, Tim. 1995. "National Myths and International Relations: Poland and Lithuania, 1989–1994." *East European Politics and Societies* 9, no. 2: 317–43.

Sørensen, Georg. 1992. "Kant and Processes of Democratization: Consequences for Neorealist Thought." *Journal of Peace Research* 29, no. 4: 397–414.

Stankevicius, Ceslovas. 1994–96. *Enhancing Security of Lithuania and Other Baltic States in 1992–94 and Future Guidelines*. Available at: http://www.nato.int/acad/fellow/94-96/stankevi/04.htm.

———. 1999. "Lithuania on Its Way to NATO." *NATO's Nations and Partners for Peace* (special issue): 79–81.

Starr, Harvey. 1992. "Democracy and War: Choice, Learning, and Security Communities." *Journal of Peace Research* 29, no. 2: 207–13.

———. 1997. "Democracy and Integration: Why Democracies Don't Fight Each Other." *Journal of Peace Research* 34, no. 2: 153–62.

Stockholm International Peace Research Institute (SIPRI). 2003. *SIPRI Yearbook 2003: Armaments, Disarmament, and International Security.* Oxford: Oxford Univ. Press.

Stranga, Aivars. 1996. "Russia and the Security of the Baltic States: 1991–1996." In *The Baltic States: Search for Security,* edited by Atis Lejins and Daina Bleiere, 141–85. Riga: Latvian Institute of International Affairs.

———. 1997. "Baltic-Russian Relations: 1995–Beginning of 1997." In *Small States in a Turbulent Environment: The Baltic Perspective,* edited by Atis Lejins and Zaneta Ozolina, 184–237. Riga: Latvian Institute of International Affairs.

Strods, Heinrihs. 1997. "Guerilla Warfare in the Baltic States during the Period of the Cold War from 1944 to 1956." In *50 Years after World War II: International Politics in the Baltic Sea Region 1945–1995,* edited by Harald Runblom, Mieczyslaw Nurek, Marceli Burdelski, Thomas Jonter, and Erik Noreen, 155–59. Gdansk, Poland: Wydawnictwo Uniwersytetu Gdanskiego.

———. 1999. "The Latvian Partisan War between 1944 and 1956." In *The Anti-Soviet Resistance in the Baltic States,* edited by Arvydas Anusauskas, 149–60. Vilnius, Lithuania: Du Ka.

Sweedler, Alan. 1994. "Security in Northern Europe and the Baltic Sea from an American Perspective." In *Common Security in Northern Europe after the Cold War,* edited by Göran Baecklund, 190–206. Stockholm: Olof Palme International Center.

Taagepera, Rein. 1993. *Estonia: Return to Independence.* Boulder, Colo.: Westview Press.

Talbott, Strobe. 1998. "The U.S. and the Baltic Region: Remarks by Strobe Talbott, 8 July 1998, Riga, Latvia." *U.S. Information and Texts,* no. 028 (15 July): 5–6.

———. 2000a. "A Baltic Home-Coming: Robert C. Frasure Memorial Lecture, Tallinn, 24 Jan." Available at: http://www.vm.ee/eng/pressreleases/speeches/2000/Talbott.html.

———. 2000b. "Press Conference at Third Annual U.S.-Baltic Partnership Commission Meeting, Tallinn, Estonia, 7 June." Available at: http://state.gov/www/policy_remarks/2000/000607_talbott_usbalticpc.html.

Tallo, Ivar. 1995. "Estonian Government and Politics." *World Affairs* 157, no. 3: 125–30.

Tammerk, Tarmu. 1994. "Treaties Leave Open Ends." *Baltic Independent* 5, no. 224, 5–11 Aug., 3.

Tamulaitis, Gintaras. 1994. *National Security and Defense Policy of the Lithuanian State.* Geneva: United Nations Institute for Disarmament Research.

Tauber, Joachim. 1997. *Die Auseinandersetzung mit der kommunistischen Vergangenheit in Litauen.* Cologne: Bundesinstitut für Ostwissenschaftliche und Internationale Studien.

Thomas, Daniel C. 2001. *The Helsinki Effect: International Norms, Human Rights, and the Demise of Communism.* Princeton, N.J.: Princeton Univ. Press.

Thompson, Kenneth W. 1958. Review of *Political Community and the North Atlantic Area,* by Karl W. Deutsch et al. *American Political Science Review* 52, no. 2: 531–33.

Tiido, Harri. 2000. "The Impossibility of a 'Baltic Sea Region Security.' " Speech at the 5th Anniversary Stockholm Conference on Baltic Sea Region Security and Cooperation, 19 Oct. Available at: http://www.usis.usemb.se/bsconf/2000/tiido.html.

Todorov, Tzvetan. 1982. *La conquête de l'Amérique: La question de l'autre.* Paris: Éditions du Seuil.

Tracevskis, Rokas M. 1998. "Americans Welcomed to Lithuanian Oil." *Baltic Times* 3, no. 129, 8–14 Oct., 3.

———. 2000a. "Defense Cuts Not Supported in Vilnius." *Baltic Times* 5, no. 205, 27 Apr.–10 May, 6.

———. 2000b. "Lithuanian Political Parties Waltz with NATO." *Baltic Times* 5, no. 194, 10 –16 Feb., 6.

Trachtenberg, Marc. 1999. *A Constructed Peace: The Making of the European Settlement, 1945–1963.* Princeton, N.J.: Princeton Univ. Press.

Trapans, Andris. 1991. "Moscow, Economics, and the Baltic Republics." In *Toward Independence: The Baltic Popular Movements,* edited by Jan Arveds Trapans, 85–98. Boulder, Colo.: Westview Press.

Trapans, Jan Arveds. 1991a. "Averting Moscow's Baltic Coup." *Orbis* 35, no. 3: 427–39.

———. 1991b. "Introduction." In *Toward Independence: The Baltic Popular Movements,* edited by Jan Arveds Trapans, 3–8. Boulder, Colo.: Westview Press.

———. 1991c. "The Sources of Latvia's Popular Movement." In *Toward Independence: The Baltic Popular Movements,* edited by Jan Arveds Trapans, 25–41. Boulder, Colo.: Westview Press.

Trenin, Dmitri. 2000. "Security Cooperation in North-Eastern Europe: A Russian Perspective." In *Russia and the United States in Northern European Security,* by Dmitri Trenin and Peter van Ham, 15–54. Helsinki: Finnish Institute of International Affairs and Institut für Europäische Politik.

Tribble, Conrad. 2000. "NEI and the Northern Dimension." In *The Northern Dimension: An Assessment and Future Development,* edited by Atis Lejins and Jörg-Dietrich Nackmayr, 61–70. Riga: Latvian Institute of International Affairs and Konrad Adenauer Stiftung.

Troebst, Stefan. 1999a. "Nordosteuropa: Geschichtsregion mit Zukunft." *Nordeuropaforum* 9, no. 1: 53–69.

———. 1999b. "Rußland und die Ostseeregion: Nordosteuropa als historisch gewachsener Kooperationsraum." In *Kooperation und Konflikt in der Ostseeregion*, edited by Christian Wellmann, 11–21. Kiel, Germany: Landeszentrale für Politische Bildung Schleswig-Holstein.

Tunander, Ola. 1989. *Cold Water Politics: The Maritime Strategy of the Northern Front.* London: Sage.

U.S.-Baltic Partnership Commission. 2000. "Joint Communiqué of the U.S.-Baltic Partnership Commission." Released following the third annual U.S.-Baltic Partnership Commission, Tallinn, Estonia, 7 June. Available at: http://www.state .gov/www/regions/eur/nei/000607_pc_commq.html.

U.S. Commission on Security and Cooperation in Europe. 1990. *Renewal and Challenge: The Baltic States 1988–1989.* Washington, D.C.: U.S. Government Printing Office.

U.S. Department of State. n.d.a. "Cooperative Security." Available at: http://www .state.gov/p/eur/rt/epine/c10610htm.

———. n.d.b. "Healthy Societies." Available at: http://www.state.gov/p/eur /rt/epine/c10611.htm.

———. Bureau of European Affairs. 2000. *Overview of the Northern Europe Initiative.* Fact sheet, released 1 May. Available at: http://www.state.gov/www/regions/eur/nei/fs_000501_nei.html.

U.S. Embassy in Stockholm. n.d. "Security and Cooperation in the Baltic and Barents Sea Region: A Priority for the United States." Available at: http://www.usemb .se/BalticSec/index.html.

U.S. Information and Texts. 1997. No. 45, 5 Nov.

———. 1998. No. 003, 21 Jan.

U.S. Senate. 2000. "Concurrent Resolution 122, 106th Congress, 2d Session, 14 June." Available at: http://www.usis.lt/ReleasesRecognition.htm.

van Ham, Peter. 1998. "The Baltic States and *Zwischeneuropa*: 'Geography Is Destiny'?" *International Relations* 14, no. 2: 47–59.

———. 2000. "Testing Cooperative Security in Europe's New North: American Perspectives and Policies." In *Russia and the United States in Northern European Security*, by Dmitri Trenin and Peter van Ham, 57–95. Helsinki: Finnish Institute of International Affairs and Institut für Europäische Politik.

van Zon, Hans. 1995. "Problems of Transitology: Towards a New Research Agenda and New Research Practice." In *Baltic Europe in the Perspective of Global Change*, edited by Antoni Kuklinski, 454–74. Warsaw: Oficyna Naukowa.

Vardys, V. Stanley. 1991. "Sajudis: National Revolution in Lithuania." In *Toward Inde-*

REFERENCES • 359

pendence: The Baltic Popular Movements, edited by Jan Arveds Trapans, 11–25.
Boulder, Colo.: Westview Press.

Vardys, V. Stanley, and Judith B. Sedaitis. 1997. Lithuania: The Rebel Nation. Boulder,
Colo.: Westview Press.

Vares, Peeter. 1993. "Dimensions and Orientations in the Foreign and Security Policies of the Baltic Countries." In New Actors on the International Arena: The Foreign Policies of the Baltic Countries, edited by Pertti Joenniemi and Peeter Vares, 3–31. Tampere, Finland: Tampere Peace Research Institute.

———. 1994. "Russia and the Baltic States: Are There Common Security Perspectives?" In Common Security in Northern Europe after the Cold War: The Baltic Sea Region and the Barents Sea Region, edited by Göran Baecklund, 139–47. Stockholm: Olof Palme International Center.

———. 1995. "The Security of the Baltic States and Russia." In Vilnius—Kaliningrad: Ideas on Cooperative Security in the Baltic Sea Region, edited by Ritva Grönick, Mia Grönqvist, and Nina Granlund, 53–64. Helsinki: Nordic Forum for Security Policy and the Finnish Committee for European Security.

———, project leader. 1997. Civil Monitoring of Security Structures in Estonia. Tallinn, Estonia: Institute of International and Social Sciences.

Vares, Peeter, and Mare Haab. 1993. "The Baltic States: Quo Vadis?" In Central and Eastern Europe: The Challenge of Transition, edited by Regina Cowen Karp, 283–307. Oxford: Oxford Univ. Press.

Väyrynen, Raimo. 1998. "Towards a Pluralistic Security Community in the Baltic Sea Region?" In And Now What? International Politics after the Cold War: Essays in Honour of Nikolaj Petersen, edited by Georg Sørensen and Hans-Henrik Holm, 149–74. Copenhagen: Politica.

———. 2000. "Stable Peace through Security Communities? Steps Towards Theory-Building." In Stable Peace among Nations, edited by Arie M. Kacowicz, Yaacov Bar-Siman-Tov, Ole Elgström, and Magnus Jerneck, 108–29. Lanham, Md.: Rowman and Littlefield.

———. 2003. "Regionalism: Old and New." International Studies Review 5, no. 1: 25–51.

Venclova, Tomas. 1991. "Letter to Czeslaw Milosz." In Beginning with My Streets: Essays and Recollections, by Czeslaw Milosz, 36–57. New York: Farrar, Straus and Giroux.

———. 1995. "A Fifth Year of Independence: Lithuania, 1922 and 1994." East European Politics and Societies 9, no. 2: 344–67.

Vesa, Unto. 1993a. "Back to Deutsch: Integration, Peaceful Change, and Security Communities." In Changes in the Northern Hemisphere, edited by Jyrki Käkönen, 159–64. Tampere, Finland: Tampere Peace Research Institute.

———. 1993b. "Environmental Security and the Baltic Sea Region." In *Cooperation in the Baltic Sea Region*, edited by Pertti Joenniemi, 87–98. London: Taylor and Francis.

Vesa, Unto, and Frank Möller. 2003. *Security Community in the Baltic Sea Region? Recent Debate and Recent Trends.* Tampere, Finland: Tampere Peace Research Institute.

Vike-Freiberga, Vaira. 2000a. "Russland ist unberechenbar." Interview. *Der Spiegel*, no. 22: 198–200.

———. 2000b. "Security Aspects of Integrating Latvia into Euro-Atlantic Structures." Address at the 5th Stockholm Conference on Baltic Sea Region Security and Cooperation, 19 Oct. Available at: http://www.mfa.gov.lv/ENG/NEWS/SPEECHES/VVF/VVF001019.htm.

Viksne, Ilmars. 1995. "Latvia and Europe's Security Structures." In *The Baltic States: Security and Defense after Independence*, edited by Peter van Ham, 61–81. Paris: Institute for Security Studies of Western European Union.

Vitas, Robert A. 1996. "Civil-Military Relations in Lithuania." In *Civil-Military Relations in the Soviet and Yugoslav Successor States*, edited by Constantine P. Danopoulos and Daniel Zirker, 73–91. Boulder, Colo.: Westview Press.

Vitkus, Gediminas. 1997. "At the Crossroads of Alternatives: Lithuanian Security Policies in 1995–1997." In *Baltic Security: Looking Towards the 21st Century*, edited by Gunnar Artéus and Atis Lejins, 53–78. Riga: Latvian Institute of Foreign Affairs and Försvarshögskolan.

Vulfsons, Mavriks. 1998. *Nationality Latvian? No, Jewish. Cards on the Table.* Riga, Latvia: Jumava.

Wæver, Ole. 1995. "Securitization and Desecuritization." In *On Security*, edited by Ronnie D. Lipschutz, 46–86. New York: Columbia Univ. Press.

———. 1997. "The Baltic Sea: A Region after Post-modernity?" In *Neo-nationalism or Regionality: The Restructuring of Political Space around the Baltic Rim*, edited by Pertti Joenniemi, 293–342. Stockholm: NordREFO.

———. 1998. "Insecurity, Security, and Asecurity in the West-European Non-war Community." In *Security Communities*, edited by Emanuel Adler and Michael Barnett, 69–118. Cambridge: Cambridge Univ. Press.

———. 2002. "Identity, Communities, and Foreign Policy: Discourse Analysis as Foreign Policy Theory." In *European Integration and National Identity: The Challenge of the Nordic States*, edited by Lene Hansen and Ole Wæver, 20–49. London: Routledge.

Walker, Martin. 2000. "Variable Geography: America's Mental Maps of a Greater Europe." *International Affairs* 76, no. 3: 459–74.

Walker, R. B. J. 1997. "The Subject of Security." In *Critical Security Studies: Concepts and Cases,* edited by Keith Krause and Michael C. Williams, 61–81. London: University College London Press.

Wallace, William. 2000. "From the Atlantic to the Bug, from the Arctic to the Tigris? The Transformation of the EU and NATO." *International Affairs* 76, no. 3: 475–93.

Wallander, Celeste. 2000. "Wary of the West: Russian Security Policy at the Millennium." *Arms Control Today* 30, no. 2: 7–12.

Wallensteen, Peter, and Karin Axell. 1994. "Conflict Resolution and the End of the Cold War, 1989–93." *Journal of Peace Research* 31, no. 3: 333–49.

Wallensteen, Peter, Kjell-Åke Nordquist, Björn Hagelin, and Erik Melander. 1994. *Towards a Security Community in the Baltic Region: Patterns of Peace and Conflict.* Uppsala: Baltic Univ.

Wallensteen, Peter, and Margareta Sollenberg. 2000. "Armed Conflict, 1989–99." *Journal of Peace Research* 37, no. 5: 635–49.

Walt, Stephen M. 1991. "The Renaissance of Security Studies." *International Studies Quarterly* 35, no. 2: 211–39.

———. 1994. *The Origins of Alliances.* 3d printing. Ithaca, N.Y.: Cornell Univ. Press.

Waltz, Kenneth N. 1979. *Theory of International Politics.* Reading, Mass.: Addison-Wesley.

Wellmann, Christian. 1996. "Russia's Kaliningrad Exclave at the Crossroads: The Interrelation Between Economic Development and Security Politics." *Cooperation and Conflict* 31, no. 2: 161–83.

Wendt, Alexander. 1992. "Anarchy Is What States Make of It: The Social Construction of Power Politics." *International Organization* 46, no. 2: 391–425.

———. 1995. "Constructing International Politics." *International Security* 20, no. 1: 71–81.

———. 1999. *Social Theory of International Politics.* Cambridge: Cambridge Univ. Press.

Wettig, Gerhard. 1993. *Der russische Truppenrückzug aus den baltischen Staaten.* Cologne: Bundesinstitut für Ostwissenschaftliche und Internationale Studien.

White House Office of the Press Secretary. 1996. "Fact Sheet: U.S. Support for Estonia, Latvia, Lithuania." Washington, D.C., 25 June. Available at: http://www.pub.whitehouse.gov.

———. 1998a. "Fact Sheet: U.S.-Baltic Relations." *U.S. Information and Texts,* no. 003 (21 Jan.): 16–17.

———. 1998b. "Remarks by the President at Charter Signing Ceremony." Washington, D.C., 16 Jan. Available at: http://www.whitehouse.gov/WH/New/html.

———. 1998c. "White House Summary on U.S.-Baltic Partnership Charter." *U.S. Information and Texts,* no. 003 (21 Jan.): 15.

Wiberg, Håkan. 2000. "Emanuel Adler, Michael Barnett, and Anomalous Northern-ers." *Cooperation and Conflict* 35, no. 3: 289–98.

———. 2004. "Nordic Peace—Another Dimension of Nordic Expectionalism?" *Kosmopolis* 34, Special Celebratory Issue for Unto Vesa's 60th Birthday: 208–17.

Williams, Michael C., and Keith Krause. 1997. "Preface: Toward Critical Security Studies." In *Critical Security Studies: Concepts and Cases,* edited by Keith Krause and Michael C. Williams, vii–xxiv. London: University College London Press.

Williams, Raymond. 1976. *Keywords: A Vocabulary of Culture and Society.* London: Fontana.

Winner, Andrew C. 1998. "Presentation." In *After Madrid and Amsterdam: Prospects for the Consolidation of Baltic Security,* edited by Paulis Apinis and Atis Lejins, 49–60. Riga: Konrad Adenauer Stiftung and Latvian Institute of International Affairs.

Wolfers, Arnold. 1962. *Discord and Collaboration: Essays on International Relations.* Baltimore: Johns Hopkins Univ. Press.

Young, Oran R. 1985–86. "The Age of the Arctic." *Foreign Policy,* no. 61: 160–79.

Zaagman, Rob. 1999. *Conflict Prevention in the Baltic States: The OSCE High Commissioner on National Minorities in Estonia, Latvia, and Lithuania.* Flensburg, Germany: European Center for Minority Issues.

Zaccor, Albert M. 1994. "Lithuania's New Army." *Journal of Slavic Military Studies* 7, no. 2: 198–217.

———. 1995a. "Problems in the Baltic Armed Forces, Part I." *Lithuania Today: Politics and Economics* 4, no. 31: 14.

———. 1995b. "Problems in the Baltic Armed Forces, Part II." *Lithuania Today: Politics and Economics* 6, no. 32: 12–14.

———. 1995c. "Problems in the Baltic Armed Forces, Part III." *Lithuania Today: Politics and Economics,* issue 7(33): 12–14.

Zakheim, Dov S. 1998. "The United States and the Nordic Countries during the Cold War." *Cooperation and Conflict* 33, no. 2: 115–29.

Zalkalns, Gunnars. 2002. "Security Concepts and the Build-up of Armed Forces in Latvia." In *The Military in Transition: Restructuring and Downsizing the Armed Forces of Eastern Europe,* edited by Andreas Heinemann-Grüder, 37–47. Bonn: Bonn International Center for Conversion.

Zaslavsky, Victor. 1993. "Success and Collapse: Traditional Soviet Nationality Policy." In *Nations and Politics in the Soviet Successor States,* edited by Ian Bremmer and Ray Taras, 29–42. Cambridge: Cambridge Univ. Press.

Zhuryari, Olga. 1994. "The Baltic Countries and Russia (1990–1993): Doomed to Good-Neighborliness?" In *The Foreign Policies of the Baltic Countries: Basic Issues,* edited by Pertti Joenniemi and Juris Prikulis, 75–86. Riga, Latvia: Center of Baltic-Nordic History and Political Studies, and Tampere Peace Research Institute.

Index